THE GREAT RIFT VALLEYS OF PANGEA IN EASTERN NORTH AMERICA

THE GREAT RIFT VALLEYS OF PANGEA IN EASTERN NORTH AMERICA

VOLUME 1

Tectonics, Structure, and Volcanism

Edited by

PETER M. LETOURNEAU AND PAUL E. OLSEN

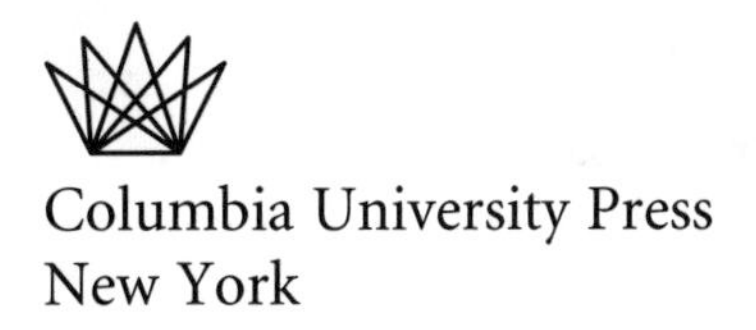

Columbia University Press
New York

Columbia University Press
Publishers Since 1893
New York Chichester, West Sussex

Library of Congress Cataloging-in-Publication Data
The great rift valleys of Pangea in eastern North America / edited by Peter M. LeTourneau
and Paul E. Olsen.
p. cm.
Includes bibliographical references and indexes.
Contents: v. 1. Tectonics, structure, and volcanism
v. 2. Sedimentology, stratigraphy, and paleontology.
ISBN 0-231-11162-2 (v. 1 : acid-free paper)
ISBN 0-231-12676-X (v. 2 : acid-free paper)
1. Geology, Stratigraphic—Triassic. 2. Geology, Stratigraphic—Jurassic. 3. Rifts (Geology)—
North America. 4. Pangaea (Geology) I. LeTourneau, Peter M.
II. Olsen, Paul Eric.

QE676 .G74 2003
551.7′6′097—dc21 2002031452

∞

Columbia University Press books are printed on permanent and durable acid-free paper.
Printed in the United States of America
c 10 9 8 7 6 5 4 3 2 1

CONTENTS

PREFACE

The 1990s were an exceptionally productive period of research in the Triassic–Jurassic age rift basins located around the central Atlantic margin (CAM). Researchers in North America, Europe, and North Africa were engaged in a broad range of projects, including sedimentology, stratigraphy, geophysics, paleontology, structural geology, and tectonics. A significant advance occurred in 1990, when the National Science Foundation made a major investment in ancient rift basin research by funding the $2.5 million Newark Basin Coring Project—the longest publicly available record of rift sedimentation and cyclical climate change and a catalyst for renewed interest in the Mesozoic rifts in eastern North America. With 6.7 km of continuous stratigraphic section spanning more than 30 million years of Earth history, the Newark cores provide unparalleled insights into the influence of climate and tectonics on rift basin stratigraphy and into the evolution and diversification of Triassic terrestrial and aquatic vertebrates. With the publication of the initial results of the Newark Basin Coring Project in the mid-1990s, a new paradigm for high-resolution stratigraphic and paleontologic analysis of the CAM rifts emerged.

In addition to a high level of research activity, significant oil and gas exploration in the CAM rifts in the mid-1980s through the early 1990s was spurred on by substantial new reserves found in rift plays in the North Atlantic, West Africa, and China. An unexpected windfall for researchers occurred in the mid-1990s, when historically low prices for oil led to a sharp decrease in wildcat plays such as the CAM rifts and industrywide downsizing that inspired some oil companies generously to release vast amounts of previously proprietary materials, including Texaco's gracious donation to the Lamont-Doherty Earth Observatory of Columbia University of more than 6 km of continuous cores and more than 7 km of test well cuttings from the Taylorsville basin.

Finally, the explosive growth of the Internet resulted in vastly increased opportunities for communication between researchers and for access to data. All these developments generated a high level of excitement among our fellow scientists and led to the realization

FIGURE P.1 Natural cast of the theropod dinosaur footprint *Eubrontes giganteus* from Dinosaur State Park. Quarter (2.4 cm) for scale. (Photograph by P. E. Olsen)

FIGURE P.2 Main exhibit area at Dinosaur State Park, Rocky Hill, Connecticut. The footprints in the forground are in situ and consist mostly of trackways of *Eubrontes giganteus.* The background is a diorama with a model of one possible trackmaker, the theropod dinosaurs *Dilophosaurus,* and a mural showing the track site during the earliest Jurassic. (Photograph by P. E. Olsen)

that multidisciplinary research of the Triassic–Jurassic rifts was indeed breaking through regional boundaries and becoming global in scope.

By 1995, we felt that time was ripe for a gathering of active workers to meet at a focused, independent conference to discuss our new activities in a congenial and stimulating atmosphere. This idea grew into the conference "Aspects of Triassic–Jurassic Rift Basin Geoscience," held at Dinosaur State Park in Rock Hill, Connecticut, on November 9 and 10, 1996. The unique setting of Dinosaur State Park, located on exposed Early Jurassic strata of the Hartford rift basin in the heart of the Connecticut Valley, provided an exciting and appropriate background for a conference on the Triassic–Jurassic rifts. The park includes an Earth history museum with spectacular in situ theropod dinosaur trackways, dioramas and murals of Triassic and Jurassic life, as well as a classroom, an auditorium, and nature trails. This conference was organized outside the regular meetings of the various geoscience societies specifically to provide more focus for researchers interested in Triassic–Jurassic geoscience and to avoid scheduling conflicts with concurrent symposia or technical sessions that often occur at national geoscience meetings. A volume of abstracts of the papers presented at the conference was published by the State Geological and Natural History Survey of Connecticut, Natural Resources Center, a division of the Connecticut Department of Environmental Protection (DEP).

Following the conference, we solicited full-length papers from the participants and other researchers to provide a lasting record of the excitement of the conference and to bring to light some of the new developments in Triassic–Jurassic rift basin geoscience described in this volume. At least one "outside" reviewer—that is, a qualified scientist who was not an author or a coauthor in this volume—reviewed each paper. The thorough peer review of the papers resulted in high-quality contributions of lasting scientific merit that are of interest to a wide audience rather than merely hastily assembled "gray" papers of passing interest. Columbia University Press expressed a strong interest in publishing these essays and was selected over several other publishers because of its enthusiasm for the project.

We are, of course, humbly and gratefully indebted to the many people and organizations that made the conference an overwhelming success and contributed to the completion of this volume. We extend our sincere thanks to all the conference participants, whose enthusiasm, excitement, and knowledge made it a special event. We especially thank Richard Krueger, director of Dinosaur State Park, and the staff and volunteers who hosted the conference, cheerfully assisted us, and anticipated our every need. We thank the staff of the Natural Resources Center of the Connecticut DEP, especially Nancy McHone, for publishing the conference abstracts. The editors at Columbia University Press—Edward Lugenbiel, Holly Hodder, and Robin Smith—were instrumental in the initial conceptualization and ultimate execution of this volume. We also thank Irene Pavitt of Columbia University Press for shepherding this project. We thank Columbia University—specifically the Department of Earth and Environmental Sciences, the Office of the Vice Provost, and the Lamont-Doherty Earth Observatory—as a major sponsor of the conference. We gratefully acknowledge TILCON Connecticut, Inc.; Greg McHone and the Graduate Liberal Studies Program of Wesleyan University; Andrew W. Lord and the Environmental Practice Group of the law firm of Murtha, Cullina, Richter, and Pinney in Hartford, Connecticut; the Connecticut Natural Gas Corporation; and the Iroquois Gas Transportation Corporation for their generous financial support of the conference. Finally, we thank the contributors and reviewers of the essays published here for their dedication and hard work in helping us to bring this necessary project to completion.

Peter M. LeTourneau
Paul E. Olsen
Palisades, N.Y.

THE GREAT RIFT VALLEYS OF PANGEA IN EASTERN NORTH AMERICA

1
Introduction

Peter M. LeTourneau and Paul E. Olsen

Rift basins of the Triassic–Jurassic age associated with the breakup of the Pangean supercontinent contain an extraordinary record of the physical and biological conditions at a critical period of Earth history. Rather than considering the rift basins as local features of limited interest, ongoing work reveals that these Triassic–Jurassic rifts should be studied in a broader context that spans the entire proto-Atlantic realm, including eastern North America, Greenland, the British Isles, and North Africa, as well as South America and central West Africa (figure 1.1). The rift province, collectively called the central Atlantic margin (CAM) system (Olsen 1977), spans more than 45° of paleolatitude and records more than 35 million years of Earth history (figure 1.2). The CAM basins are of broad appeal to researchers interested in topics as diverse as extensional tectonics, the global magnetostratigraphic timescale, the evolution of early mammals, the appearance and diversification of dinosaurs, rift to drift crustal dynamics, astronomical forcing of climate, and models for the formation and occurrence of economic mineral and fossil fuel deposits.

The Late Triassic and Early Jurassic was a critical time in Earth history, representing one of the fundamental end members of the Earth System. The breakup of the most recent supercontinent, the emplacement of what may be the largest igneous province on Earth, the sedimentary record of the largest known system of rifts, and the pattern of biotic change during the second largest mass extinction of the Phanerozoic (figure 1.3) are fundamental problems of Earth System

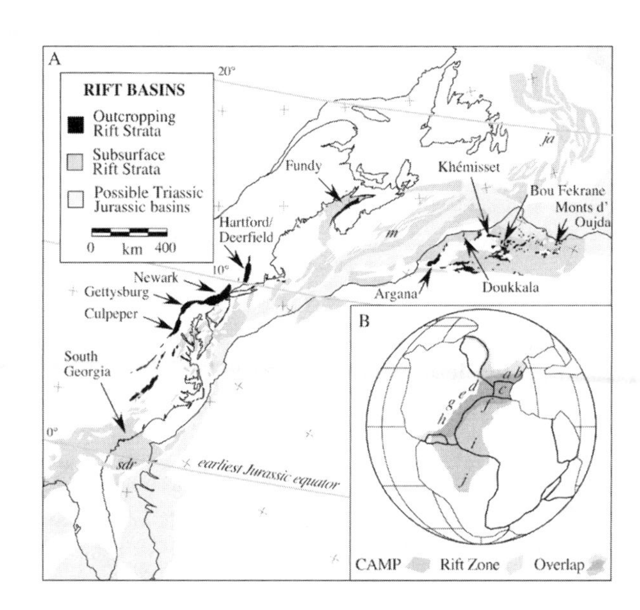

FIGURE 1.1 Distribution of rift basins in eastern North America and Morocco and the distribution of the CAMP). (*A*) Eastern North American rift basins; *ja*, Jeanne d'Arc basin of the Grand Banks area; *m*, Mohican basin on the Scotian Shelf; *sdr*, seaward-dipping reflectors on the southeastern United States continental margin. (*B*) Pangea in the earliest Jurassic, showing the distribution of rifts, the CAMP, and areas discussed in the text; *a*, pyroclastics and ?sills of the Aquitaine basin of southwestern France; *b*, alkali basalt pyroclastics and flows in Provence and the Ecrins-Pelvoux of the external massif of the Alps, France; *c*, flows in Iberia, including the Pyrenees; *d*, flows in the basins on the Grand Banks region, Canada; *e*, Fundy basin area of the Maritime Provinces, Canada; *f*, flows in Morocco, Tunisia, and Algeria; *g*, flows in the major rifts in the eastern United States; *h*, flows in the South Georgia rift and the offshore seaward-dipping reflectors; *i*, ultrabasic layered plutons and dikes of Liberia, Mali, and Senegal; *j*, CAMP flows of Brazil.

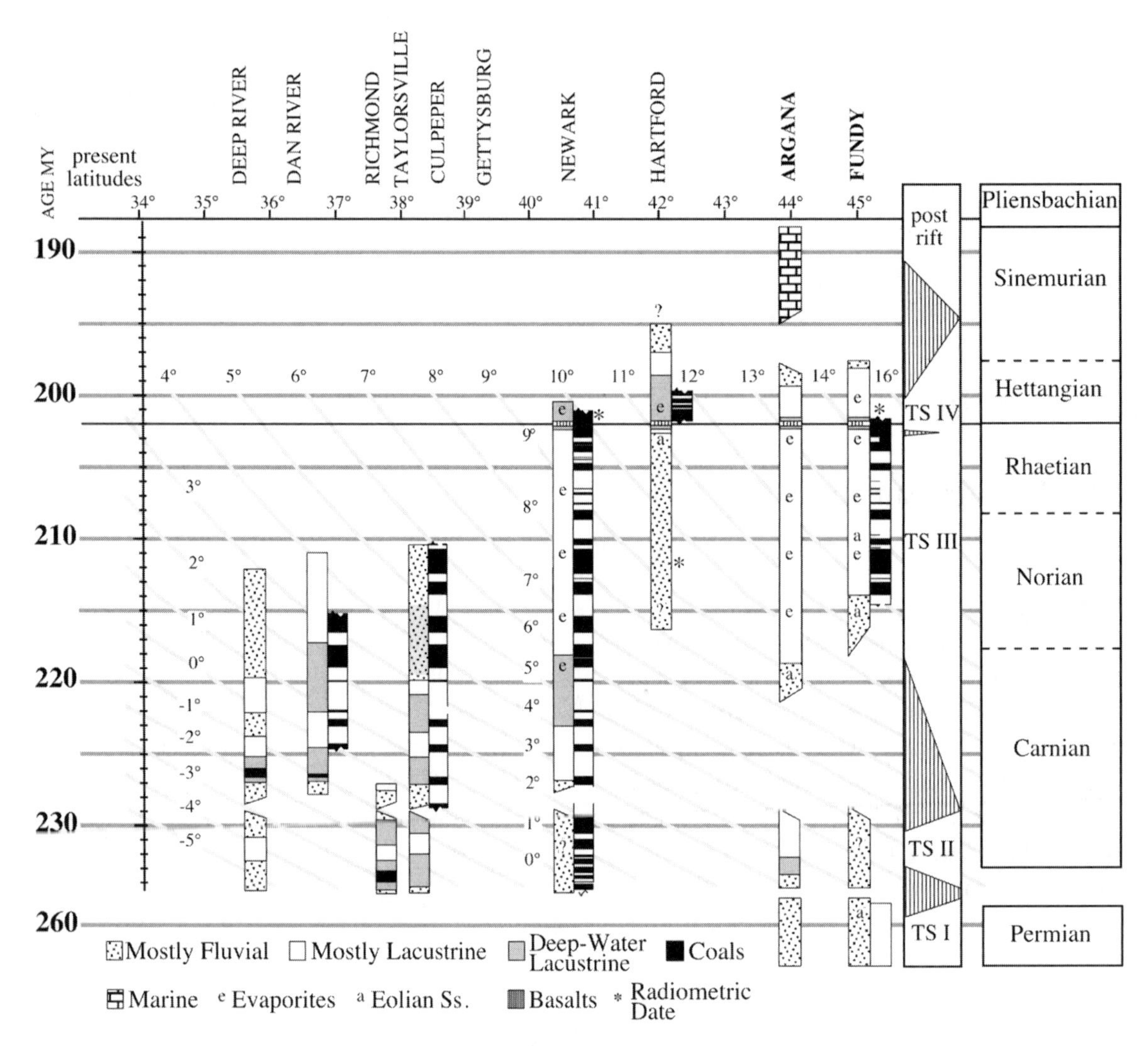

FIGURE 1.2 Spatiotemporal matrix of tectonostratigraphic sequences of CAM basins. Darker-gray tones represent facies deposited under wetter climates. (Data from Olsen 1977; Kent, Olsen, and Witte 1995; Olsen and Kent 1999; LeTourneau 1999; and Kent and Olsen 1997, 2000)

studies that require innovative approaches and high-resolution records for their analysis. Fortunately, CAM basins preserve a spectacular sedimentary and igneous record of Triassic–Jurassic tectonic, climatic, and biotic events. The North American components of the CAM rifts are the focus of the two volumes of *The Great Rift Valleys of Pangea in Eastern North America.*

Volume 1, *Tectonics, Structure, and Volcanism,* addresses recent advances in structural geology, tectonics, and volcanism of the Central Atlantic Magmatic Province (CAMP). Volume 2, *Sedimentology, Stratigraphy, and Paleontology,* presents new interpretations of Triassic–Jurassic rift sedimentology and stratigraphy and new findings in vertebrate and invertebrate paleontology.

Each volume is further divided into two parts that reflect the main concepts that these essays address. Each part, in turn, opens with an introductory chapter, which is followed by the separate thematic papers. The introductory chapters set the thematic papers in a larger context for the general reader. Although we wanted to integrate the chapters as much as possible into a larger context to benefit general readers, we also recognized that it would be most useful to both authors and readers if the papers could be freestanding. Thus there is some unavoidable duplication of introductory material and figures in the papers.

The chapters in part I of volume 1, "Tectonics and Structure of Supercontinent Breakup," establish the tectonic and structural context of these rifts and of the Newark Supergroup in particular, ranging from the

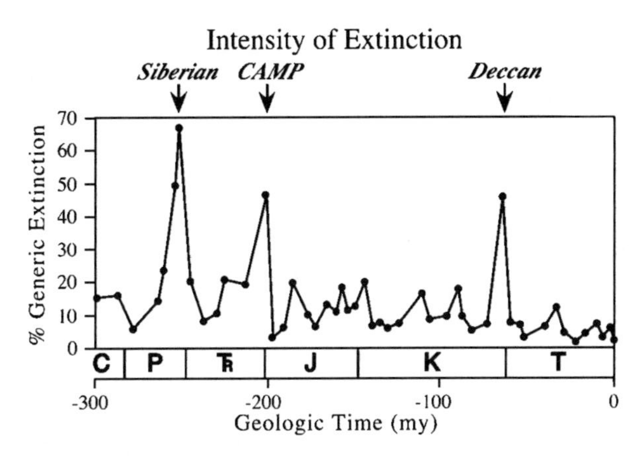

FIGURE 1.3 Generic-level extinctions of marine, shelly organisms during the past 300 million years and the distribution of giant flood basalt provinces (*italics*). Modified from Sepkoski (1997), with timescale modified according to Kent and Olsen (1999), to reflect more recent estimates of the age of the Triassic–Jurassic boundary.

scale of global plate tectonics (chapter 3) to microscale structural analysis (chapter 8). The regional- to basin-scale structure and tectonic development of eastern North American rifts are described in chapters 4 to 7.

Aspects of what may have been the largest known igneous event in Earth history—the 6,000 km diameter CAMP event, which may also mark the emplacement of the earliest Atlantic Ocean seafloor—are analyzed in part II: "The Central Atlantic Large Igneous Province." After an overview of the global context of the CAMP (chapter 9), the North American components of CAMP are dealt with in more detail in chapter 10. Chapters 11 and 12 explore the spatial and temporal trends of the geochemistry of mafic igneous rocks in the CAMP and in other continental large igneous provinces, and chapter 13 describes evidence for the mode of emplacement of this large igneous province.

Literature Cited

Kent, D. V., and P. E. Olsen. 1997. Paleomagnetism of Upper Triassic continental sedimentary rocks from the Dan River–Danville rift basin (eastern North America). *Geological Society of America Bulletin* 109:366–377.

Kent, D. V., and P. E. Olsen. 2000. Magnetic polarity stratigraphy and paleolatitude of the Triassic–Jurassic Blomidon Formation in the Fundy basin (Canada): Implications for early Mesozoic tropical climate gradients. *Earth and Planetary Science Letters* 179:311–324.

Kent, D. V., P. E. Olsen, and W. Witte. 1995. Late Triassic–earliest Jurassic geomagnetic polarity sequence and paleolatitudes from drill cores in the Newark rift basin, eastern North America. *Journal of Geophysical Research* 100:14965–14998.

LeTourneau, P. M. 1999. Depositional history and tectonic evolution of Late Triassic age rifts of the U.S. central Atlantic margin: Results of an integrated stratigraphic, structural, and paleomagnetic analysis of the Taylorsville and Richmond basins. Ph.D. diss., Columbia University.

Olsen, P. E. 1977. Stratigraphic record of the early Mesozoic breakup of Pangea in the Laurasia–Gondwana rift system. *Annual Review of Earth and Planetary Science* 25:337–401.

Olsen, P. E., and D. V. Kent. 1999. Long-period Milankovitch cycles from the Late Triassic and Early Jurassic of eastern North America and their implications for the calibration of the early Mesozoic timescale and the long-term behavior of the planets. *Philosophical Transactions of the Royal Society of London,* part A 357: 1761–1786.

Sepkoski, J., Jr. 1997. Biodiversity: Past, present, and future. *Journal of Paleontology* 71:533–539.

PART I

TECTONICS AND STRUCTURE OF SUPERCONTINENT BREAKUP

2
Introduction

Martha Oliver Withjack and Roy W. Schlische

The chapters in part I provide new and valuable information about the tectonic and structural evolution of eastern North America during the breakup of the Pangean supercontinent. Specifically, these chapters reveal a remarkably consistent story about the synrift and postrift stages of development of this passive continental margin. This consistency is especially noteworthy because the authors employ different approaches applied at different scales along virtually the entire length of the Mesozoic rift system.

In chapter 3, Dennis V. Kent and Giovanni Muttoni use paleomagnetic data to discuss the likely plate-tectonic configurations of Pangea from Permian to Middle Jurassic time. In eastern North America, this time interval corresponds to the prerift, synrift, and early postrift stages. Robert J. Altamura in chapter 5; Dave Goldberg, Tony Lupo, Michael Caputi, Colleen Barton, and Leonardo Seeber in chapter 7; and Rolf V. Ackermann, Roy W. Schlische, Lina C. Patiño, and Lois A. Johnson in chapter 8 use the orientation of small-scale structures to define the NW–SE extension direction in eastern North America during the synrift stage. Altamura's work defines the extension direction specifically for Middle Triassic time (238 Ma). This information, together with previously published studies of Early Jurassic age diabase dikes (e.g., McHone 1988) and sediment-filled fissures (Schlische and Ackermann 1995), indicates that a NW–SE extension direction was temporally consistent, at least in the northeastern United States and maritime Canada. In chapter 4, Roy W. Schlische complements these studies of small-scale structures by integrating structural and stratigraphic data at the basin scale. He describes a broad outline of diachronous rifting and drifting in eastern North America, synthesizing this study with work he did previously with Martha O. Withjack and Paul E. Olsen (Withjack, Olsen, and Schlische 1995; Withjack, Schlische, and Olsen 1998). Specifically, rifting ended earlier and drifting began earlier in the south. Given the diachronous nature of the large-scale breakup of Pangea, it is not surprising that the central North Atlantic margin also had a diachronous history. As discussed in chapter 6, MaryAnn Love Malinconico's work with vitrinite reflectance data from the Taylorsville basin, Virginia, supports the concept that the cessation of rifting was diachronous in eastern North America. Her work suggests that rifting ceased in the south during the latest Triassic and earliest Jurassic.

Schlische (chapter 4) reviews the field and seismic evidence for postrift shortening and inversion in eastern North America. (Basin inversion is a two-stage tectonic process in which a contractional phase follows the extensional phase [Buchanan and Buchanan 1995].) Malinconico (chapter 6) provides independent support for inversion with her vitrinite reflectance study in the Taylorsville basin. Particularly noteworthy is her observation that the amount of postrift erosion is highest over inversion-related anticlines. Although we have known that most rift basins of eastern North America have experienced considerable postrift erosion, Malinconico was the first to demonstrate that at least part of this erosion is controlled directly by structures produced during basin inversion. Fracture studies by Goldberg and colleagues (chapter 7) and by Ackermann and co-workers (chapter 8) corroborate the theory that postrift shortening occurred in eastern North America. These authors (along with Withjack, Olsen, and Schlische 1995; Withjack, Schlische, and Olsen 1998; and Schlische [chapter 4]) conclude that the postrift shortening direction was NW–SE. Previous studies, however, support different shortening directions. For example, structural analyses in the Newark basin indicate a N–S (Lomando and Engelder 1980) or NE–SW (Lucas, Hull, and Manspeizer 1988) shortening direction. The postrift shortening direction is generally difficult to constrain because the geometries of the preexisting extensional structures strongly influence the geometries of the inversion structures. Moreover, the timing of postrift shortening is difficult to constrain because of an absence of growth strata. We believe that additional work is needed to define conclusively the timing and direction of postrift shortening in eastern North America. The modern state of stress and strain, defined by borehole breakouts and quarry buckles, is NE–SW compression and shortening (e.g., Goldberg et al. [chapter 7]; Ackermann et al. [chapter 8]).

The chapters in part I also provide insights into some general aspects of extensional tectonics and passive-margin development. The Mesozoic rift basins of eastern North America can (and should) serve, therefore, as models for interpreting other less well studied or less well understood rift basins and passive margins.

Preexisting Fabric and Oblique Rifting

In chapter 5, Altamura reports that the Lantern Hill fault, a reactivated Paleozoic contractional structure, was oblique to the Triassic extension direction, resulting in oblique-slip faulting. Oblique extension as applied to the Hartford basin is described in Clifton and colleagues (2000). Schlische (chapter 4) describes its application to the Fundy basin and to the Narrow Neck.

Fault-Population Systematics

The Solite Quarry, described in chapter 8, has been useful in understanding the scaling relationship between fault length and displacement (Schlische et al. 1996), the deflection geometry associated with normal faults (Gupta and Scholz 1998), and the mechanics of fault interaction (Gupta and Scholz 2000b). The Solite faults have shown that fault populations do not always follow a simple power-law distribution of fault sizes (Ackermann and Schlische 1997), as is conventionally assumed (e.g., Marrett and Allmendinger 1991; most papers in Cowie, Knipe, and Main 1996). Experimental models (e.g., Ackermann, Schlische, and Withjack 2001) and field data from Afar (Gupta and Scholz 2000a) show that fault populations change from power-law to exponential-size distributions with increasing strain. Modeling work by Ackermann, Schlische, and Withjack (2001) has further demonstrated that this transition is governed by mechanical layer thickness. In chapter 8, Ackermann and co-workers further describe the importance of mechanical stratigraphy in controlling fracture populations. The lack of self-similar (power-law) fault growth under certain conditions poses problems for the very simple basin-filling models described by Schlische (1991, chapter 4) and by Contreras, Scholz, and King (1997).

Fault Spatial Relationships

In chapter 5, Altamura highlights the fact that more faults were active during rifting than those that now preserve synrift units in their hanging-wall blocks. This is especially the case during the early stages of rifting. Data from the Taylorsville basin (LeTourneau 1999) and from other rift systems indicate that faulting is distributed originally on many small faults but that it localizes on a few large faults. Localization occurs because of stress-enhancement and stress-reduction zones (e.g., Cowie 1998; Gupta et al. 1998). Stress-enhancement zones encourage the along-strike linkage of colinear fault segments, whereas stress-reduction zones deactivate noncolinear but parallel segments. Ackermann and colleagues (chapter 8) describe small-scale examples of stress-reduction zones in the Solite Quarry (see also Ackermann and Schlische 1997).

Basin Inversion on Passive Margins

Like many other passive margins (as described in Hill et al. 1995; Doré and Lundin 1996; Vågnes, Gabrielsen, and Haremo 1998; Withjack and Eisenstadt 1999), the rift basins of eastern North America have undergone basin inversion. However, unlike most other inverted basins, many of those in eastern North America are exposed. Thus field and seismic data allow us to determine more accurately the three-dimensional geometry of inversion structures and related smaller-scale fractures, which is essential for constraining the shortening direction (or directions) involved with the inversion and the cause (or causes) of this shortening in a passive-margin setting. Possible mechanisms for producing basin inversion in this tectonic setting are reviewed by Withjack, Olsen, and Schlische (1995); Withjack, Schlische, and Olsen (1998); Vågnes, Gabrielsen, and Haremo (1998); and Schlische (chapter 4). Finally, basin inversion may be even more widespread than currently assumed. Experimental models (Eisenstadt and Withjack 1995) indicate that some inversion structures may be difficult to recognize. Unless the magnitude of shortening exceeds the magnitude of extension, inversion structures may have a subtle, anticlinal expression and may be overlooked easily. These low-amplitude anticlines would be the first to be removed during the erosion associated with uplift.

Literature Cited

Ackermann, R. V., and R. W. Schlische. 1997. Anticlustering of small normal faults around larger faults. *Geology* 25:1127–1130.

Ackermann, R. V., R. W. Schlische, and M. O. Withjack. 2001. The geometric and statistical evolution of normal fault systems: An experimental study of the effects of mechanical layer thickness on scaling laws. *Journal of Structural Geology* 23:1803–1819.

Buchanan, J. G., and P. G. Buchanan, eds. 1995. *Basin Inversion*. Geological Society Special Publication, no. 88. London: Geological Society.

Clifton, A. E., R. W. Schlische, M. O. Withjack, and R. V. Ackermann. 2000. Influence of rift obliquity on fault-population systematics: Results of clay modeling experiments. *Journal of Structural Geology* 22:1491–1509.

Contreras, J., C. H. Scholz, and G. C. P. King. 1997. A model of rift basin evolution constrained by first order stratigraphic observations. *Journal of Geophysical Research* 102:7673–7690.

Cowie, P. A. 1998. A healing-reloading feedback control on the growth rate of seismogenic faults. *Journal of Structural Geology* 20:1075–1088.

Cowie, P. A., R. Knipe, and I. G. Main. 1996. Introduction to *Scaling Laws for Fault and Fracture Populations: Analyses and Applications.* Special issue of *Journal of Structural Geology* 18:v–xi.

Doré, A. G., and E. R. Lundin. 1996. Cenozoic compressional structures on the NE Atlantic margin: Nature, origin, and potential signficance for hydrocarbon exploration. *Petroleum Geoscience* 2:299–311.

Eisenstadt, G., and M. O. Withjack. 1995. Estimating inversion: Results from clay models. In J. G. Buchanan and P. G. Buchanan, eds., *Basin Inversion,* pp. 119–136. Geological Society Special Publication, no. 88. London: Geological Society.

Gupta, S., P. A. Cowie, N. H. Dawers, and J. R. Underhill. 1998. A mechanism to explain rift-basin subsidence and stratigraphic patterns through fault-array evolution. *Geology* 26:595–598.

Gupta, A., and C. H. Scholz. 2000a. A brittle strain regime in the Afar Depression: Implications for fault growth and seafloor spreading. *Geology* 28:1087–1090.

Gupta, A., and C. H. Scholz. 2000b. A model of normal fault interaction based on observations and theory. *Journal of Structural Geology* 22:865–879.

Hill, K. C., K. A. Hill, G. T. Cooper, A. J. O'Sullivan, P. B. O'Sullivan, and M. J. Richardson. 1995. Inversion around the Bass basin, SE Australia. In J. G. Buchanan and P. G. Buchanan, eds. *Basin Inversion,* pp. 525–548. Geological Society Special Publication, no. 88. London: Geological Society.

LeTourneau, P. M. 1999. Depositional history and tectonic evolution of Late Triassic age rifts of the U.S. central Atlantic margin: Results of an integrated stratigraphic, structural, and paleomagnetic analysis of the Taylorsville and Richmond basins. Ph.D. diss., Columbia University.

Lomando, A. J., and T. Engelder. 1984. Strain indicated by calcite twinning: Implications for deformation of the early Mesozoic northern Newark basin, New York. *Northeast Geology* 6:192–195.

Lucas, M., J. Hull, and W. Manspeizer. 1988. A foreland-type fold and related structures of the Newark rift basin. In W. Manspeizer, ed., *Triassic–Jurassic Rifting: Continental Breakup and the Origin of the Atlantic Ocean and the Passive Margins,* pp. 307–332. Developments in Geotectonics, no. 22. Amsterdam: Elsevier.

Marrett, R., and R. W. Allmendinger. 1991. Estimates of strain due to brittle faulting: Sampling of fault populations. *Journal of Structural Geology* 13:735–738.

McHone, J. G. 1988. Tectonic and paleostress patterns of Mesozoic intrusions in eastern North America. In W. Manspeizer, ed., *Triassic–Jurassic Rifting: Continental Breakup and the Origin of the Atlantic Ocean and the Passive Margins*, pp. 607–620. Developments in Geotectonics, no. 22. Amsterdam: Elsevier.

Schlische, R. W. 1991. Half-graben basin filling models: New constraints on continental extensional basin evolution. *Basin Research* 3:123–141.

Schlische, R. W., and R. V. Ackermann. 1995. Kinematic significance of sediment-filled fissures in the North Mountain basalt, Fundy rift basin, Nova Scotia, Canada. *Journal of Structural Geology* 17:987–996.

Schlische, R. W., S. S. Young, R. V. Ackermann, and A. Gupta. 1996. Geometry and scaling relations of a population of very small rift-related normal faults. *Geology* 24:683–686.

Vågnes, E., R. H. Gabrielsen, and P. Haremo. 1998. Late Cretaceous–Cenozoic intraplate contractional deformation at the Norwegian continental shelf: Timing, magnitude, and regional implications. *Tectonophysics* 300:29–46.

Withjack, M. O., and G. Eisenstadt. 1999. Structural history of the Northwest Shelf, Australia: An integrated geological, geophysical, and experimental approach. *American Association of Petroleum Geologists Annual Meeting, Abstracts* 8:A151.

Withjack, M. O., P. E. Olsen, and R. W. Schlische. 1995. Tectonic evolution of the Fundy basin, Canada: Evidence of extension and shortening during passive-margin development. *Tectonics* 14:390–405.

Withjack, M. O., R. W. Schlische, and P. E. Olsen. 1998. Diachronous rifting, drifting, and inversion on the passive margin of eastern North America: An analog for other passive margins. *American Association of Petroleum Geologists Bulletin* 82:817–835.

3

Mobility of Pangea: Implications for Late Paleozoic and Early Mesozoic Paleoclimate

Dennis V. Kent and Giovanni Muttoni

Several recent analyses of paleomagnetic data support the concept of Pangea, the late Paleozoic and early Mesozoic assemblage of most of the world's continents, that was mobile in terms of large-scale internal deformation and with respect to paleolatitude. The main feature of internal deformation involved the transformation from a Pangea B–type configuration in the late Paleozoic, with northwestern South America adjacent to eastern North America, to a more traditional Pangea A–type configuration in the early Mesozoic, with northwestern Africa adjacent to eastern North America. Pangea B thus seems to coincide in time with extensive low-latitude coal deposition and high-southern-latitude Gondwana glaciations, whereas Pangea A coincides with generally drier conditions over the continents and no polar ice sheets. Although the configuration of Pangea may have been more stable as an A-type configuration in the early Mesozoic prior to breakup, the paleomagnetic evidence suggests that there was appreciable latitudinal change of the assembly. Such changing tectonic boundary conditions emphasize the practical importance of age registry of paleoclimate data in making valid comparisons with model results. A simple zonal climate model coupled with the geocentric axial dipole hypothesis for establishing paleolatitudes in precisely controlled paleogeographic reconstructions can explain many of the climate patterns in both the late Paleozoic and the early Mesozoic, but it cannot explain the presence or absence of continental ice sheets.

The supercontinent of Pangea included most of the continents over the late Paleozoic and the early Mesozoic (~300 Ma to ~175 Ma). Over its existence, Pangea experienced a wide range of global climate conditions, from extensive Gondwana glaciations in the Permo-Carboniferous to generally drier continental climates and the absence of evidence for high-latitude ice sheets in the Triassic and Jurassic (Frakes 1979). If there was indeed little change in continental positions from the late Paleozoic to the early Mesozoic, as is often supposed, then the role of paleogeography in explaining the large contrast of climates across these intervals is not very obvious (Hallam 1985).

However, some recent analyses of paleomagnetic data support the concept of a more mobile Pangea, in which the relative arrangement of the main continental elements and the paleolatitudinal setting changed appreciably with time. The concept of a mobile Pangea expands opportunities for testing the sensitivity of tectonic boundary conditions in climate simulations; however, the changing tectonic boundary conditions also emphasize the importance of precise age correlation of paleoclimate data in making valid paleogeographic reconstructions. This chapter highlights some recent developments in our understanding of the evolution of Pangea in the late Paleozoic and early Mesozoic and calls attention to their possible paleoclimate implications.

Paleomagnetic Data

Paleomagnetism provides a record of past locations of the geographic axis and therefore can place critical constraints independent of paleoclimatology on the latitudinal distribution of landmasses. Descriptions of the methodology and assumptions of the subject are discussed extensively elsewhere (for a recent general treatment, see Butler 1992, and for a modern and comprehensive analysis of tectonic applications, see Van der Voo 1993). Some comments are offered here to remind the reader of the scope and limitations of paleomagnetic data when such data are used for paleocontinental reconstructions.

A central assumption of paleomagnetism is that the Earth's magnetic field, when averaged over some thousands of years, will be approximated closely by that of a geocentric axial dipole (GAD). Thus an observed mean inclination, I, can be related to geographic latitude, L, according to the dipole formula ($\tan I = 2 \tan L$). Paleomagnetic observations for the past few million years, where accumulated plate motions are small or can be taken into account, indicate that the GAD model is an accurate representation of the paleofield to within a few degrees (e.g., Opdyke and Henry 1969; Schneider and Kent 1990b). The detailed configuration of the paleofield is known less precisely for earlier eras (e.g., Livermore, Vine, and Smith 1984; Schneider and Kent 1990a), and evidence for the general validity of the GAD hypothesis comes to rely mainly on the degree of congruence with the inferred latitudinal dependence of various climate indicators (e.g., Blackett 1961; Briden and Irving 1964). It is therefore useful to maintain a distinction between estimates of latitude obtained from paleomagnetic data and those obtained from paleoclimate data so that the nature of any discrepancies (e.g., those possibly arising from more complicated magnetic fields, regionally nonzonal climate, or simply poor or undiagnostic data) can be better understood.

According to the GAD hypothesis, paleomagnetic directions of similar age from any locality on a rigid tectonic plate should give the same paleomagnetic pole and hence the location of that plate with respect to the geographic axis. In practice, there will be some scatter in such determinations, and the resulting uncertainty in the mean pole position is usually expressed by the radius of the circle of 95% confidence (A95), according to Fisher statistics (Fisher 1953). The uncertainty in paleopole determinations, where A95 is typically a few degrees and higher, is thought to be caused mainly by experimental and observational errors rather than by complexities in the paleofield (for a caveat on Paleozoic and Precambrian data, see Kent and Smethurst 1998).

A temporal sequence of paleopoles or an apparent polar wander (APW) path is a convenient way to represent the motion of a tectonic plate with respect to the rotation axis. Van der Voo (1993) has introduced a set of seven minimum reliability criteria for evaluating published paleopoles. A low overall quality factor is grounds for rejection, but relatively few results satisfy all the threshold criteria for reliability. Van der Voo (1993) suggested that only those results that pass three or more reliability criteria should be retained as the most suitable candidates for documenting polar wander. Acceptable paleopoles from specific rock units are typically averaged over intervals, usually something such as a geologic epoch or 10 to 30 million years depending on the availability of data, in order to enhance the coherence of the APW path. Paleomagnetic data are not uniform in quality or abundance, so there is considerable variability in the definition and reliability of APW paths for different plates or time intervals. Paleocontinental reconstructions accordingly are not uniformly constrained and may change as better data become available.

Relative motions between tectonic plates or their continental proxies are detected by systematic differences in their respective APW paths. It is also possible that the whole Earth rotated with respect to the rotation axis (e.g., Goldreich and Toomre 1969), a phenomenon referred to as true polar wander (TPW). For the late Mesozoic and Cenozoic, where plate motions are constrained by evidence from the seafloor and by a hot-spot reference frame more or less fixed to the solid Earth, the amount of any TPW has been estimated to have been much smaller than plate motions in accounting for polar wander (e.g., Courtillot and Besse 1987; Gordon 1987). Some researchers have suggested episodes of larger TPW for some earlier intervals, such as the Permo-Triassic (Marcano, Van der Voo, and Mac Niocaill 1999), but TPW will remain a moot issue for paleoclimatology provided that the GAD hypothesis remains valid; that is, the dynamo in the fluid outer core always responds to the geographic or rotation axis so that the dipole field tends to remain aligned along it.

There have been occasional suggestions, although they are not well posed as testable hypotheses, that the paleomagnetic and geographic axes were significantly decoupled for long intervals. For example, Donn (1982, 1989) attempted to explain apparently large disagreements between paleomagnetic data and other paleolatitudinal indicators for the Triassic to Eocene with what might be referred to as paleomagnetic polar wander. Some discrepancies between paleomagnetic and paleoclimatic estimates of paleolatitude have been resolved with better paleomagnetic data—for example, in the Devonian of North America (Heckel and Witzke 1979; Miller and Kent 1988; Stearns, Van der Voo, and Abrahamsen 1989)—or have been taken as an indication of a significant climate problem, as in the case of high-latitude warmth in the Cretaceous and Eocene (Barron, Thompson, and Schneider 1981; Barron 1987). But the underlying nature of other discrepancies is debated still—for example, the evidence for low-latitude glaciation in the late Precambrian (Crowley and Baum 1993; Meert and Van der Voo 1994; Schmidt and Williams 1995; Sohl, Christie-Blick, and Kent 1999). Here we assume the validity of the GAD hypothesis as a basis for determining paleolatitudes in assessing paleoclimate implications.

Finally, it should be noted that the axial symmetry of the dipole model of the field means that a paleomagnetic pole can used to determine the position of a tectonic plate only with respect to lines of paleolatitude, leaving paleolongitude indeterminate. For the late Mesozoic and younger time, marine magnetic anomalies and other signatures of seafloor spreading provide a precise measure of relative paleolongitudes for most tectonic plates. For early Mesozoic and earlier times, the matching of geological or biogeographic features or provinces, in conjunction with minimum motion arguments, must be relied on to provide some constraints on the relative longitudinal distribution of tectonic plates.

Pangean Configurations in the Permian, Triassic, and Jurassic

Although a variety of data have demonstrated decisively the existence of Pangea in the late Paleozoic and early Mesozoic, uncertainty remains on its precise configuration. For example, the positions of the various tectonic blocks in eastern Asia may not have become assembled with the rest of Laurasia until sometime in the Jurassic (Enkin et al. 1992). However, a first-order problem considered here is the position of the southern continental assembly of Gondwana with respect to the northern continental assembly of Laurasia. Two general models have been proposed: (1) a supercontinent that maintained a more or less static Pangea A–type configuration over the late Paleozoic and early Mesozoic; and, largely on the basis of paleomagnetic results, (2) a supercontinent that experienced appreciable internal deformation and evolved from a Pangea B–type to a Pangea A–type configuration over this interval.

The Pangea A–type model is the traditional Wegenerian configuration, whose salient feature for this discussion is the placement of northwestern Africa adjacent to eastern North America. A widely used version (Pangea A-1) is Bullard, Everett, and Smith's (1965) reconstruction, which is based on a computerized best fit of the present circum-Atlantic continental margins. Pangea A-1 should thus approximate the relative position of the major continents just prior to the opening of the Atlantic Ocean sometime in the Jurassic.

It has long been recognized, however, that Permian and Triassic paleomagnetic poles from Gondwana (mainly from Africa and South America) with respect to Laurasia (mainly from North America and Europe) do not agree well with the Bullard fit for Pangea, so various alternative reconstructions have been offered to explain the systematic discordance between the respective APW paths. Van der Voo and French (1974) proposed a modification of Pangea A-1 that requires a further approximately 20° clockwise rotation of Gondwana with respect to Laurasia. This adjustment (Pangea A-2) brings the northwestern margin of South America much closer to the Gulf Coast margin of North America but does not fully account for the differences in Permian poles.

To bring Permian paleomagnetic poles into even better mutual agreement, Gondwana has to be rotated clockwise as well as northward. However, this rotation results in an unacceptable overlap with Laurasia unless Gondwana is also shifted appreciably eastward, taking advantage of the longitude indeterminacy of the paleomagnetic method. The Pangea B reconstruction proposed by Irving (1977) and by Morel and Irving (1981) is the best-known model of this type (see also Smith, Hurley, and Briden 1980; Livermore, Smith, and Vine 1986). An important distinguishing feature of Pangea B is that northwestern South America is placed against eastern North America, whereas northwestern Africa is positioned south of Europe. This would require approximately 3,500 km of subsequent right-lateral relative motion between Gondwana and the northern continents because the Atlantic Ocean opened from a Pangea A–type configuration. The reality of Pangea B thus has been much debated on the basis of geologic and paleomagnetic data (e.g., Hallam 1983; Van der Voo, Peinado, and Scotese 1984; Smith and Livermore 1991; see summary in Van der Voo 1993).

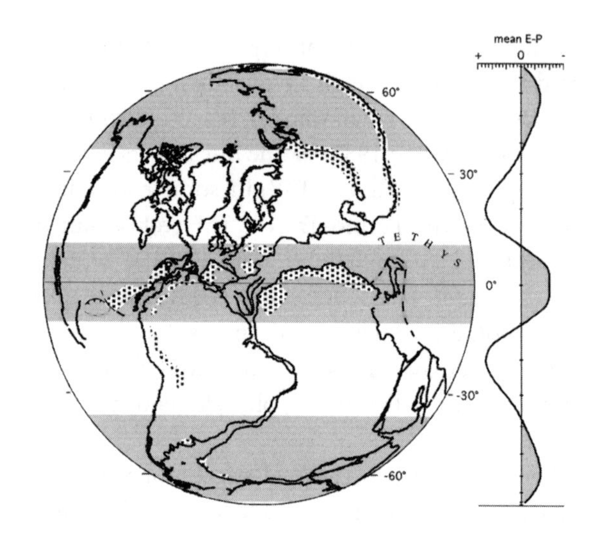

re-
construction that resembles Pangea B. The Gondwana and Laurasia continents were assembled by using reconstruction parameters from Lottes and Rowley (1990) and from Bullard, Everett, and Smith (1965), respectively, and were oriented with respect to the paleolatitudinal grid according to mean paleopoles summarized by Muttoni, Kent, and Channell (1996), with Gondwana translated longitudinally to avoid overlap with Laurasia. The distribution of the Alleghanian–Hercynian–Variscan orogenic belt (*gray*) is from Morel and Irving (1981); tectonic lines are from Arthaud and Matte (1977). Tethyan tectonics are highly diagrammatic. Shown for reference is the mean zonal variation in evaporation minus precipitation (E − P) for the modern land plus ocean surface (Crowley and North 1991), which has been further averaged over the Southern and Northern Hemispheres. (From Muttoni, Kent, and Channell 1996)

The Permo-Triassic APW path of Gondwana generally is poorly documented, which contributes to the uncertainty in Pangea configurations. To augment the definition of APW in this interval, Muttoni, Kent, and Channell (1996) included paleomagnetic data from the southern Alps in Italy as a proxy for Gondwana and generated a new tectonic model for the evolution of Pangea from the Early Permian to the Early Jurassic by integrating these data with the recently published APW paths representative of Gondwana and Laurasia (Van der Voo 1993). According to this model, the reconstruction that satisfies Early Permian paleopoles is virtually the same as Morel and Irving's (1981) Pangea B (figure 3.1). This conclusion is not strongly dependent on the paleomagnetic data from the southern Alps.

For the Late Permian–Early Triassic and into the Middle–Late Triassic, the Gondwana-proxy data from the southern Alps allow a reconstruction resembling a Pangea A-2 model (figure 3.2). According to Muttoni, Kent, and Channell's (1996) analysis, the major transformation from a Pangea B–type to a Pangea A–type configuration thus occurred effectively by the end of the Permian. This timing allows the large strike-slip motion associated with the transformation to be connected with the final stages of the Variscan orogeny (Arthaud and Matte 1977). The transition from Pangea B to Pangea A was suggested previously by Irving (1977), by Morel and Irving (1981), and more recently by Torcq et al. (1997) to have occurred during the Triassic. We believe that the Triassic is more likely associated with the tectonically more modest shift from Pangea A-2 to Pangea A-1. Unconformity-bound tectonostratigraphic units preserved in the early Mesozoic rift basins along the central Atlantic margins (Olsen 1997) may reflect such regional transtensional tectonic activity.

Early to Middle Jurassic paleopoles representative of Laurasia and Gondwana are reasonably concordant

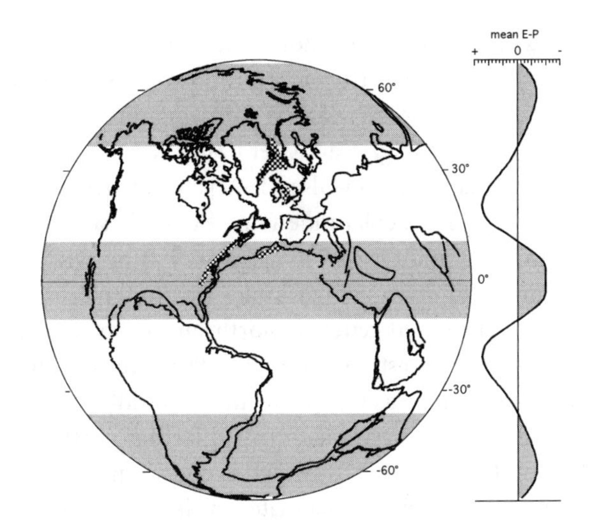

en-
tal reconstruction that resembles Pangea A-2. Notes and abbreviations are as given in figure 3.1. (From Muttoni, Kent, and Channell 1996)

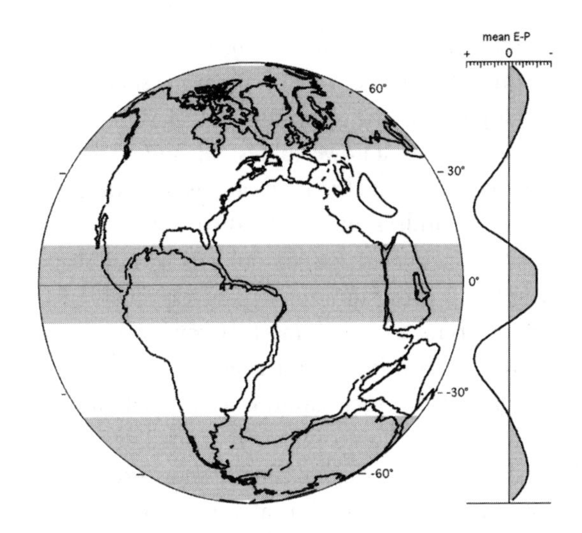

re-
construction that resembles Pangea A-1 (for just prior to the opening of the Atlantic Ocean), with a high-latitude option for Jurassic APW. Notes and abbreviations are as given in figure 3.1. The position of Gondwana versus Laurasia (i.e., Africa versus North America) is from Bullard, Everett, and Smith (1965). The Pangea assembly was oriented with respect to the paleolatitude grid by using the approximately 175 Ma Newark B paleopole of Witte and Kent (1991), which agrees with the synthetic APW path for North America of Courtillot, Besse, and Theveniaut (1994).

with a Pangea A-1 fit (Van der Voo 1993; Muttoni, Kent, and Channell 1996) from which the Atlantic Ocean began opening at approximately 175 Ma, or during the middle Middle Jurassic (Klitgord and Schouten 1986) (figure 3.3). As indicative of the amount of APW over the Jurassic as a whole, there is approximately 30° of great-circle distance separating Early Jurassic and Early Cretaceous poles for North America, the continent with the most abundant and reliable paleopoles for this interval (Van der Voo 1993). However, because of possible remagnetizations and local tectonic rotations, it is uncertain whether the Jurassic APW track for North America follows more or less along the 60th parallel (e.g., May and Butler 1986) or reaches high present-day latitudes of approximately 75° or more in the Middle Jurassic (e.g., Van Fossen and Kent 1990; for reviews and further analysis, see also Hagstrum 1993 and Van der Voo 1992, 1993). Courtillot, Besse, and Theveniaut (1994) constructed synthetic APW paths by transferring data from other Atlantic-bordering continents according to various reconstruction parameters. They concluded that high-latitude APW for North America was the option most compatible with the global paleopole database, as Van der Voo (1993) also found. The high-latitude APW model for North America results in a paleogeographic position of Pangea in the Middle Jurassic, as shown in figure 3.3.

Pangean Climates

Pangea A–type reconstructions have been used in climate simulations of the late Paleozoic and the early Mesozoic (Crowley, Hyde, and Short 1989; Kutzbach and Gallimore 1989; Chandler, Rind, and Ruedy 1992; Klein 1994). However, rather than a single configuration over the Permian and Triassic, an internally mobile Pangea suggested by Irving (1977) is supported in manner, if not in exact timing, by several more recent analyses of paleomagnetic data (Muttoni, Kent, and Channell 1996; Torq et al. 1997; but see dissent in Van der Voo 1993). If corroborated by further work to exclude artifacts from remagnetizations, inclination error, and even nondipole fields, the internal mobility of Pangea provides an additional source of paleogeographic variability that might be relevant to changes in global climate. In general, the Pangea B configuration seems to coincide temporally with extensive low-latitude coal deposits and high-southern-latitude Gondwana glaciations in the late Paleozoic. In contrast, a Pangea A paleogeography seems to have ushered in the apparently drier continentality of the early Mesozoic when there were no polar ice sheets.

Analogous to the geocentric axial dipole model, which constitutes a powerful testable model for paleomagnetism, a zonal climate model is a useful null hypothesis for understanding the distribution of climate proxies. The zonal-mean annual averages of evaporation (E) and precipitation (P) and especially of their relative variations (E-P rate) (Crowley and North 1991:fig. 2.3) are important elements of climate and hydrology, just as inclination is an important observable element of the geomagnetic field. The measured E-P values in modern-day climate show excess of precipitation over evaporation at middle and high latitudes as well as in the equatorial zone between 10° north and 10° south, whereas a deficit of precipitation is found in the subtropical regions between approximately 10 and 35° latitude. These latitudinal bands of relative aridity and humidity are superposed on the Pangea reconstructions (figures 3.1–3.3). It should be noted that this zonal configuration of relative humidity and aridity need not have been the same in the past. For example, many analyses have suggested that there was a desertlike equatorial belt in the Triassic (e.g., Ziegler et al. 1993; Wilson et al. 1994), which, if true, would constitute a first-order discrepancy with the modern atmospheric circulation pattern. Climate proxies should also represent the characteristic state over several million years to minimize aliasing from Milankovitch climate cyclicity (Olsen and Kent 1996).

The difference between Pangea B and Pangea A is mainly in the relative paleolongitudes of Laurasia versus Gondwana; hence the latitudinal distribution of land areas is not very different in the two alternative reconstructions. However, Pangea as a whole gradually shifted northward over the Permian and Triassic and into the Jurassic so that, for example, European sites of near-equatorial coal deposition in the late Paleozoic (figure 3.1) had migrated into the arid belt by the early Mesozoic (figure 3.2). Although Late Triassic paleoclimate indicators from eastern North America are often regarded as the result of local orographic effects (e.g., Manspeizer 1982; Hay and Wold 1998), they are actually very consistent, when correlated precisely using magnetic polarity stratigraphy, with a latitudinal variation in the balance of evaporation to precipitation not that different from today's mean zonal regime. In contrast to many previous analyses for the early Mesozoic, the existence of an equatorial humid belt is clearly shown in the Late Triassic by the progression from coal-bearing deposits near the paleoequator in eastern North America (e.g., Dan River basin, North Carolina and Virginia [Kent and Olsen 1997]) to eolian deposits at around 10° paleolatitude (e.g., Fundy basin, Nova Scotia [Kent and Olsen 2000]), as well as by evidence of a major through-going Chinle–Dockum paleoriver system that flowed generally northward from Texas in the U.S. Southwest (Riggs et al. 1996). By the time Pangea breakup was under way in the Middle Jurassic, eastern North America had drifted farther north into the arid belt (figure 3.3), which is consistent with the presence of extensive evaporites of likely Jurassic age under the North American east coast margin (Poppe and Poag 1993) and of major eolian deposits such as the Navajo Sandstone in the U.S. Southwest.

A zonal climatic model applied to a mobile Pangea configuration therefore can account for the comings and goings of a variety of climatic conditions. Furthermore, some major paleogeographic differences between Pangea B and Pangea A may have important, nonzonal paleoclimate implications. Pangea B is characterized by a narrow Tethys Sea, open ocean to the south of North America, and an orogenic belt (Alleghanian–Hercynian–Variscan) along the equatorial zone (figure 3.1). Otto-Bliesner (1993) suggested that significant topography along the equator can enhance humidity in the tropical zone and hence the formation of coals. In contrast, the Pangea A–type reconstruction has a wide Tethys Sea, mostly landmass to the south of North America, and the eventual development of rift basins with a more meridional trend along the central Atlantic margins (figures 3.2 and 3.3).

An important issue is the ability to separate actual nonzonal features of climate from artifacts of poor spatiotemporal resolution or from errors in continental reconstructions. Even with very slow latitudinal drift, which characterized North America in the Late Triassic (~0.3°/million years [Kent and Witte 1993]), the customary mapping of paleoclimate data in time slices of 10 to 15 million years (e.g., the Carnian or Norian–Rhaetian [Hay et al. 1982; Wilson et al. 1994]) is likely to obscure evidence for narrow climate zones with steep latitudinal gradients, such as the tropical humid belt. Distortion of climate patterns is exacerbated for faster latitudinal drift (e.g., 1° per million years in the Jurassic [Van Fossen and Kent 1990]),

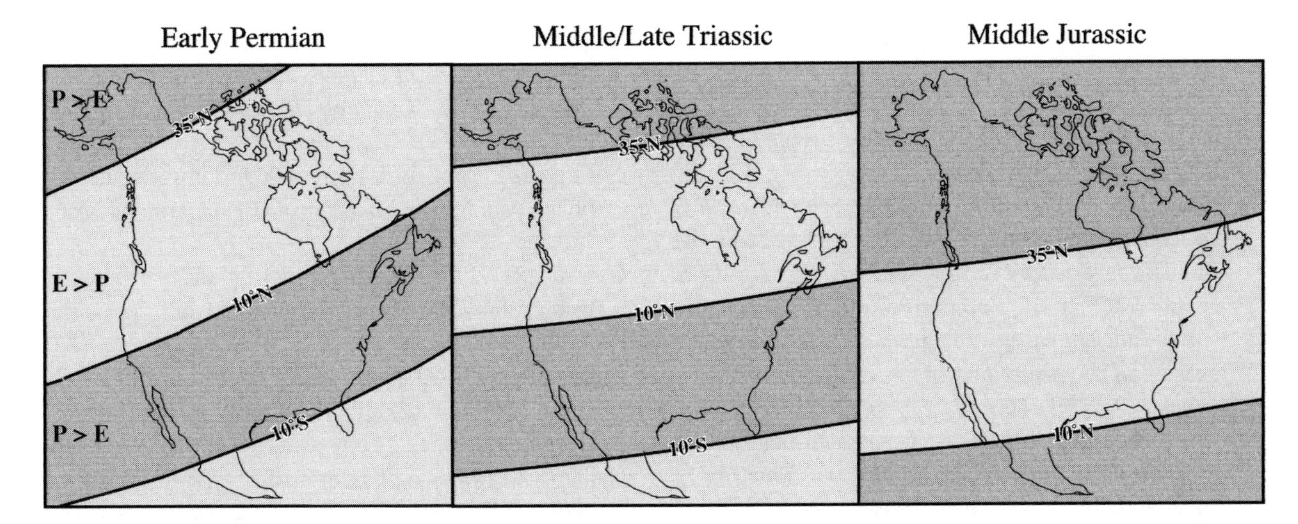

FIGURE 3.4 Map of present-day North America with inferred zonal bands of arid (E > P) and humid (P > E) climate for the Early Permian, Middle–Late Triassic, and Middle Jurassic. Paleopoles used to construct paleolatitudes for North America are 46°N 124°E for Early Permian (~275 Ma), 51°N 96°E for Middle–Late Triassic (~225 Ma) (both poles from Muttoni, Kent, and Channell 1996:table 3), and 74°N 96°E for Middle Jurassic (~175 Ma) (Witte and Kent 1991).

where there is also the greater likelihood of poor age registry and systematic offsets between the paleoclimate indicators and the paleomagnetic observations that are inevitably needed for the paleolatitudinal framework. Artifacts of such loose cataloging provide a testable alternative explanation to nonzonal effects for what are sometimes interpreted as contradictory geographic arrays of climate indicators (e.g., Chandler, Rind, and Ruedy 1992; Hay and Wold 1998).

Conclusions

The mobility of Pangea has important but largely underestimated implications for understanding the significance of late Paleozoic and early Mesozoic paleoclimate indicators. Pangea paleogeographic reconstructions from the Early Permian to the Middle Jurassic (Muttoni, Kent, and Channell 1996; this study) have been compared with modern-day climatic indicators such as the relative rate of evaporation and precipitation under the null hypothesis of a zonal climate model. Climatic variations from arid (E > P) to humid (P > E) conditions across large portions of the Pangea-forming continents can be accounted for by internal or total Pangea mobility or by both. A large portion of North America, for example, was characterized in the Early Permian by humid (P > E) conditions, which correspond on modern-day Earth to the savanna–rain forest equatorial belt, whereas by the Middle Jurassic the same areas fell within the arid belt (E > P), which currently characterizes tropical deserts such as the Sahara (figure 3.4). Reliable paleomagnetic-derived paleogeographic reconstructions under the hypothesis of a zonal climate model therefore can explain a wide variety of climatic conditions over late Paleozoic and early Mesozoic Pangea. At the present level of analysis, it is not even obvious that the widths of the humid and arid belts in the late Paleozoic and early Mesozoic were markedly different from those of today's world, despite the major differences in paleogeography.

Acknowledgments

We thank Paul Olsen for sharing stimulating ideas and valuable information on many aspects of early Mesozoic history, the reviewers of an earlier version of this chapter for their many constructive comments, and Bill Lowrie for a critical reading of this version. Support for this work was provided by the Climate Dynamics Program of the U.S. National Science Foundation (grant ATM93–17227). This chapter is Lamont-Doherty Earth Observatory contribution no. 6101.

Literature Cited

Arthaud, F., and P. Matte. 1977. Late Paleozoic strike-slip faulting in southern Europe and northern Africa: Result of a right-lateral shear zone between the Appalachians and the Urals. *Geological Society of America Bulletin* 88:1305–1320.

Barron, E. J. 1987. Eocene equator-to-pole surface ocean temperatures: A significant climate problem? *Paleoceanography* 2:729–739.

Barron, E. J., S. L. Thompson, and S. H. Schneider. 1981. An ice-free Cretaceous? Results from climate model simulations. *Science* 212:501–508.

Blackett, P. M. S. 1961. Comparison of ancient climates with the ancient latitudes deduced from rock magnetic measurements. *Proceedings of the Royal Society of London,* part A 263:1–30.

Briden, J. C., and E. Irving. 1964. *Palaeolatitude Spectra of Sedimentary Palaeoclimatic Indicators: Problems in Palaeoclimatology.* New York: Interscience.

Bullard, E. C., J. E. Everett, and A. G. Smith. 1965. A symposium on continental drift. IV. The fit of the continents around the Atlantic. *Philosophical Transactions of the Royal Society of London,* part A 258:41–51.

Butler, R. F. 1992. *Paleomagnetism: Magnetic Domains to Geologic Terranes.* Oxford: Blackwell Scientific.

Chandler, M. A., D. Rind, and R. Ruedy. 1992. Pangaean climate during the Early Jurassic: GCM simulations and the sedimentary record of paleoclimate. *Geological Society of America Bulletin* 104:543–559.

Courtillot, V., and J. Besse. 1987. Magnetic field reversals, polar wander, and core-mantle coupling. *Science* 237:1140–1147.

Courtillot, V., J. Besse, and H. Theveniaut. 1994. North American Jurassic apparent polar wander: The answer from other continents? *Physics of the Earth and Planetary Interiors* 82:87–104.

Crowley, T. J., and S. K. Baum. 1993. Effect of decreased solar luminosity on Late Precambrian ice extent. *Journal of Geophysical Research* 98:16723–16732.

Crowley, T. J., W. T. Hyde, and D. A. Short. 1989. Seasonal cycle variations on the supercontinent of Pangaea. *Geology* 17:457–460.

Crowley, T. J., and G. R. North. 1991. *Paleoclimatology.* New York: Oxford University Press.

Donn, W. L. 1982. The enigma of high-latitude paleoclimate. *Palaeogeography, Palaeoclimatology, Palaeoecology* 40:199–212.

Donn, W. L. 1989. Paleoclimate and polar wander. *Palaeogeography, Palaeoclimatology, Palaeoecology* 71: 225–236.

Enkin, R. J., Z. Yang, Y. Chen, and V. Courtillot. 1992. Paleomagnetic constraints on the geodynamic history of the major blocks of China from the Permian to the present. *Journal of Geophysical Research* 97:13953–13989.

Fisher, R. A. 1953. Dispersion on a sphere. *Proceedings of the Royal Society of London,* part A 217:295–305.

Frakes, L. A. 1979. *Climates Throughout Geologic Time.* New York: Elsevier.

Goldreich, P., and A. Toomre. 1969. Some remarks on polar wandering. *Journal of Geophysical Research* 74:2555–2567.

Gordon, R. G. 1987. Polar wandering and paleomagnetism. *Annual Review of Earth and Planetary Science* 15:567–593.

Hagstrum, J. T. 1993. North American Jurassic APW: The current dilemma. *Eos: Transactions of the American Geophysical Union* 74:65–69.

Hallam, A. 1983. Supposed Permo-Triassic megashear between Laurasia and Gondwana. *Nature* 301:499–502.

Hallam, A. 1985. A review of Mesozoic climates. *Journal of the Geological Society of London* 142:433–445.

Hay, W. W., J. F. Behensky Jr., E. J. Barron, and J. L. I. Sloan. 1982. Late Triassic–Liassic paleoclimatology of the proto-central North Atlantic rift system. *Palaeogeography, Palaeoclimatology, Palaeoecology* 40: 13–30.

Hay, W. W., and C. N. Wold. 1998. The role of mountains and plateaus in a Triassic climate model. In T. J. Crowley and K. C. Burke, eds. *Tectonic Boundary Conditions for Climate Reconstructions,* vol. 39, pp. 116–143. New York: Oxford University Press.

Heckel, P. H., and B. J. Witzke. 1979. Devonian world palaeogeography determined from distribution of carbonates and related lithic palaeoclimatic indicators. In C. T. Scrutton and M. G. E. Bassett, eds., *The Devonian System,* pp. 9–123. Special Papers in Palaeontology, no. 23. London: Palaeontological Association.

Irving, E. 1977. Drift of the major continental blocks since the Devonian. *Nature* 270:304–309.

Kent, D. V., and P. E. Olsen. 1997. Paleomagnetism of Upper Triassic continental sedimentary rocks from the Dan River–Danville rift basin (eastern North America). *Geological Society of America Bulletin* 109:366–377.

Kent, D. V., and P. E. Olsen. 2000. Magnetic polarity stratigraphy and paleolatitude of the Triassic–Jurassic Blomidon Formation in the Fundy basin (Canada): Implications for early Mesozoic tropical climate gradients. *Earth and Planetary Science Letters* 179:311–324.

Kent, D. V., and M. A. Smethurst. 1998. Shallow bias of paleomagnetic inclinations in the Paleozoic and Precambrian. *Earth and Planetary Science Letters* 160: 391–402.

Kent, D. V., and W. K. Witte. 1993. Slow apparent polar wander for North America in the Late Triassic and large Colorado Plateau rotation. *Tectonics* 12:291–300.

Klein, G. D., ed. 1994. *Pangea: Paleoclimate, Tectonics, and Sedimentation During Accretion, Zenith, and Breakup of a Supercontinent.* Geological Society of America Special Paper, no. 288. Boulder, Colo.: Geological Society of America.

Klitgord, K. D., and H. Schouten. 1986. Plate kinematics of the central Atlantic. In P. R. Vogt and B. E. Tucholke, eds., *The Western North Atlantic Region*, pp. 351–378. Vol. M of *The Geology of North America.* Boulder, Colo.: Geological Society of America.

Kutzbach, J. E., and R. G. Gallimore. 1989. Pangaean climates: Megamonsoons of the megacontinent. *Journal of Geophysical Research* 94:3341–3357.

Livermore, R. A., A. G. Smith, and F. J. Vine. 1986. Late Palaeozoic to early Mesozoic evolution of Pangaea. *Nature* 322:362–365.

Livermore, R. A., F. J. Vine, and A. G. Smith. 1984. Plate motions and the geomagnetic field. II. Jurassic to Tertiary. *Geophysical Journal of the Royal Astronomical Society* 79:939–961.

Lottes, A. L., and D. B. Rowley. 1990. Reconstruction of the Laurasian and Gondwanan segments of Permian Pangaea. In W. S. McKerrow and C. R. Scotese, eds., *Palaeozoic Palaeogeography and Biogeography*, pp. 383–395. Geological Society Memoir, no. 12. London: Geological Society.

Manspeizer, W. 1982. Triassic–Liassic basins and climate of the Atlantic passive margins. *Geologische Rundschau* 71:895–917.

Marcano, M. C., R. Van der Voo, and C. Mac Niocaill. 1999. True polar wander during the Permo-Triassic. *Journal of Geodynamics* 28:75–95.

May, S. R., and R. F. Butler. 1986. North American Jurassic apparent polar wander: Implications for plate motion, paleogeography, and Cordilleran tectonics. *Journal of Geophysical Research* 91:11519–11544.

Meert, J. G., and R. Van der Voo. 1994. The Neoproterozoic (1000–540 Ma) glacial intervals: No more snowball Earth? *Earth and Planetary Science Letters* 123:1–13.

Miller, J. D., and D. V. Kent. 1988. Paleomagnetism of the Siluro-Devonian Andreas Redbeds: Evidence for an Early Devonian supercontinent? *Geology* 16:195–198.

Morel, P., and E. Irving. 1981. Paleomagnetism and the evolution of Pangea. *Journal of Geophysical Research* 86:1858–1987.

Muttoni, G., D. V. Kent, and J. E. T. Channell. 1996. Evolution of Pangea: Paleomagnetic constraints from the southern Alps, Italy. *Earth and Planetary Science Letters* 140:97–112.

Olsen, P. E. 1997. Stratigraphic record of the early Mesozoic breakup of Pangea in the Laurasia–Gondwana rift system. *Annual Review of Earth and Planetary Science* 25:337–401.

Olsen, P. E., and D. V. Kent. 1996. Milankovitch climate forcing in the tropics of Pangea during the Late Triassic. *Palaeogeography, Palaeoclimatology, Palaeoecology* 122:1–26.

Opdyke, N. D., and K. W. Henry. 1969. A test of the dipole hypothesis. *Earth and Planetary Science Letters* 6:139–151.

Otto-Bliesner, B. L. 1993. Tropical mountains and coal formation: A climate model study of the Westphalian (306 Ma). *Geophysical Research Letters* 20:1947–1950.

Poppe, L. J., and C.W. Poag.1993. Mesozoic stratigraphy and paleoenvironments of the Georges Bank basin: A correlation of exploratory and COST wells. *Marine Geology* 113:147–162.

Riggs, N. R., T. M. Lehman, G. E. Gehrels, and W. R. Dickinson. 1996. Detrital zircon link between headwaters and terminus of the Upper Triassic Chinle–Dockum paleoriver system. *Science* 273: 97–100.

Schmidt, P. W., and G. E. Williams. 1995. The Neoproterozoic climatic paradox: Equatorial palaeolatitude for Marinoan glaciation near sea level in South Australia. *Earth and Planetary Science Letters* 134:107–124.

Schneider, D. A., and D. V. Kent. 1990a. Testing models of the Tertiary paleomagnetic field. *Earth and Planetary Science Letters* 101:260–271.

Schneider, D. A., and D. V. Kent. 1990b. The time-averaged paleomagnetic field. *Reviews of Geophysics* 18:71–96.

Smith, A. G., A. M. Hurley, and J. C. Briden. 1980. *Phanerozoic Paleocontinental World Maps.* Cambridge: Cambridge University Press.

Smith, A. G., and R. A. Livermore. 1991. Pangea in Permian to Jurassic time. *Tectonophysics* 187:135–179.

Sohl, L. E., N. Christie-Blick, and D. V. Kent. 1999. Paleomagnetic polarity reversals in Marinoan (ca. 600 Ma) glacial deposits of Australia: Implications for the duration of low-latitude glaciation in Neoproterozoic times. *Geological Society of America Bulletin* 111:1120–1139.

Stearns, C., R. Van der Voo, and N. Abrahamsen. 1989. A new Siluro-Devonian paleopole from early Paleozoic rocks of the Franklinian basin, north Greenland fold belt. *Journal of Geophysical Research* 94:10669–10683.

Torcq, F., J. Besse, D. Vaslet, J. Marcoux, L. E. Ricou, M. Halawani, and M. Basahel. 1997. Paleomagnetic results from Saudi Arabia and the Permo-Triassic Pangea configuration. *Earth and Planetary Science Letters* 148:553–567.

Van der Voo, R. 1992. Jurassic paleopole controversy: Contributions from the Atlantic-bordering continents. *Geology* 20:975–978.

Van der Voo, R. 1993. *Paleomagnetism of the Atlantic, Tethys, and Iapetus Oceans.* Cambridge: Cambridge University Press.

Van der Voo, R., and R. B. French. 1974. Apparent polar wandering for the Atlantic-bordering continents: Late Carboniferous to Eocene. *Earth-Science Reviews* 10: 99–119.

Van der Voo, R., J. Peinado, and C. R. Scotese. 1984. A paleomagnetic reevaluation of Pangea reconstructions. *American Geophysical Union Geodynamics Series* 12:11–26.

Van Fossen, M. C., and D. V. Kent. 1990. High-latitude paleomagnetic poles from Middle Jurassic plutons and Moat Volcanics in New England and the controversy regarding Jurassic apparent polar wander for North America. *Journal of Geophysical Research* 95:17503–17516.

Wilson, K. M., D. Pollard, W. W. Hay, S. L. Thompson, and C. N. Wold. 1994. General circulation model simulations of Triassic climates: Preliminary results. In G. D. Klein, ed., *Pangea: Paleoclimate, Tectonics, and Sedimentation During Accretion, Zenith, and Breakup of a Supercontinent,* pp. 91–116. Geological Society of America Special Paper, no. 288. Boulder, Colo.: Geological Society of America.

Witte, W. K., and D. V. Kent. 1991. Tectonic implications of a remagnetization event in the Newark basin. *Journal of Geophysical Research* 96:19569–19582.

Ziegler, A. M., J. M. Parrish, Y. Jiping, E. D. Gyllenhaal, D. B. Rowley, J. T. Totman, N. Shangyou, A. Bekker, and M. L. Hulver. 1993. Early Mesozoic phytogeography and climate. *Philosophical Transactions of the Royal Society of London,* part B 341:297–305.

4

Progress in Understanding the Structural Geology, Basin Evolution, and Tectonic History of the Eastern North American Rift System

Roy W. Schlische

Five key developments have contributed significantly to our understanding of the structural geology, basin evolution, and tectonic history of the eastern North American rift system:

1. *Acquisition of new data.* Over the past two decades, regional and local geologic mapping, drilling and coring, and seismic-reflection profiling have increased vastly our structural and tectonic database. It is now clear that these basins are predominantly half-graben, with generally synthetic intrabasinal faults and fault-perpendicular folds that in many cases are related to fault segmentation.

2. *Role of preexisting structures.* The rift system is located within the Appalachian orogen, and thus the border-fault systems of the rift basins consist of reactivated structures. The attitude of the reactivated faults with respect to the rift-related extension direction controlled the nature of the reactivation (dip-slip dominated versus strike-slip dominated), which affected the amount of basin subsidence and types of associated structures. The uniform dip direction of preexisting faults over large areas accounts for the lack of half-graben polarity reversals within rift zones (e.g., the Newark–Gettysburg–Culpeper rift zone).

3. *Application of fault-population studies.* In the past 10 years, considerable progress has been made in our understanding of the geometry and scaling relationships of populations of normal-fault systems. This information is directly applicable to rift basin structural geology in that half-graben are large, normal-fault–bounded basins. The most relevant features of normal-fault systems to basin geometry are: (a) Displacement is greatest at or near the center of a normal fault and decreases systematically to the fault tips; displacement also decreases with distance perpendicular to the fault. (b) Normal-fault systems are segmented, and many fault segment boundaries are areas of (at least temporary) displacement deficits. (c) As displacement builds up on a normal fault, the fault increases in length. Consequently, rift basins consist of scoop-shaped depressions that grow longer, wider, and deeper through time. In segmented border-fault systems, the scoop-shaped subbasins are separated by intrabasinal highs.

4. *Integration of stratigraphy and structural geology.* The sedimentary deposits of half-graben are influenced by basin geometry; consequently, stratigraphy can be used to infer aspects of basin evolution and structural geology. On a local scale, thickness variations of fixed-period Milankovitch cycles are particularly useful for

assessing variations in basin subsidence and for determining whether structures formed syndepositionally. On a regional scale, (a) the lack of Jurassic strata in the southern basins likely indicates that they stopped subsiding before the northern basins did; (b) high accumulation rates in Early Jurassic strata in the northern rift basins indicate accelerated basin subsidence during eastern North American magmatism; and (c) the presence of a tripartite stratigraphy (basal fluvial unit, middle deep-water lacustrine unit, and upper shallow-lacustrine and fluvial unit) in most basins indicates that they share a similar evolutionary trend, most likely related to the infilling of basins growing larger through time. All or parts of a tripartite stratigraphy comprise unconformity-bounded sequences associated with extensional pulses.

5. *Recognition of inversion structures.* Although postrift contractional structures have long been recognized, recent work shows that the magnitude of postrift shortening was greater than previously thought and that the initiation of shortening and basin inversion was diachronous. In particular, shortening in the southern basins began after synrift deposition and before the eastern North America magmatic event (~202 Ma), whereas rifting and subsidence continued in the northern basins. Inversion in the northern basins occurred between early Middle Jurassic and Early Cretaceous time. Postrift shortening is attributed to ridge-push forces and to continental resistance to plate motion during the initiation of seafloor spreading, which itself was diachronous along the North American margin.

Over the past three decades, interest in extensional basins has been increasing steadily in both academia and industry, and with good reason: the basins are the best recorders of the history of crustal extension. Extensional basins contain not only the structures formed during extension but also the dateable strata influenced by those structures. In addition, extensional basins are the sites of active hydrocarbon exploration and exploitation (e.g., Harding 1984; Lambiase and Bosworth 1995).

Much of the research on extensional tectonics has been conducted in regions where extension is currently active or was recently active (e.g., Basin and Range, Aegean Sea, Gulf of Suez, and East African rifts). Although it is possible to study extension-related earthquakes or recent fault scarps in these basins, the extensional basins themselves—those recorders of the history of extension—are buried by their youngest fill. Therefore, most of the record is not available for direct observation. Even worse, some extensional basins are covered by hundreds of meters of water (e.g., North Sea, Gulf of Suez, Aegean Sea, and East African rift lakes). Drilling and geophysical methods provide the only (and commonly ambiguous) clues to unraveling that much-sought-after extensional history.

Eastern North America provides an excellent opportunity to examine the 40-million-year extensional history of the early fragmentation of Pangea locked within the basins containing the rocks of the early Mesozoic Newark Supergroup (Froelich and Olsen 1984). The exposed basins are erosionally dissected, permitting much of the stratigraphic section to be scrutinized. The largely nonmarine record of the exposed basins means that sea level is not a factor in influencing depositional patterns. However, the largely lacustrine strata are commonly sensitive recorders of climate change as well as potential hydrocarbon source rocks. In addition, these basins are related to continental separation that led to the initiation of seafloor spreading in the North Atlantic and are the synrift counterparts to the postrift deposits of the U.S. continental margin, perhaps the most intensely studied and certainly the type passive margin (Sheridan and Grow 1988).

Following an overview of points of contention, this chapter explores five key areas that have contributed most to our understanding of the structural geology, basin evolution, and tectonic history of the eastern North American rift system in the past 15 years: (1) acquisition of new data, which have defined more specifically the geometry of the basins themselves and the geologic structures present within them; (2) recognition of the role of preexisting structures, which can account for some of the first-order structural differences among the various basins; (3) application of fault-population studies, which has provided a framework for a better understanding of a suite of normal-fault–related structures; (4) integration of structural geology and stratigraphy, which has allowed us to date many geologic structures as well as to make strides toward unraveling the effects of tectonics, climate, and sediment supply on stratigraphic architecture; and

(5) recognition of inversion structures, which indicate that the classical rift-drift models of passive margin evolution are overly simplistic.

This review focuses principally on the eastern North American rift system, although many of the concepts apply to other extensional settings. Lorenz (1988) provides an excellent historical summary of research on the Mesozoic basins, and Schlische (1990) gives a brief historical review focusing on the structural aspects of the rift system.

Points of Contention

Considerable debate has centered around three questions:

1. Is the large suite of structures contained in the eastern North American rift system attributable to a uniform stress field reactivating variably oriented Paleozoic structures or to a history of changing stress regimes?
2. Are the major structures bounding and contained within the basins syndepositional or postdepositional?
3. What are the causes of the major stratigraphic transitions (e.g., fluvial sedimentation to lacustrine sedimentation) within the exposed basins?

In the following paragraphs, the various competing hypotheses related to these three questions are outlined. Progress toward resolving these questions is then described in subsequent sections. The current status of these questions is summarized in the discussion section.

The extensional basins of the Newark Supergroup are host to a variety of structures, many of which generally are not associated with continental extension. Structures include normal (e.g., Barrell 1915), strike-slip (e.g., Manspeizer 1980; de Boer and Clifton 1988), and thrust faults (e.g., Wise 1988); dikes (e.g., King 1971) and joints (e.g., Wise 1982); and folds (e.g., Wheeler 1939), some of which contain an axial planar cleavage (Lucas, Hull, and Manspeizer 1988). This association of structures has led some workers to conclude that not all structures are related to rifting but reflect a historical sequence of changing stress regimes (Sanders 1963; Faill 1973, 1988; Swanson 1982; de Boer and Clifton 1988; de Boer et al. 1988; Wise 1988; Manning and de Boer 1989). In this scenario, the normal faults, joints, and dikes are related to early Mesozoic extension, with the strike-slip faults and folds related to postrift episodes of shearing and compression. Other workers maintain that a large suite of the structures can be explained as a consequence of a relatively uniform and protracted period of extension reactivating variously oriented preexisting structures, which then controlled the formation of new structures (Ratcliffe and Burton 1985; Schlische and Olsen 1988; Olsen, Schlische, and Gore 1989; Olsen and Schlische 1990). A third group argues that the structures can be explained as a consequence of a regional shear couple acting on North America during the early Mesozoic (Manspeizer 1980, 1981, 1988; Manspeizer and Cousminer 1988; Manspeizer et al. 1989). The latter two groups do not discount the notion that postrift deformation affected the basins and their fill, but they also argue that many of the structures ascribed by the first group to postrift deformation are in actuality synrift.

The timing of border faulting with respect to sedimentation has been controversial. One group maintains that the basins initially formed as synformal downwarps, with border faulting occurring late in the history of rifting (e.g., Faill 1973, 1988; de Boer and Clifton 1988; de Boer et al. 1988; Root 1988, 1989). Others have followed Barrell's (1915) line of thought and argued that border faulting was contemporaneous with sedimentation (e.g., Sanders 1963; Ratcliffe 1980; LeTourneau 1985; Nadon and Middleton 1985; LeTourneau and McDonald 1988; Schlische and Olsen 1988; Olsen and Schlische 1990). Manspeizer (1988) suggested that basin-bounding faulting became active when sedimentation switched from fluvial to lacustrine.

Although a wide variety of depositional environments have been identified within Newark Supergroup basins, two broad types predominate: fluvial and lacustrine (each has associated marginal facies). Changes from fluvial to lacustrine sedimentation typically have been interpreted to reflect a change in tectonic activity: the border fault moved, the basin deepened, and lacustrine sedimentation began (Manspeizer 1988; Manspeizer et al. 1989; Lambiase 1990). But Schlische and Olsen (1990) proposed that the simple infilling of a basin growing in size through time can also explain the transition and its noncontemporaneity in unconnected basins.

Acquisition of New Data

New geologic mapping at a variety of scales has contributed to our understanding of the structural geology of the eastern North American rift system. On the largest scale, Williams's (1978) map of the Appalachians clearly shows the strong influence of Appalachian structures on basin geometry (figure 4.1a). At the next lower scale range are the state and provincial geologic maps that include all or parts of the exposed rift basins; many of these maps have been updated in the past 20 years (see references in the caption to figure 4.1). Olsen, Schlische, and Gore (1989) and Schlische (1990) recompiled the geologic maps of the major exposed basins, which were then published using uniform lithologic and structural symbols by Schlische (1993). Simplified versions of these maps, all published for the first time at a common map scale, are presented in figure 4.1b. At the next scale are regional maps prepared for a specific purpose—for example, quadrangle-scale maps of the areas surrounding the Newark Basin Coring Project (NBCP) drill sites (Olsen et al. 1996). Perhaps the smallest-scale maps are those showing the spatial distribution of normal faults in the Solite Quarry of the Danville basin (Ackermann et al., chapter 8 in this volume).

Cross sections of the various basins are based in part on projecting surface features to depth and subsurface data, which include seismic-reflection profiles as well as core- and drill holes. In the past 15 years, there has been an explosion in the availability of seismic-reflection data of variable quality (data available prior to 1988 are summarized in Unger 1988). The United States Geological Survey has acquired the majority of the offshore seismic-reflection data (summaries in Hutchinson, Klitgord, and Detrick 1986; Benson and Doyle 1988; Hutchinson and Klitgord 1988a, 1988b; and Hutchinson et al. 1988). The recent EDGE project obtained high-quality data for the offshore Norfolk basin (Musser 1993; Sheridan et al. 1993). Seismic-reflection data for a variety of onshore basins in the United States (summarized in Costain and Coruh 1989) were acquired by the U.S. Geological Survey and by other federal and state agencies, various academic groups (including COCORP [McBride, Nelson, and Brown 1989]), and industry. Industry-acquired data have become increasingly available in the past decade. Reynolds (1994) interpreted multiple seismic-reflection profiles across the southeastern Newark basin released by Exxon. Withjack, Olsen, and Schlische (1995) described only a few of the scores of seismic-reflection profiles acquired by industry in the Fundy basin. Industry-acquired seismic-reflection data for the Taylorsville basin are described by LeTourneau (1999, chapter 3 in volume 2 of *The Great Rift Valleys of Pangea*) and by Malinconico (chapter 6 in this volume).

Drill-hole and core data also have become increasingly available, especially in the past few years. Notable examples include data from Army Corps of Engineers cores from the Connecticut Valley basin (e.g., Pienkowski and Steinen 1995) and the Newark basin (e.g., Olsen, Schlische, and Fedosh 1996); NBCP cores from the Newark basin (Kent, Olsen, and Witte 1995; Olsen and Kent 1996; Olsen et al. 1996); Department of Energy cores from the Dunbarton basin (Thayer and Summer 1996); industry cores from the Taylorsville basin (Letourneau 1999, chapter 3 in volume 2; Malinconico, chapter 6 in this volume); and a long

→

FIGURE 4.1 (*a*) Eastern North American rift basins (modified from Schlische 1993 and Olsen 1997). Basins discussed in the text are labeled. (*b*) Geologic maps of selected exposed basins, all shown at the same scale. The Newark, Gettysburg, and Culpeper basins are shown in their true spatial arrangement, and the same applies to the Pomperaug and Connecticut Valley basins; *C*, Chalfont fault; *LP*, longitudinal pinchout; *RB*, rider block; *RR*, relay ramp. Seismic lines 81-79 and 81-47 are shown in figure 4.14. TS II to TS IV refer to tectonostratigraphic sequences of Olsen (1997) (figure 4.12). For enlarged versions of the portions of the Newark basin enclosed within the rectangles, see figures 4.3 and 4.10a; for an enlarged version of the portion of the Minas subbasin enclosed within the rectangle, see figure 4.6 (maps simplified from Schlische 1993, based on the following sources: Fundy [Keppie 1979; Donohoe and Wallace 1982; Olsen and Schlische 1990; Withjack, Olsen, and Schlische 1995]; Connecticut Valley and Pomperaug [Zen et al. 1983; Rodgers 1985]; Newark [Berg 1980; Olsen 1980; Ratcliffe et al. 1986; Lyttle and Epstein 1987; Parker, Houghton, and McDowell 1988; Ratcliffe and Burton 1988; Schlische and Olsen 1988; Schlische 1992; Olsen et al. 1996]; Gettysburg [Berg 1980; Root 1988]; Culpeper [Leavy, Froelich, and Abram 1983]; Richmond [Olsen, Schlische, and Gore 1989]; Danville–Dan River [Meyertons 1963; Thayer 1970; Kent and Olsen 1997]; Deep River [Bain and Harvey 1977; Brown et al. 1985]). (*c*) Longitudinal cross sections of the basins shown in *b* (modified from Schlische 1993).

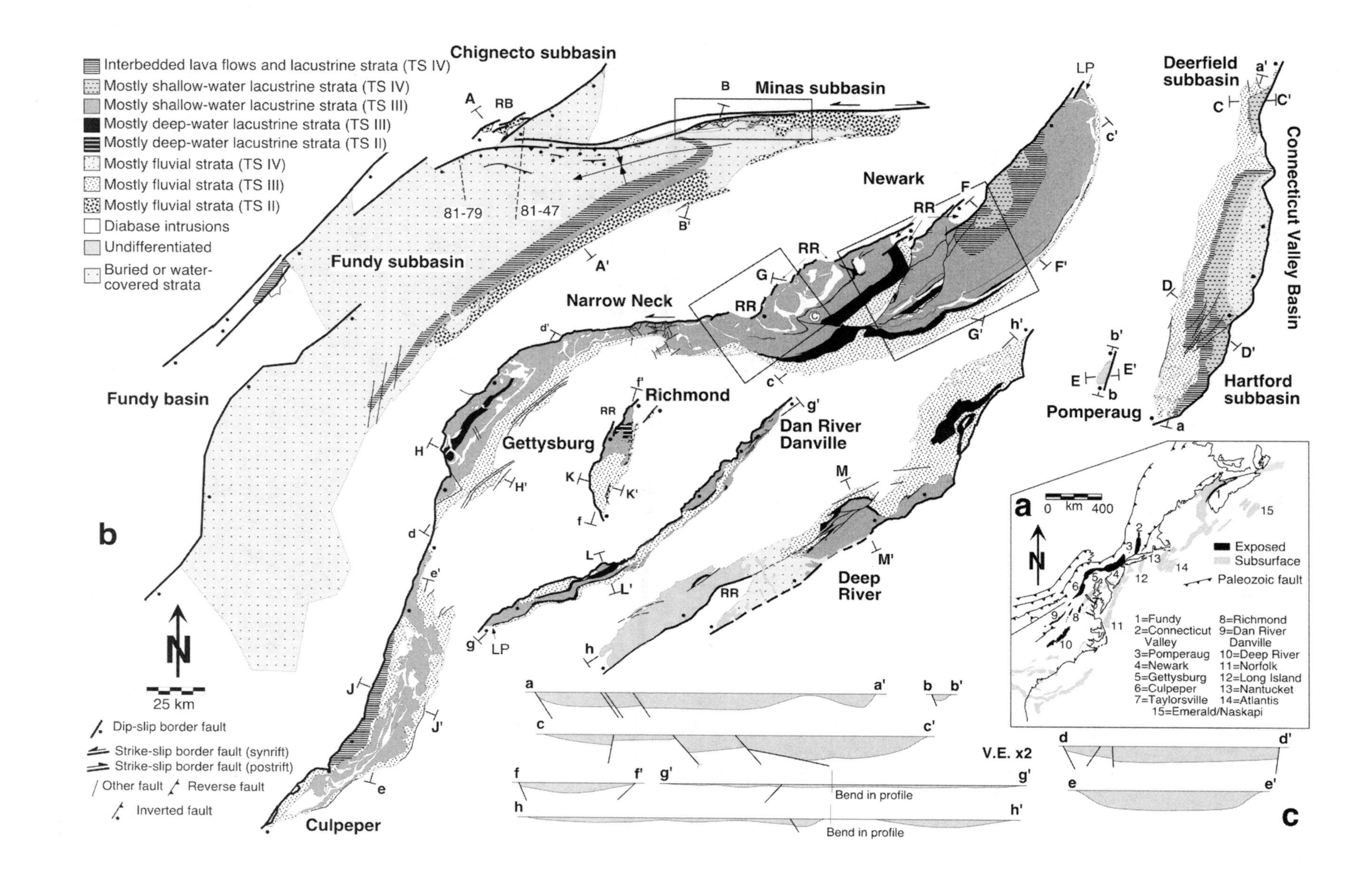

Interbedded lava flows and lacustrine strata (TS IV)
Mostly shallow-water lacustrine strata (TS IV)
Mostly shallow-water lacustrine strata (TS III)
Mostly deep-water lacustrine strata (TS III)
Mostly deep-water lacustrine strata (TS II)
Mostly fluvial strata (TS IV)
Mostly fluvial strata (TS III)
Mostly fluvial strata (TS II)
Diabase intrusions
Undifferentiated
Buried or water-covered strata
Chignecto subbasin
Minas subbasin
Fundy subbasin
Fundy basin
81-79
81-47
Newark
Narrow Neck
Richmond
Gettysburg
Dan River Danville
Deep River
Culpeper
Deerfield subbasin
Connecticut Valley Basin
Hartford subbasin
Pomperaug
b
25 km
N
Dip-slip border fault
Strike-slip border fault (synrift)
Strike-slip border fault (postrift)
Other fault
Reverse fault
Inverted fault
a
0 km 400
Exposed
Subsurface
Paleozoic fault
1=Fundy
2=Connecticut Valley
3=Pomperaug
4=Newark
5=Gettysburg
6=Culpeper
7=Taylorsville
8=Richmond
9=Dan River Danville
10=Deep River
11=Norfolk
12=Long Island
13=Nantucket
14=Atlantis
15=Emerald/Naskapi
V.E. x2
Bend in profile
Bend in profile
c

core obtained by the mining industry from the Fundy basin (Kent and Olsen 2000). The core data are most relevant in integrating stratigraphy with basin analysis, which is addressed later in the chapter.

Dating the activity of geologic structures requires good age control for the affected rock units. Age control in the exposed Mesozoic basins traditionally has been based on biostratigraphy (e.g., Cornet 1977; Cornet and Olsen 1985; Olsen, Schlische, and Gore 1989). In recent years, a more robust chronostratigraphic control has been achieved through isotopic dating of igneous rocks, cyclostratigraphy, and magnetostratigraphy. The age of quartz-normative tholeiitic magmatism in basins from the Culpeper basin to the north (hence referred to as the northern basins) is 202 ± 2 Ma (Sutter 1988; Dunning and Hodych 1990; Hodych and Dunning 1992); this age likely applies to the entire eastern North American magmatic event, including subsurface lava flows in the southeastern United States and diabase dikes of various orientations (Olsen, Schlische, and Gore 1989; Ragland, Cummins, and Arthur 1992; Olsen, Schlische, and Fedosh 1996). Interbasinal correlations based on basalt geochemistry and cyclostratigraphy indicate that the duration of extrusive igneous activity in the Culpeper, Newark, Connecticut Valley, and Fundy basins was less than 600,000 years (Olsen, Schlische, and Fedosh 1996). Using the age of 202 Ma for the lava flows in the Newark basin, Olsen, Schlische, and Fedosh (1996) used core-based cyclostratigraphy to apply absolute ages to the Newark basin stratigraphic section. Cyclostratigraphy also provides data on accumulation rates (discussed in depth later in the chapter) and constrains the ages of magnetic polarity reversals for the Newark basin magnetic polarity section (Kent, Olsen, and Witte 1995). Kent and Olsen (1997) demonstrated that several correlative polarity reversal boundaries are present in the Danville basin; LeTourneau (chapter 3 in volume 2) is developing a magnetostratigraphy for the Taylorsville basin. Interbasinal paleomagnetic correlations with the Newark basin section potentially provide excellent chronostratigraphic control even in strata lacking Milankovitch cycles.

Structural Synthesis

This review of the structural geology of the eastern North American rift system is based mainly on map views and longitudinal cross sections of the exposed basins shown in figure 4.1b and c and on the transverse cross sections presented in figure 4.2. Additional information is available in the syntheses of Olsen, Schlische, and Gore (1989) and Schlische (1990, 1993) and in the references listed in the figure captions for this chapter.

The majority of the rift basins in eastern North America are half-graben, which are asymmetrical fault-bounded basins roughly triangular in cross section (figure 4.2). The more steeply inclined basin margin consists of a network of predominantly normal-slip faults comprising the border-fault system (BFS), which generally trends NNE to NE. The more gently inclined margin is known as the ramping margin and consists of the contact between prerift and synrift rocks. This contact may be disrupted by intrabasinal faults. Both the contact and the overlying synrift strata dip toward the BFS. The dip angle of the synrift strata commonly steepens somewhat in proximity to the BFS; this geometry is known as *reverse drag* or *rollover* (e.g., Hamblin 1965; Gibbs 1983; Dula 1991; Xiao and Suppe 1992; Schlische 1995).

The BFSs are commonly segmented in plan view as well as in cross section. The most common type of plan-view segmentation consists of a number of fault segments with a relay geometry, meaning that the individual faults have the same strike as the fault system. Relay ramps are present between the overlapping fault segments (figures 4.1b and 4.3a); these regions are marked by the prerift-synrift contact that strikes obliquely with respect to the fault segments and that dips in the general direction of the mutual footwall. A less common type of BFS segmentation is best developed in the Hartford subbasin. In this case, longer NNE-striking segments are connected by shorter NE-striking segments; the NE-striking segments are oriented similarly to intrabasinal faults. The relay ramp style of segmentation is an example of soft linkage in that faults transfer displacement from one segment to another without being physically connected, at least in map view; the Hartford subbasin style is an example of hard linkage in that the segments are physically connected (Walsh and Watterson 1991; Davison 1994).

In cross section, the BFS segmentation is expressed as a series of fault riders that bound rider blocks along which the prerift-synrift contact progressively steps down (figures 4.2 and 4.3c). As a result, the deepest

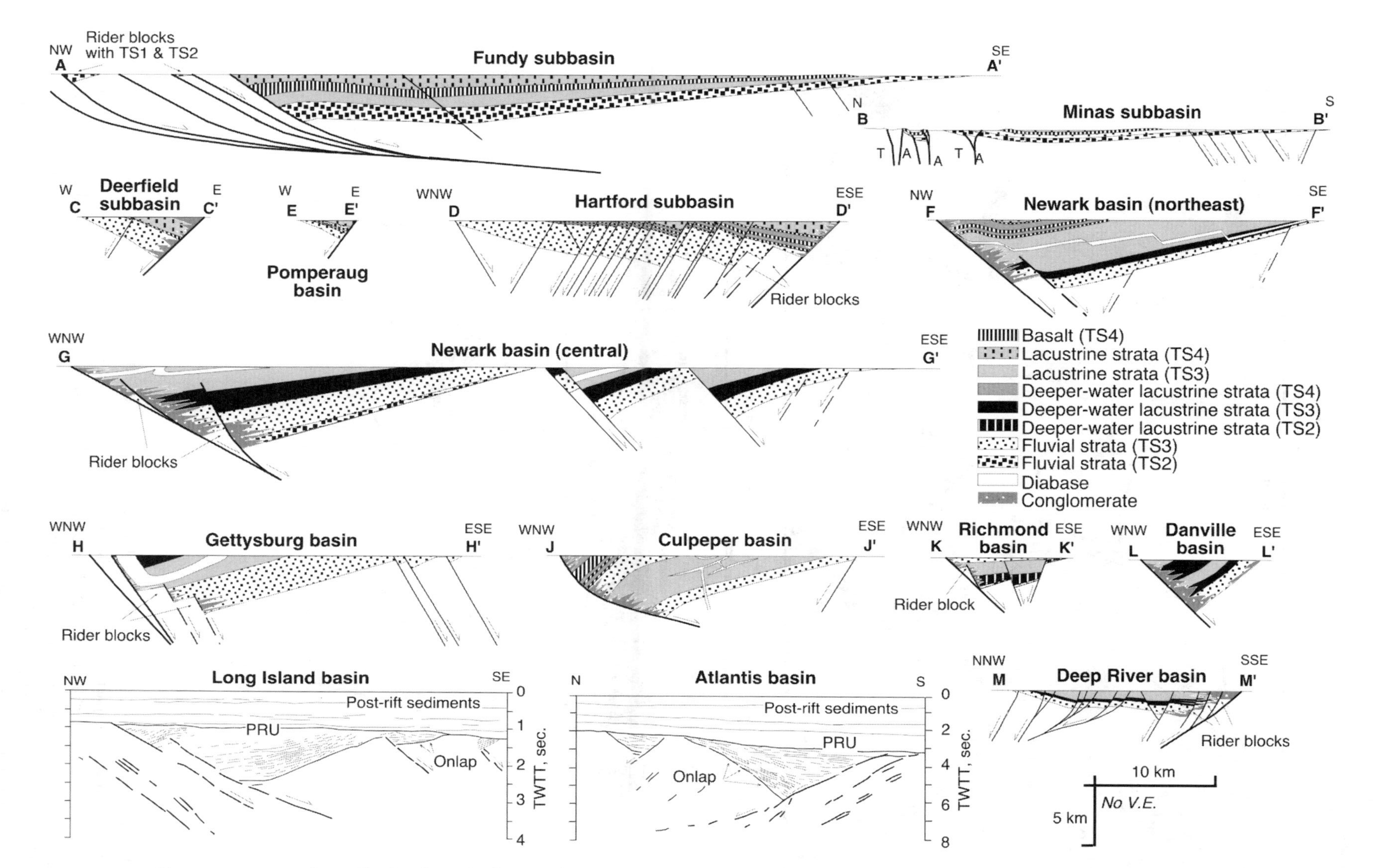

FIGURE 4.2 Transverse cross sections of basins shown in figure 4.1b and line drawings of seismic-reflection profiles across the Long Island and Atlantis basins. For location, see figure 4.1a. (Cross sections modified from Schlische 1993, based on the following additional sources: Fundy [Olsen and Schlische 1990]; Newark [Schlische 1992]; Gettysburg [Root 1988]; Culpeper [seismic profile in Manspeizer et al. 1989]; Deep River [Olsen et al. 1990]. Line drawings reinterpreted from Hutchinson et al. 1986)

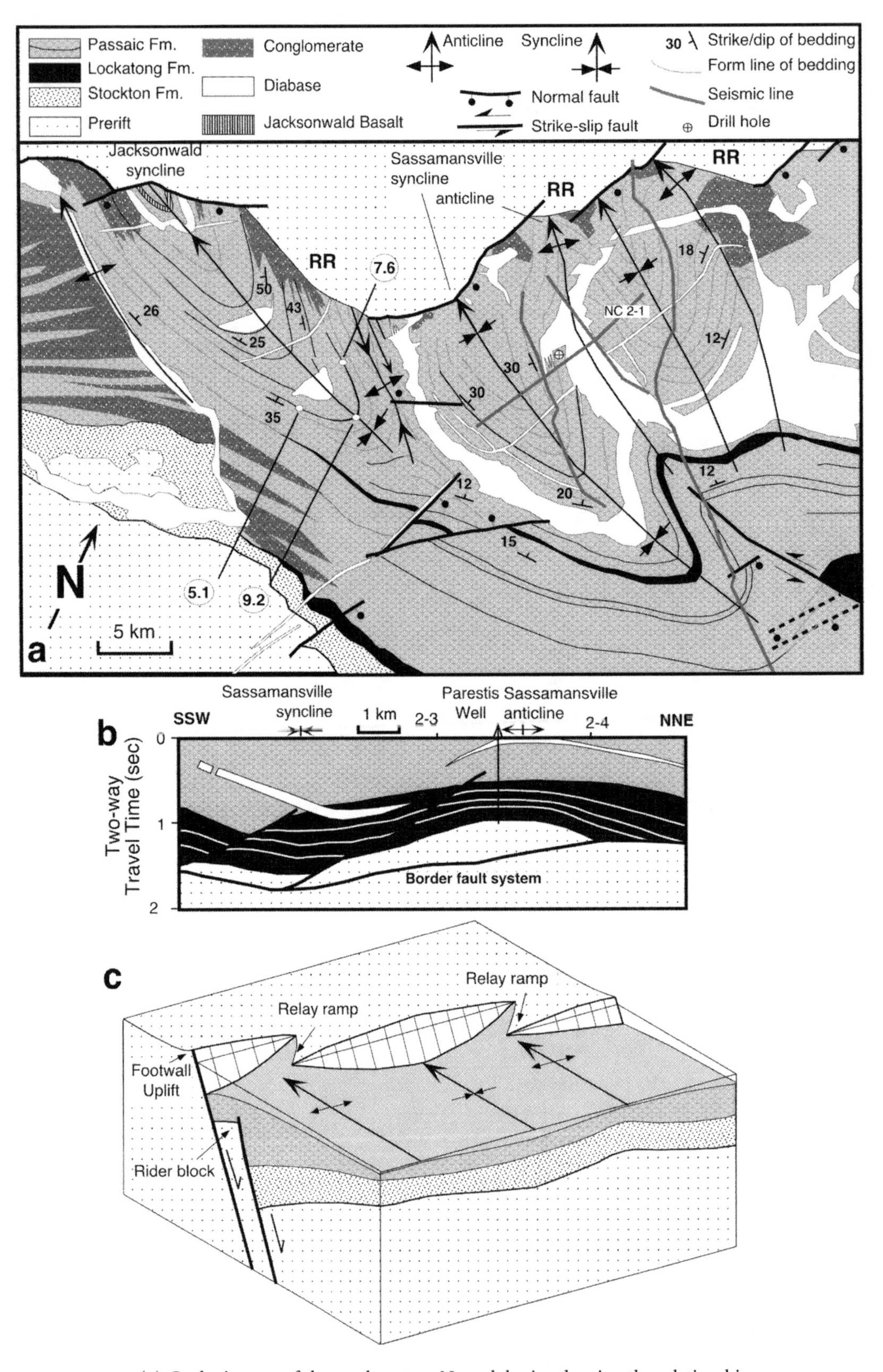

FIGURE 4.3 (*a*) Geologic map of the southwestern Newark basin, showing the relationships among a segmented border-fault system, transverse folds, and relay ramps (*RR*) and the seismic line and drill hole shown in *b* (modified from Schlische 1992 and Reynolds 1994). For location, see figure 4.1b. Circled numbers refer to thicknesses in meters of a marker unit. (*b*) Geologic interpretation of longitudinal seismic line NC 2-1 (modified from Reynolds 1994). (*c*) Block diagram illustrating the geometric relationships among a segmented normal-fault system, relay ramps, rider blocks, and transverse folds (modified from Schlische 1993).

part of the basin is shifted hingeward from the surface trace of the BFS. Faulting appears to have propagated preferentially into the footwall block (Schlische 1992, 1993; Reynolds 1994). This progressive footwall incisement is most likely attributable to gravitational collapse of the uplifted footwall blocks (Gibbs 1984), particularly in cases where the uplift rate exceeded the erosion rate. Footwall uplift for basin-bounding faults (e.g., Zandt and Owens 1980; King and Ellis 1990; Anders and Schlische 1994) occurs in response to absolute upward motion during fault displacement (e.g., Stein and Barrientos 1985), coupled with isostatic unloading of the footwall blocks (Jackson and McKenzie 1983). The gravitational collapse process may have been aided by preexisting fault structures in the footwall block (see the next section). This model for footwall incisement contrasts with the rolling-hinge model (Buck 1988), which predicts hanging-wall incisement. As illustrated in figure 4.3, border-fault segmentation, relay ramps, and rider blocks are related features.

The dip angle of border faults is quite variable (figure 4.2), and the geometry of the border faults at depth is a matter of some controversy. Several workers have inferred that border faults are moderately to strongly listric, soling into detachments at midcrustal depths (e.g., Brown 1986; Bell, Karner, and Steckler 1988; Crespi 1988; de Boer and Clifton 1988; Hutchinson et al. 1988; Manspeizer 1988; Manspeizer and Cousminer 1988; Ressetar and Taylor 1988; Root 1988, 1989; Manspeizer et al. 1989). In many instances, these inferences are based on reverse-drag or rollover stratal geometries, which are not necessarily diagnostic of listric faults (Shelton 1984; Barnett et al. 1987; Gibson, Walsh, and Watterson 1989), or on unmigrated seismic-reflection profiles. Unger (1988) cautioned that unmigrated or time-migrated seismic-reflection profiles over Mesozoic rift basins commonly display a pronounced kink where the basal unconformity intersects the fault trace; this kink comes about because the higher seismic velocities recorded in basement rocks result in a velocity pull-up from the true position of the border fault (see also Withjack and Drickman-Pollock 1984). Seismological studies (Stein and Barrientos 1985; Jackson 1987) indicate that most active normal faults dip at moderate to moderately steep angles and remain essentially planar to the base of the seismogenic crust. For these reasons, the cross sections shown in figure 4.2 depict listric faults only in cases where there is good evidence of listricity based on depth-migrated seismic-reflection profiles.

The map-view geometry of the basins falls into two end-member types:

1. Simple basins—such as the Pomperaug, Newark, Gettysburg, Culpeper, and Richmond basins—are shaped somewhat like brazil nuts, with basin width generally being greatest at or near the center of the basin and decreasing toward the ends of the basin. Longitudinal sections reveal a synclinal or faulted synclinal geometry (figure 4.1c); maximum basin depth occurs at or near the centers of the basins.
2. Composite basins—such as the Connecticut Valley, Danville, and Deep River basins—are composed of multiple subbasins; basin width generally decreases at subbasin boundaries. Longitudinal cross sections reveal relatively broad synclinal subbasins separated by narrower anticlinal features, referred to as intrabasinal highs (Anders and Schlische 1994; Schlische and Anders 1996). All subbasins exhibit a half-graben geometry in transverse cross section. Individual half-graben within a composite basin all exhibit the same polarity (dip direction of BFS).

Intrabasinal faults are overwhelmingly synthetic to the BFS (figure 4.2), in part because this architecture decreases the likelihood that the fault systems will cross at depth, allowing both fault systems to accommodate the extension. In some instances, the intrabasinal faults are oblique to the BFS, suggesting that the orientation of the BFS was controlled by a preexisting weakness, whereas the orientation of the intrabasinal faults—at least in the synrift section—was controlled by the Mesozoic state of stress. Antithetic faults, typically with far less displacement than the synthetics, are also present, particularly along the hinged margin of the basin (figure 4.2); these faults appear to have formed to relieve the bending stresses associated with the regional dip of the hanging wall toward the BFS (see models in Gibbs 1984). The majority of the intrabasinal faults generally strike in a NE direction and have significant dip-slip components. A notable exception is the Chalfont fault in the Newark basin (figure 4.1b, C), which has been proposed to be a major ESE-striking intrabasinal strike-slip fault; it formed as a transfer structure to accommodate differential extension within the basin and formed subparallel to the extension direction (Schlische and Olsen 1988). Diabase dikes generally parallel the basin margin or are slightly oblique to it in the northern basins but are perpendicular or highly oblique to the basin margins in the southern basins (figure 4.4). The significance of the dike trends is discussed in the section on postrift deformation and basin inversion.

Folds occur at a variety of scales in the Mesozoic rifts, are generally fault related, and can be classified as either transverse, with fold axis perpendicular to associated fault, or longitudinal, with fold axis parallel to associated fault (Schlische 1995) (figure 4.5). In all the rifts, the folds are preferentially expressed in the hanging-wall blocks. The largest-scale transverse folds are the basin-scale synclines and subbasin-scale synclines separated by intrabasin highs (anticlines) (figure 4.1c). Smaller transverse folds typically are associated with BFS segmentation at a scale smaller than the basin or subbasin scale (Schlische 1992) (figure 4.3a) as well as with fault-line deflections (Wheeler 1939). These folds decrease in amplitude away from the BFS and do not affect basin width (unlike the basin-scale or subbasin-scale transverse folds). In the case of segmented BFSs, hanging-wall synclines occur at or near the centers of fault segments, whereas hanging-wall anticlines occur at segment boundaries (figure 4.3). In the case of fault-line deflections, synclines form at recesses and anticlines form at salients. Longitudinal folds include rollover folds, reverse-drag folds, drag folds (forced folds or fault-propagation folds), and inversion-related folds (figure 4.5).

The structural features described here apply to those rift basins whose border faults underwent significant dip slip. As further explored in the next section, a different suite of structures developed in basins whose border faults underwent significant strike slip. The nature of slip along the BFS (dip-slip dominated or strike-slip dominated) is a function of the orientation of synrift extension direction with respect to the orientation of reactivated faults.

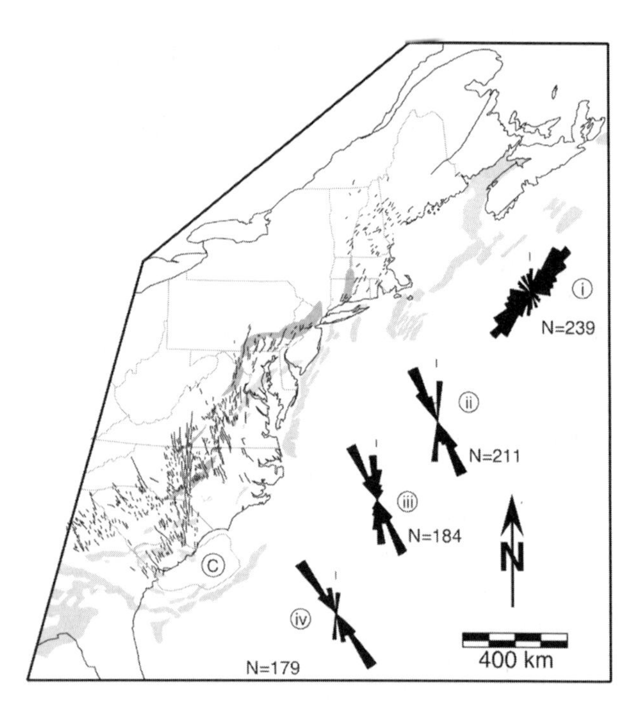

FIGURE 4.4 Early Jurassic age diabase dikes (*thin black lines*) in northeastern North America. Rift basins are stippled; *C*, extent of Clubhouse Crossroads Basalt (Oh et al. 1995). Rosettes indicate dike orientations (small tick marks indicate north) for the following regions: *i*, maritime Canada, New England, and New Jersey; *ii*, Pennsylvania, Maryland, and Virginia; *iii*, North Carolina; and *iv*, South Carolina and Georgia. (Modified from McHone 1988)

Role of Preexisting Structures

Even a cursory examination of the distribution of early Mesozoic basins (figure 4.1) shows that they parallel many of the key structural elements of the Appalachian orogen (e.g., Williams 1978). It is thus no surprise that workers on the Mesozoic basins have long suggested that they were profoundly influenced by preexisting structural controls (e.g., Lindholm 1978). For example, Swanson (1986) summarized the evidence that the border faults of the Danville, Richmond, Taylorsville, Culpeper, Gettysburg, Newark, Connecticut Valley, and Fundy basins were inherited from the Paleozoic orogenies.

Ratcliffe and Burton (1985) presented a generalized fault reactivation model that predicts the nature of slip and the relative amounts of throw on faults oriented at various angles to the extension direction. In cases where the preexisting zone of weakness is oriented perpendicular to the extension direction, the fault is reactivated in pure dip-slip mode, and the amount of throw on the fault is maximized. In cases where the preexisting zone of weakness is oriented parallel to the extension direction, the fault is reactivated in strike-

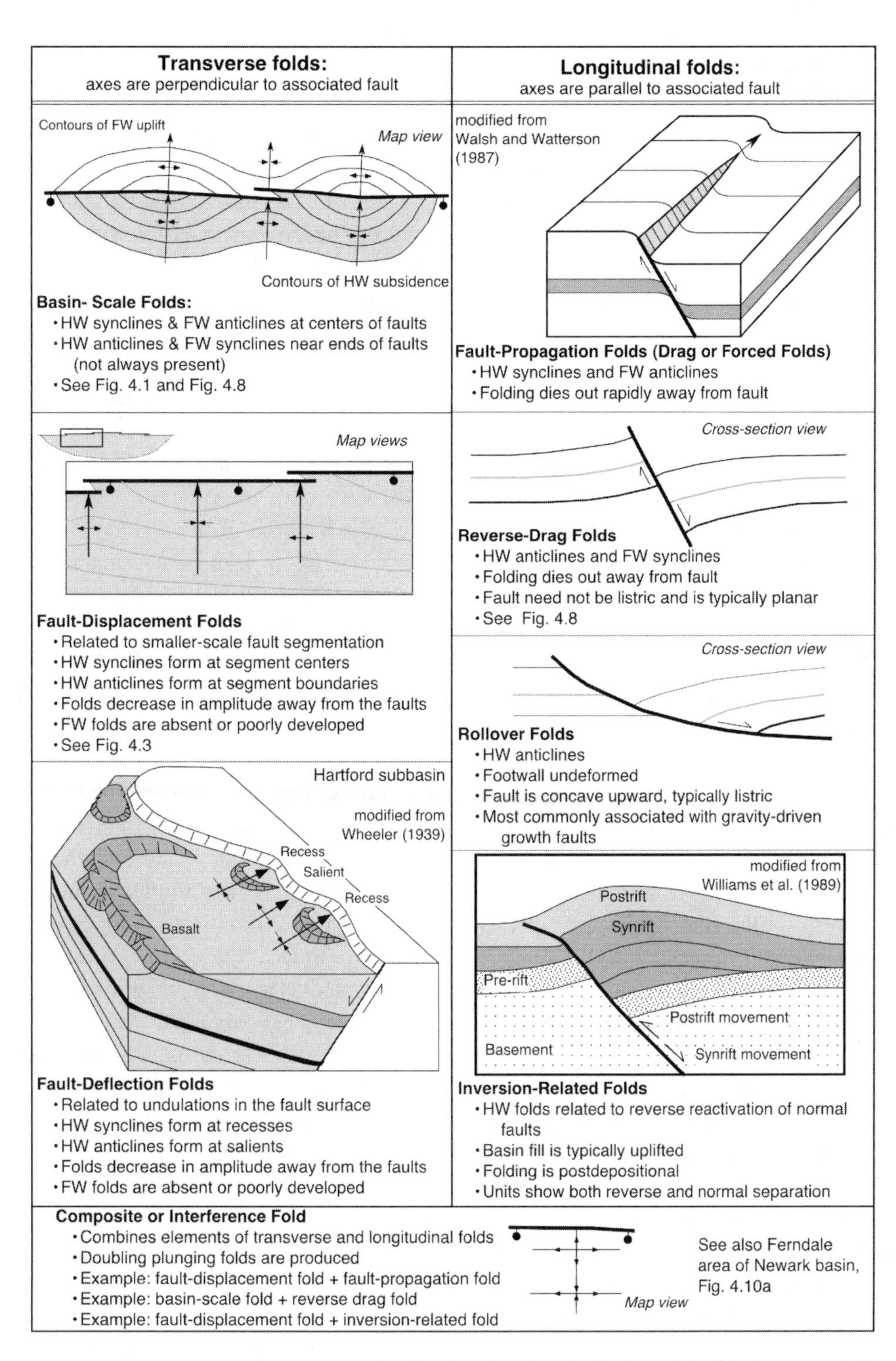

FIGURE 4.5 Geometry, features, and classification (based on Schlische 1995) of fault-related folds in rift basins.

slip mode, and the amount of throw is minimized. Of course, these examples are only two end-member extremes. Schlische and Olsen (1988) noted that the amount of throw is also strongly dependent on the dip angle of the fault, increasing as the dip angle increases, assuming all other factors are equal. After a brief digression on what is meant by extension direction and how it can be constrained, I present two case studies of the Ratcliffe–Burton model of fault reactivation.

Estimating the extension direction is not a trivial task, and it is an undertaking made more difficult through an often ambiguous use of the term "extension direction." When used strictly, "extension" refers to strain. When used in conjunction with rift basins, "extension direction" refers to the direction of maximum elongation, most typically the direction of maximum horizontal elongation. The extension direction is parallel to the minimum principal stress direction (σ_3) only if the deformation is coaxial (e.g., pure shear) or infinitesimal. S_{Hmin} refers to the horizontal minimum stress direction. Most published estimates of the extension direction are actually estimates of σ_3 or S_{Hmin}.

Dikes form perpendicular to σ_3 (Anderson 1942) and are likely the most reliable paleostress indicators for the eastern North American rift system (e.g., McHone 1988; de Boer 1992). If all or most of the dikes in eastern North America were intruded at the same time (~200 Ma), as now seems likely, then σ_3 was clearly oriented differently in the southern basins than in the northern basins (the significance of this difference is discussed later in the chapter). Stress directions also can be estimated using fault orientation and slip direction data (e.g., Angelier 1994). This method is applied most often to very small faults, which are commonly difficult to date, although crosscutting relationships and overprinting of slickensides assist in establishing a relative chronology (e.g., de Boer and Clifton 1988; de Boer 1992). For the northern basins, the estimates of σ_3 based on small normal faults are in good agreement with those based on the diabase dikes (de Boer and Clifton 1988; de Boer 1992). Furthermore, the dikes and the majority of intrabasinal faults are subparallel (Schlische 1993), indicating that the faults formed as predominantly normal-slip faults during the synrift period. However, both the dikes and the intrabasinal normal faults are commonly oblique to BFSs (figure 4.4), suggesting that their orientation is not a reliable indicator of σ_3. Less widely available kinematic indicators include sediment-filled fissures in basalt and clastic dikes, which in the Fundy basin yield extension directions that are in agreement with diabase dikes (Schlische and Ackermann 1995). These kinematic indicators are particularly useful because they can be dated more accurately.

The first case study of the Ratcliffe–Burton model involves the Newark–Gettysburg–Culpeper rift system, which exhibits the same curvature as that of the Pennsylvania reentrant/New York promontory (terminology used in Williams 1978) (figure 4.1a). The boundary faults of the Newark basin (Ratcliffe 1971, 1980; Ratcliffe and Burton 1985; Ratcliffe et al. 1986), the Gettysburg basin (Root 1988, 1989), and the Culpeper basin (Lindholm 1978; Volckmann and Newell 1978) have been interpreted to be reactivated structures. On the basis of the regional orientation of diabase dikes (figure 4.4), intrabasinal normal faults, and the ESE-striking Chalfont transfer fault, the extension direction, or σ_3, is inferred to have been oriented WNW–ESE. The Ratcliffe–Burton model predicts predominantly dip slip on the NNE-striking boundary faults of the Culpeper and Gettysburg basins and mostly dip slip along the generally NE-striking boundary faults of the Newark basin (figure 4.6). Indeed, all three basins are half-graben (figure 4.2). In contrast, the E-striking boundary fault of the Narrow Neck, connecting the Newark and Gettysburg basins, is predicted to have a significant component of left-lateral strike slip. Indeed, the narrow width of the appropriately named Narrow Neck and the reduced stratigraphic thickness compared with that of the dip-slip–dominated basins are consistent with reduced dip slip (especially reduced heave) along the BFS. Border-fault geometry in the Narrow Neck consists of a mosaic of E- and NE-striking faults (figure 4.1b), the latter of which are probably normal faults. The E-striking faults are likely left-lateral or left-oblique-slip faults based on the predictions of the Ratcliffe–Burton model, the mapped left-lateral strike-slip faults at the east end of the Narrow Neck (Lucas, Hull, and Manspeizer 1988), and the WNW-plunging folds that intersect ESE-striking faults at dihedral angles of 10 to 15° (McLaughlin 1963). Transverse folds are absent from the Narrow Neck.

The second case study focuses on the Fundy basin, which consists of three structurally distinct subbasins:

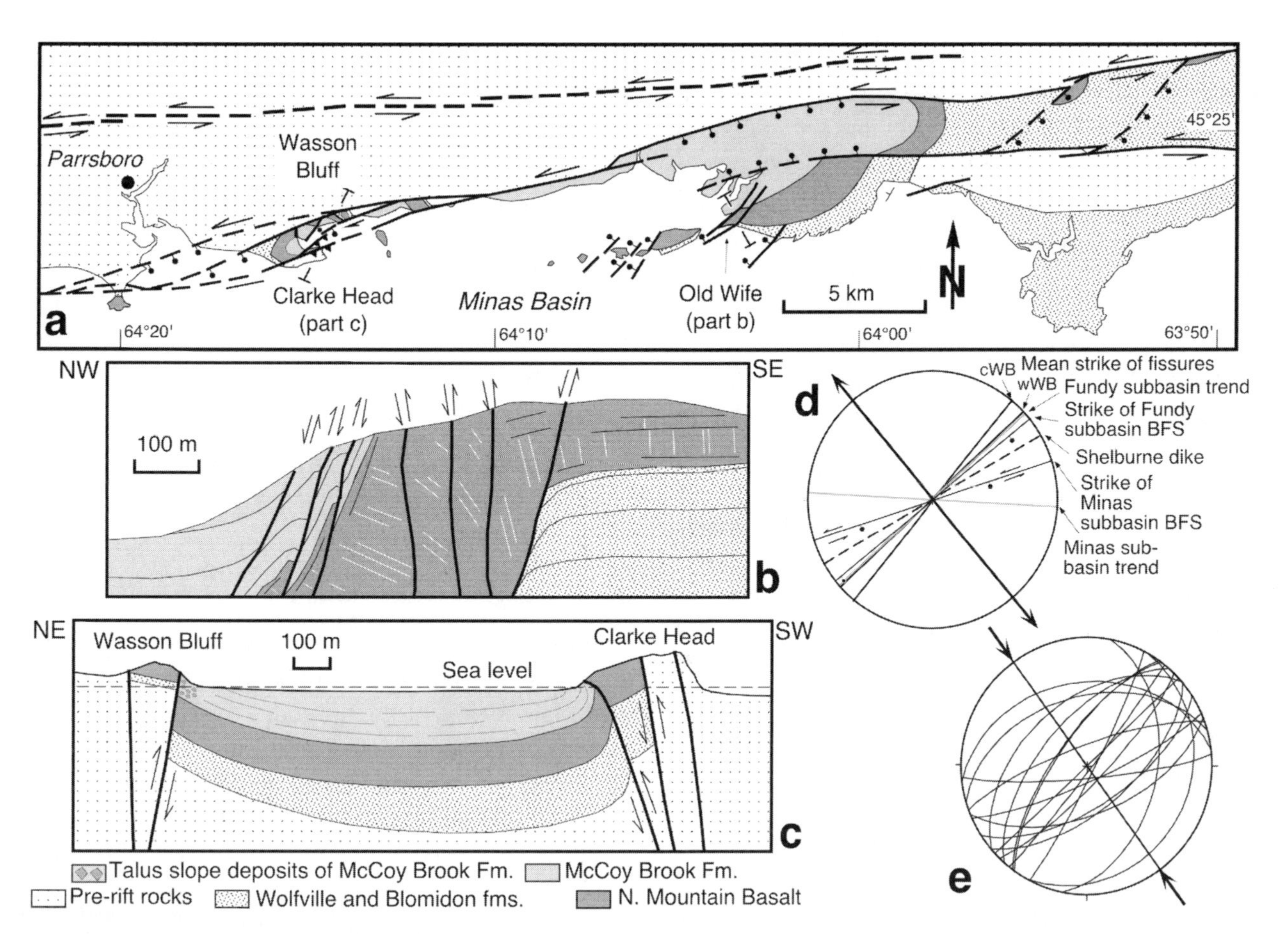

FIGURE 4.6 (*a*) Geologic map of the Five Islands region, northern part of Minas subbasin (modified from Olsen and Schlische 1990). For location, see figure 4.1b. Note the mosaic of dip-slip and strike-slip faults. Only synrift movement sense is shown on faults. Because the pre-Jurassic stratigraphy of the Fundy basin is being revised (Olsen et al. 2000), the Wolfville and Blomidon Formations are mapped as a single unit to avoid confusion between the old and new nomenclatures. (*b*) Geologic cross section through Old Wife Point (for location, see *a*), illustrating localized reactivation of predominantly normal-slip fault zones (modified from Withjack, Olsen, and Schlische 1995). Reverse faults at the northwestern end have associated drag folds and a considerable component of strike-slip movement. White lines in basalt depict cooling joints. (*c*) Geologic cross section from Wasson Bluff to Clarke Head (for location, see *a*) (modified from Withjack, Olsen, and Schlische 1995). The thickening of units toward the Clarke Head fault zone indicates that it likely originated as a NW-dipping normal-fault zone and was rotated and reactivated during inversion. (*d*) Summary of geologic evidence that the extension direction was oriented NW–SE during synrift phase. cWB and wWB refer to central and western Wasson Bluff, where the orientations of sediment-filled fissures in the North Mountain Basalt (for location, see *a*) were measured (Schlische and Ackermann 1995). (*e*) Equal-area projection, showing the attitude of faults with reverse separation and associated right-oblique-slip faults from the northern margin of the Minas subbasin (from Withjack, Olsen, and Schlische 1995). The shortening direction is normal to the mean strike of the faults.

the NE-trending Fundy and Chignecto subbasins and the E-trending Minas subbasin (figure 4.1b). The boundary faults of all subbasins are reactivated faults (Keppie 1982; Plint and van de Poll 1984; Mawer and White 1987; Withjack, Olsen, and Schlische 1995). Based on the attitude of the Shelburne dike, sediment-filled fissures in basalt, and clastic dikes in the Blomidon Formation (Olsen and Schlische 1990; Schlische and Ackermann 1995) (figure 4.6d), the extension direction, or σ_3, was oriented NW–SE. The Ratcliffe–Burton model therefore predicts predominantly synrift dip slip along the border faults of the Fundy and Chignecto basins and synrift left-oblique slip along the boundary fault (Minas fault zone) of the Minas subbasin. Olsen and Schlische (1990) verified this prediction by assembling data that showed that (1) the Fundy and Chignecto basins are half-graben, are wider, and contain a much thicker stratigraphic section than the

Minas subbasin; (2) the northern margin of the Minas subbasin is an extensional strike-slip duplex made up of a network of E-striking left-oblique-slip faults and NE-striking normal faults bounding a series of syndepositional graben and half-graben (figure 4.6a); and (3) dip-slip faulting and strike-slip faulting were coeval. The fault pattern along the northern margin of the Minas subbasin is also remarkably similar to that produced in clay models of oblique extension (Schlische, Withjack, and Eisenstadt 2002). Schlische (1990) suggested that the Minas fault zone is linked physically to and transfers its displacement to the BFS of the Fundy subbasin and thus marks the boundary between the Fundy and Chignecto subbasins, a scenario largely confirmed by Withjack, Olsen, and Schlische (1995).

The Paleozoic fabric of eastern North America consists largely of generally NE-striking thrust faults and E-striking strike-slip faults (e.g., Arthaud and Matte 1977). The NE-striking faults were reactivated as primarily dip-slip faults during Mesozoic extension and are associated with half-graben–type sedimentary basins, segmented border faults, rider blocks, relay ramps, and transverse folds. The E-striking faults were reactivated as left-oblique-slip faults during Mesozoic extension and are associated with complex mosaics of dip-slip and strike-slip faults, an absence of transverse folds, and reduced basin width and depth compared with dip-slip–dominated basins. A large suite of quite different structures therefore can form, depending on the orientation of the reactivated fault with respect to the extension direction (Olsen and Schlische 1990).

Fault reactivation likely accounts for the lack of half-graben polarity reversals within the eastern North American rift system. As noted earlier, individual half-graben units within composite basins (Connecticut Valley and Deep River basins) or rift basins within a larger rift zone (Newark–Gettysburg–Culpeper rift zone) do not alternate asymmetry along strike, as is characteristic of the East African rift system (Rosendahl 1987). This notable difference stems from the localization of North American basins along preexisting structures that generally dip in the same direction over large areas (Burgess et al. 1988; Reynolds and Schlische 1989). Adjacent half-graben in eastern North America generally are not linked by accommodation zones, which are a common occurrence in East Africa (see also Morley 1995). The larger number of accommodation zones in the East African rift system may be related to the closer spacing of and more frequent polarity reversals between adjacent half-graben in East Africa (Schlische 1993). Some type of accommodation structure transferring displacement from one boundary fault to another is required between two oppositely dipping, closely overlapping BFS segments. Large-scale hard-linkage structures are not required where half-graben are sufficiently widely spaced, border faults dip in the same direction, and displacement goes to zero at the fault tips—a feature discussed in greater detail in the next section.

Application of Fault-Population Studies

The past decade has experienced an explosion of research in fault-population studies (e.g., Cowie, Knipe, and Main 1996). In this section, I focus on those aspects of fault-population studies most relevant to the structural geology and basin evolution of rifts: fault-displacement geometry, length-displacement scaling, and fault segmentation. Additional aspects of this topic are explored in chapter 8 of this volume.

Features of Normal Faults

Fault displacement is greatest at or near the center of blind, isolated faults (figure 4.7a) and decreases to zero at the fault tips (Chapman, Lippard, and Martyn 1978; Muraoka and Kamata 1983; Barnett et al. 1987; Walsh and Watterson 1987; Dawers, Anders, and Scholz 1993) (figure 4.7b–d). For an initially horizontal surface, this displacement geometry results in a synclinal fold in the hanging-wall block (figure 4.7d) and in an anticlinal fold in the footwall block. For any given point on the fault surface, displacement is greatest at the fault surface and decreases away from the fault, resulting in a reverse-drag geometry in both the hanging wall and the footwall (Barnett et al. 1987) (figure 4.7c). The reverse-drag geometry is in most instances associated with planar normal faults, and thus Barnett et al. (1987) and Gibson, Walsh, and Watterson (1989) cautioned that reverse drag does not necessarily indicate the presence of listric normal faults, as is commonly assumed in many two-dimensional fault kinematic models (e.g., Gibbs 1983; Dula 1991). The displacement geometries described here are character-

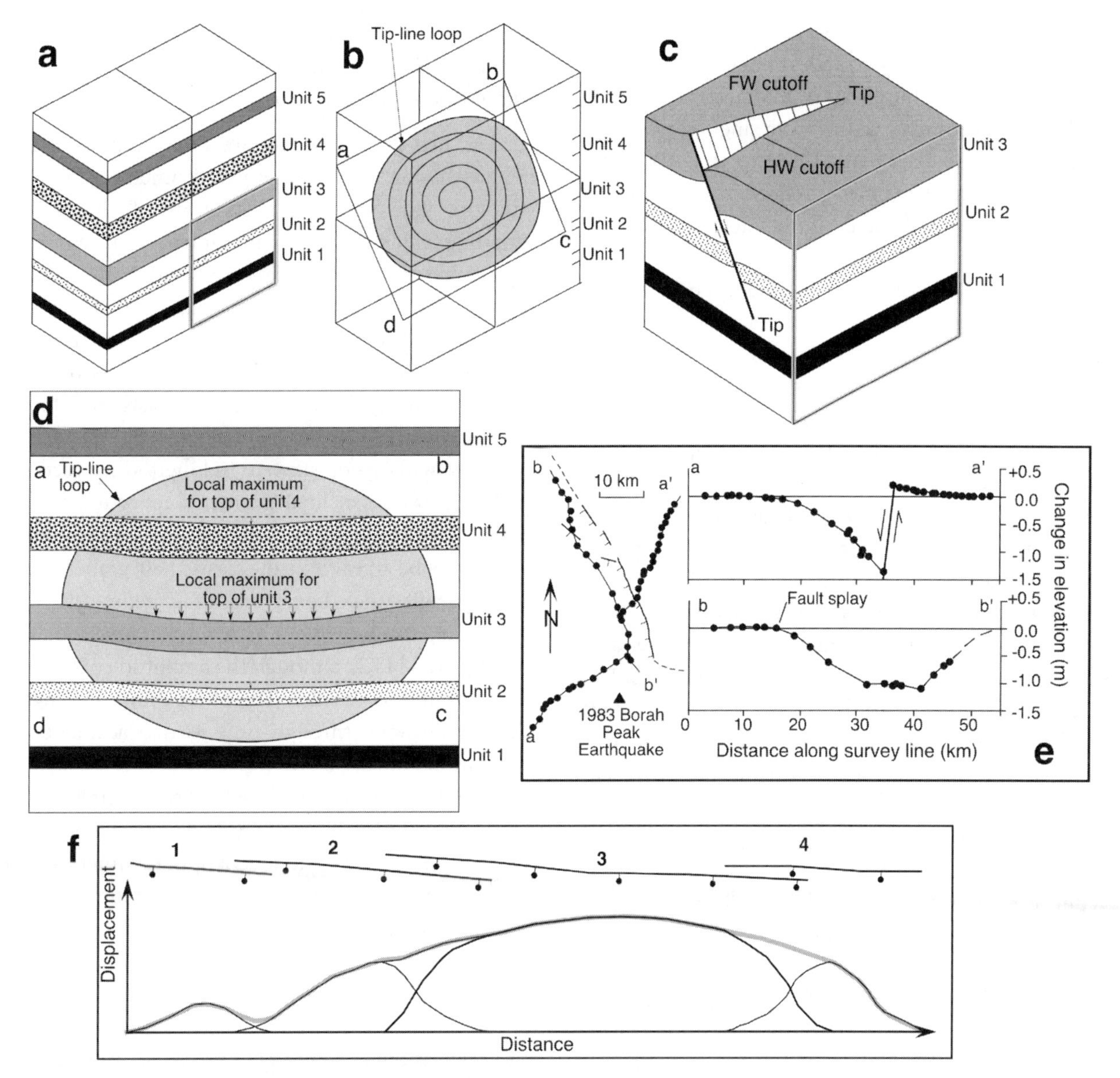

FIGURE 4.7 Geometric features of normal-fault systems. (*a*) Block with five marker units containing a blind normal fault. (*b*) "Transparent" version of block in *a*, showing the fault surface (*shaded*). The tip-line loop separates faulted rocks from unfaulted rocks and corresponds to the zero-displacement contour. Displacement increases everywhere from the tip-line loop toward the center of the fault. (*c*) Block diagram of the lower-right quarter of the block in *a*. The top surface of unit 3 is cut by the fault. The distance between the footwall and hanging-wall cutoffs decreases toward the fault tip. In the vertical cross section perpendicular to the fault, displacement at the fault surface decreases toward the tip. Displacement also decreases away from the fault, producing reverse-drag folds in both the hanging-wall and footwall blocks. (*d*) Fault-surface projection of the fault in *b*, showing the tip-line loop and the hanging-wall cutoffs of units 1 to 5. Along-strike variations in displacement produce hanging-wall synclines (shown) and footwall anticlines (not shown), both of which are fault-displacement folds. Parts *c* and *d* are at the same scale and are shown twice the size of *a* and *b;* no absolute scale is shown because these features are scale invariant. (*e*) Changes in ground-surface elevation following the 1983 Borah Peak earthquake on the Lost River normal fault, Idaho (epicenter marked by triangle) (modified from Barrientos, Stein, and Ward 1987). Black circles indicate survey points; tick marks indicate parts of the fault that developed prominent scarps following the earthquake. Note the well-developed reverse-drag geometry in section a-a′; b-b′ shows a hanging-wall fault-displacement syncline. (*f*) Geologic map (*top*) and displacement-distance plot (*bottom*) of a segmented fault system (inspired by data in Dawers and Anders 1995). The shaded line reflects the summed displacement profile. Segments 2, 3, and 4 sum to a smooth profile, indicating that they are linked kinematically. A displacement deficit is evident for segments 1 and 2, which are incompletely linked. No scale is shown because these features are scale invariant.

istic of normal faults over scales ranging from centimeters to tens and hundreds of kilometers. A similar displacement geometry results from single-slip seismic events (Stein and Barrientos 1985) (figure 4.7e). The finite displacement geometry reflects the cumulative effect of multiple slip events (King, Stein, and Rundle 1988).

It is generally agreed that there is a positive relationship between maximum displacement, D, and fault length, L, of the form $D = cL^n$, where c is related to rock properties and n is the scaling exponent (e.g., Watterson 1986). Although the value of n has been controversial in the literature (compare Walsh and Watterson 1988; Marrett and Allmendinger 1991; Cowie and Scholz 1992; Gillespie, Walsh, and Watterson 1992), high-quality data sets that span a sufficient scale range indicate that $n = 1$; that is, the relationship between D and L is linear (e.g., Dawers, Anders, and Scholz 1993; Schlische et al. 1996; see also Ackermann et al., chapter 8 in this volume). The length-displacement scaling relation indicates that small faults with small displacements grow into large faults with large displacements during successive slip events.

Normal-fault systems at a variety of scales commonly are segmented (e.g., Larsen 1988; Jackson and White 1989; Peacock and Sanderson 1991; Gawthorpe and Hurst 1993; Anders and Schlische 1994; Faulds and Varga 1998) (figure 4.7f). Segment boundaries may be recognized by significant changes in fault strike, fault overlaps and offsets, reduced displacement (especially in the hanging-wall block), and differences in the age of faulting across the segment boundary (e.g., Zhang, Slemmons, and Mao 1991). Segmented faults typically are linked kinematically (Walsh and Watterson 1991): summing the displacements of individual faults within the system results in a cumulative profile that resembles that of an isolated fault (e.g., Trudgill and Cartwright 1994; Dawers and Anders 1995), although displacement deficits may indicate regions of incomplete linkage (Dawers and Anders 1995) (figure 4.7f). At least during the early stages of kinematic linkage, relay ramps form between closely spaced fault segments. These relay ramps may become breached during continued fault-segment propagation (Peacock and Sanderson 1991; Childs, Watterson, and Walsh 1995).

Effect on Basin Geometry and Associated Structures

Because half-graben basins are fault-bounded sedimentary basins, the displacement geometry associated with the faults and the evolution of the basin-bounding fault system strongly control basin geometry (e.g., Schlische 1991). The fault- and basin-growth models shown in figure 4.8 are based on the three-dimensional displacement field associated with normal faults, the *L-D* scaling relation, and aspects of fault segmentation. The simplest case consists of a basin bounded by single border fault (figure 4.8a). The basin itself is roughly scoop shaped. In longitudinal section, the basin has a synclinal geometry reflecting along-strike variations in fault displacement. In transverse section, the basin exhibits the classic half-graben morphology reflecting the reverse-drag geometry. The (uneroded) footwall block is a mirror image of the hanging-wall block, although the amplitude of the uplift is commonly smaller. The basin grows in length, width, and depth through time as displacement accrues on the border fault.

Basins bounded by multiple fault segments are considerably more complex (figure 4.8b–d). All begin as isolated subbasins. For nonoverlapping synthetic faults (figure 4.8b, Case 1) and closely overlapping synthetic faults (figure 4.8c, Case 2), the subbasins eventually merge. In the former case, the intrabasin high that separates the subbasins eventually subsides. In the latter case, the intrabasin high persists because fault displacement is divided among multiple, active fault segments that distribute basin subsidence (Anders and Schlische 1994) (see transverse cross sections in figure 4.8c). For widely overlapping fault segments (figure 4.8d, Case 3), the subbasins may never merge. The Connecticut Valley basin is an example of a Case 1 basin (figure 4.8b) in which faulting stopped prior to complete linkage. The Deep River basin consists of both Case 1 and Case 2 subbasins. All the other half-graben–type basins appear to be "simple" basins bounded by a single border fault or by a fault that effectively acts like a single fault, although it may be composed of several geometric segments. The gross geometry of these basins does not indicate whether they grew from a single border fault (figure 4.8a) or through the consolidation and complete merger of

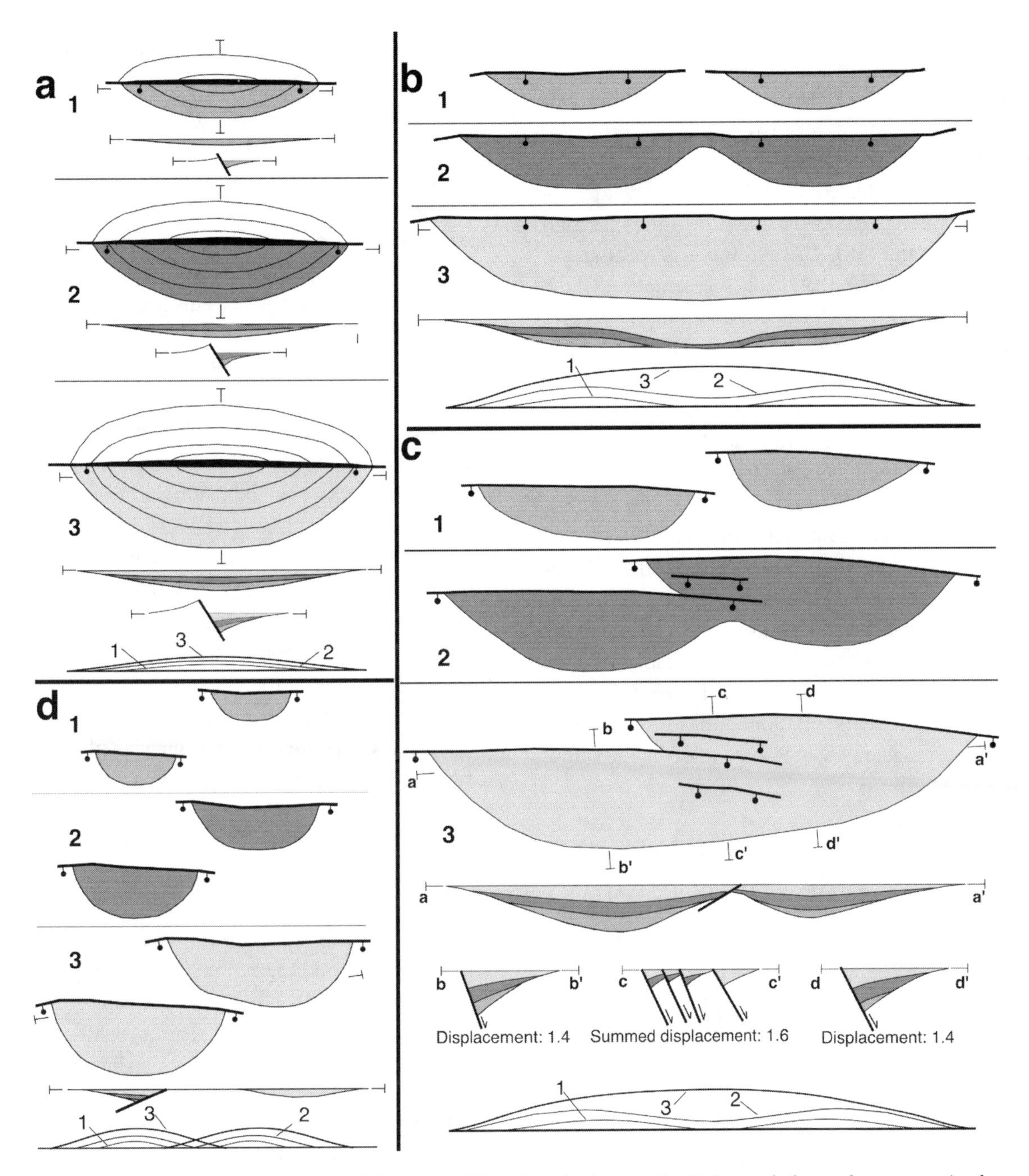

FIGURE 4.8 Evolution of extensional basins resulting from basin growth. Each panel shows three stages in the evolution of an isolated basin or subbasin system. Normal faults are indicated by bar-and-ball symbols. The lower diagram in each panel shows the longitudinal geometry of footwall uplift after each of the three stages. (*a*) Single fault (thin lines represent contours of footwall uplift and hanging-wall subsidence; these lines are omitted for simplicity in the other parts of this figure). (*b*) Nonoverlapping synthetic fault segments. (*c*) Closely spaced overlapping synthetic fault segments. (*d*) Widely spaced overlapping synthetic fault segments. (Modified from Schlische 1995 and Schlische and Anders 1996)

subbasins (figure 4.8b). The currently segmented nature of many of the border faults suggests the latter. Stratal geometry in longitudinal section also may be of help (cf. figure 4.8a and b, Stage 3).

Fault-displacement geometry, growth, and segmentation also account for nearly all the folds present in rift basins (Schlische 1995) (figure 4.5). The largest-scale transverse folds are the basin-scale synclines and subbasin-scale synclines separated by anticlinal highs (figures 4.1c and 4.8). Other transverse folds are associated with smaller-scale fault segmentation (figure 4.3), with hanging-wall synclines forming in areas of locally higher fault displacement (segment centers and recesses), and with hanging-wall anticlines forming in areas of locally lower fault displacement (segment boundaries and salients). Reverse-drag folds are longitudinal folds that reflect the decrease in displacement away from the fault surface. Drag folds are longitudinal folds that form as a consequence of fault propagation into the monoclinally flexed strata commonly surrounding the fault tips. These folds thus more properly are called fault-propagation folds and commonly are superimposed on the much longer wavelength reverse-drag folds. The majority of the fault-related folds formed during sedimentation and thus controlled thickness and facies distributions, which is the subject of the next section.

Integration of Stratigraphy with Structural Geology and Basin Analysis

The first-order control on basin geometry is the deformation field associated with the basin-bounding fault system. In turn, the architecture of the basin fill is strongly influenced by basin geometry. For example, as shown in figure 4.8, the basin fill thickens toward the centers of subbasins; lacustrine facies would be expected to deepen in a similar manner. These relationships suggest that the thickness and facies patterns in the basin fill can be used to infer aspects of basin geometry. Furthermore, they can be used to determine whether or not faults (and folds) formed syndepositionally. This section explores these topics as well as the significance of vertical facies transitions for the tectonostratigraphic evolution of extensional basins.

Syndepositional Faulting and Folding

Evidence of syndepositional border faulting includes (1) alluvial-fan conglomerates deposited adjacent to the border fault; (2) an increase in the thickness of strata in the hanging wall as the normal fault is approached with thinning toward the fault in the footwall (assuming sedimentation within the footwall block); (3) evidence of growth (i.e., decreasing amounts of fault displacement in younger rocks); and (4) a progressive decrease in dip of younger strata (figure 4.9a). Syndepositional intrabasinal faulting is more difficult to assess. For example, as shown in figure 4.9a, strata thicken toward fault B in the hanging-wall block and thin toward fault B in the footwall block, but these thickness changes are exclusively attributable to syndepositional movement on the border fault. Fault A, on the contrary, is a syndepositional intrabasinal fault because there are dramatic changes in thickness of a given unit across the fault, and thickening rates change across the intrabasinal fault. Schlische (1990, 1993) reviewed the evidence for syndepositional border faulting in the major exposed rift basins. In this section, I focus on the evidence for syndepositional faulting and folding in the Newark basin.

Conglomerates are present along virtually the entire length of the BFS of the Newark basin and occur in all sedimentary formations; conglomerates are also inferred to be present adjacent to the BFS at depth on the basis of reflection characteristics on the NORPAC seismic line (Schlische 1992). Although the pattern of stratal dip is complex as a result of intrabasinal faulting, transverse and longitudinal folding, and basin inversion, there are some structurally simple areas—for example, the Delaware River region (Schlische 1992) (figure 4.8)—that exhibit decreasing dips in progressively younger strata.

The NBCP cores also provide information on regional variations in thickness and facies. The offset coring technique (Olsen et al. 1996) produced overlap sections for six pairs of stratigraphically adjacent cores (figure 4.10a); correlations are based on cyclostratigraphy (Olsen et al. 1996) and magnetostratigraphy (Kent, Olsen, and Witte 1995). The overlap section of the Rutgers and Somerset cores, for example, shows that equivalent units thicken by approximately 12% between the Rutgers and Somerset sites (figure 4.10b).

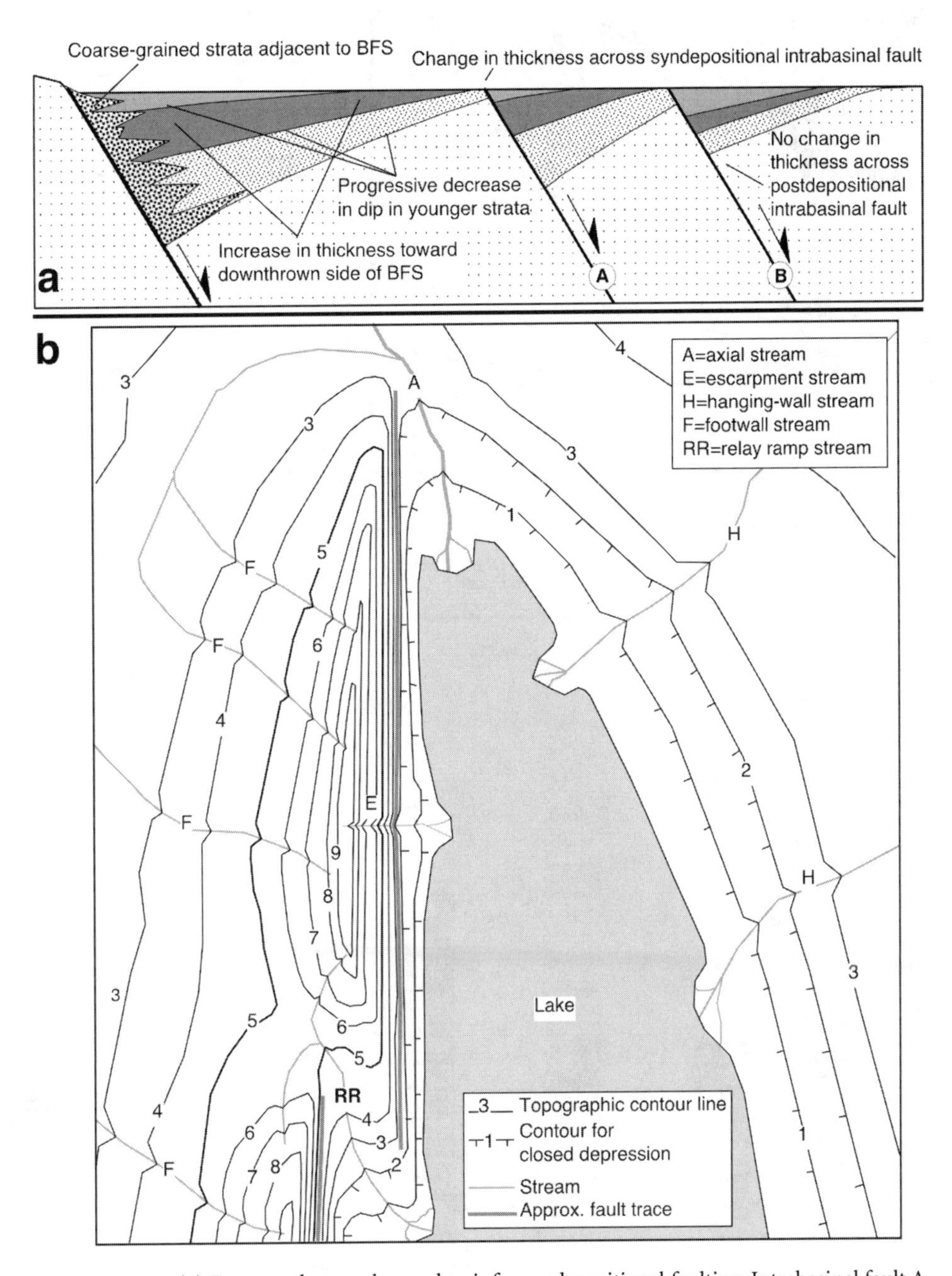

FIGURE 4.9 (*a*) Features that can be used to infer syndepositional faulting. Intrabasinal fault A is syndepositional; fault B is postdepositional. The border fault is syndepositional. (*b*) Highly schematic diagram of the topography associated with the northern end of a basin-bounding fault system. The stream patterns are based on data from the East African rifts (Cohen 1990). For further discussion, see the text. The topographic contours are inspired by surface-deflection profiles measured by Gupta and Scholz (1998) on small normal faults from the Danville basin (see also Schlische et al. 1996 and Ackermann et al., chapter 8 in this volume).

Furthermore, lacustrine facies are deeper in the Somerset core than in the Rutgers core. All overlap sections are thicker in the core located closer to the center of the basin or to the BFS or to the Hopewell intrabasinal fault system. Syndepositional faulting is therefore the most likely cause for these variations in thickness and facies.

Stratigraphic intervals recovered in the NBCP cores may also be correlated to outcrop sections (Silvestri 1994, 1997; Olsen et al. 1996). For example, the Perkasie

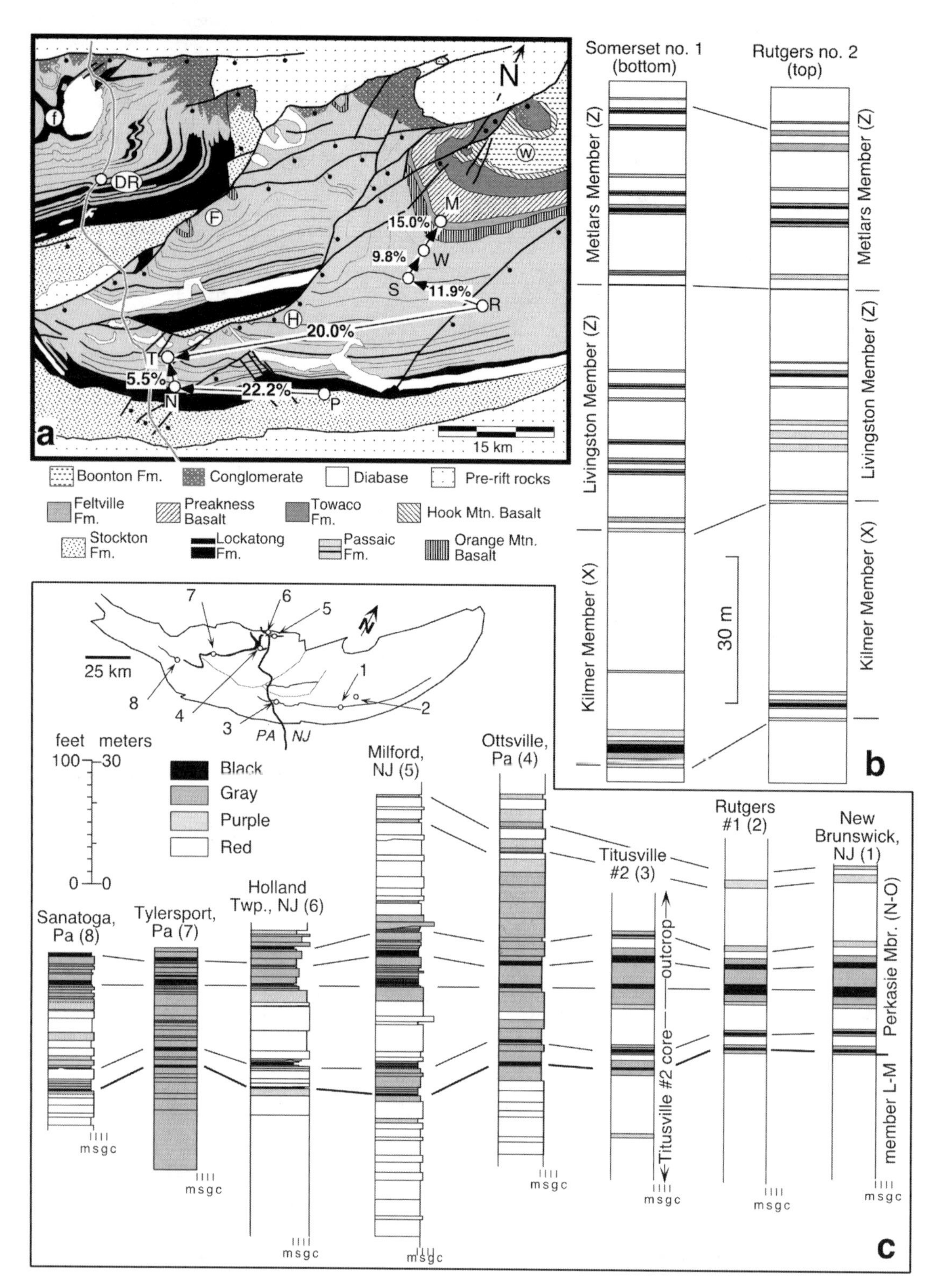

FIGURE 4.10 (*a*) Geologic map of the north-central Newark basin, showing the locations of the seven NBCP drill sites: *M*, Martinsville; *N*, Nursery; *P*, Princeton; *R*, Rutgers; *S*, Somerset; *T*, Titusville; *W*, Weston. Arrows indicate the amount of thickening between overlap sections of stratigraphically adjacent cores, based on correlations in Olsen et al. (1996); *DR*, Delaware River outcrops equivalent to units in overlap section of the Nursery and Titusville cores; *F*, Flemington fault; *f*, Ferndale folds; *H*, Hopewell fault; *w*, Watchung syncline. (*b*) Overlap section of the Rutgers and Somerset cores, showing a pronounced increase in thickness and proportion of deeper-water mudstones (*gray and black units*; for legend, see *c*) from Rutgers to Somerset (based on data in Olsen et al. 1996). (*c*) Basinwide correlation of the Perkasie Member, showing variations in thickness and facies (modified from Olsen et al. 1996).

Member, which was recovered in the bottom of the Rutgers core and in the top of the Titusville core, has been correlated to outcrop sections that have been traced for more than 125 km across the Newark basin (figure 4.10c). The units generally thicken toward the center of the basin (Rutgers core to Titusville core) and from the hinged margin toward the BFS (e.g., Titusville core to Milford section). The variations in thickness suggest that the Newark basin is a plunging syncline in longitudinal section; this configuration is consistent with syndepositional BFS displacement being highest near its center and decreasing toward its lateral ends (Schlische 1992). These interpretations also are supported by the marked increase in thickness of the Lockatong Formation toward the center of the basin (figure 4.1b).

Core-to-outcrop correlations provide constraints on the timing of activity of some of the intrabasinal faults. Different thickening rates in different fault blocks led Schlische (1992) to conclude that the Hopewell and Flemington fault systems were syndepositionally active. In addition, Jones (1994) and Schlische (1995) showed that some transverse folds in the hanging walls of the Hopewell and Flemington fault systems were syndepositionally active; if the folds are fault related, then the intrabasinal fault systems were syndepositionally active.

By far the best evidence for syndepositional folding along the BFS comes from the southwestern part of the Newark basin. Outcrop, drill-hole, and seismic-reflection data (Schlische 1992, 1995; Reynolds 1994) show that units thicken toward synclinal hinges and thin toward anticlinal hinges (figure 4.3a and b). In addition, the concordant diabase intrusions restricted to the hinge of the Jacksonwald syncline (figure 4.3a) are interpreted as phacoliths (Manspeizer 1988; Schlische and Olsen 1988), which suggests that intrusion occurred during or after at least some folding. Thus transverse folds from the scale of the Newark basin itself to the Jacksonwald syncline formed syndepositionally. Geometrically similar transverse folds in other basins also likely formed syndepositionally.

Onlap

Hanging-wall onlap refers to a stratal geometry in which progressively younger synrift strata progressively onlap prerift rocks of the hanging-wall block (Schlische and Olsen 1990; Schlische 1991). Transverse onlap is observed in sections normal to the BFS, whereas longitudinal onlap is observed in sections parallel to the BFS. Hanging-wall onlap is readily produced in basin-growth models in which the basin is completely filled to its lowest outlet following each increment of extension (figure 4.11c, stage 4). Both types of onlap should be readily observable on seismic lines (but see Morley 1999), but they are difficult to detect in map view. Longitudinal pinchout (figure 4.8) is a map-view manifestation of longitudinal onlap. Transverse onlap has been observed in outcrop in the Fundy basin (figure 4.11b) and on a transverse seismic line across the Newark basin, and it also has been inferred from drill-hole data in the Richmond basin (figure 4.11a) as well as on seismic lines across the Atlantis, Long Island, and Nantucket basins (Schlische and Olsen 1990) (figure 4.2). Longitudinal onlap has been inferred from longitudinal pinchout in the Newark and Danville basins (Schlische 1993) (figure 4.1b). Hanging-wall onlap is the primary stratigraphic evidence that the basins have grown wider and longer through time (Leeder and Gawthorpe 1987; Gibson, Walsh, and Watterson 1989; Schlische and Olsen 1990; Schlische 1991, 1993). Internal onlap refers to a stratal geometry in which younger synrift strata onlap older synrift strata; this phenomenon is discussed later in the chapter in relation to tectonostratigraphic sequences (Olsen 1997).

Hanging-wall onlap has important implications for estimates of the age of the synrift deposits. Although the rift basins of eastern North America are erosionally dissected, the oldest synrift deposits are not exposed at the surface (figure 4.11c, stage 4). Thus age estimates based exclusively on surface outcrops underestimate the age of the oldest synrift deposits.

Vertical Transitions and Basin-Filling Models

The typical Triassic stratigraphy (Olsen, Schlische, and Gore 1989; Schlische and Olsen 1990; Olsen 1997) consists of a basal fluvial unit overlain by lacustrine strata, with the deepest lakes occurring near the base of the lacustrine succession and then generally shoaling upward (figure 4.12). In some basins, the shallow-water lacustrine deposits are capped by fluvial strata. This stratigraphic architecture is referred to as a *tripartite stratigraphy.* In the northern basins, this Triassic tripartite sequence is overlain by an Early Jurassic age package of lava flows and intercalated lacustrine (com-

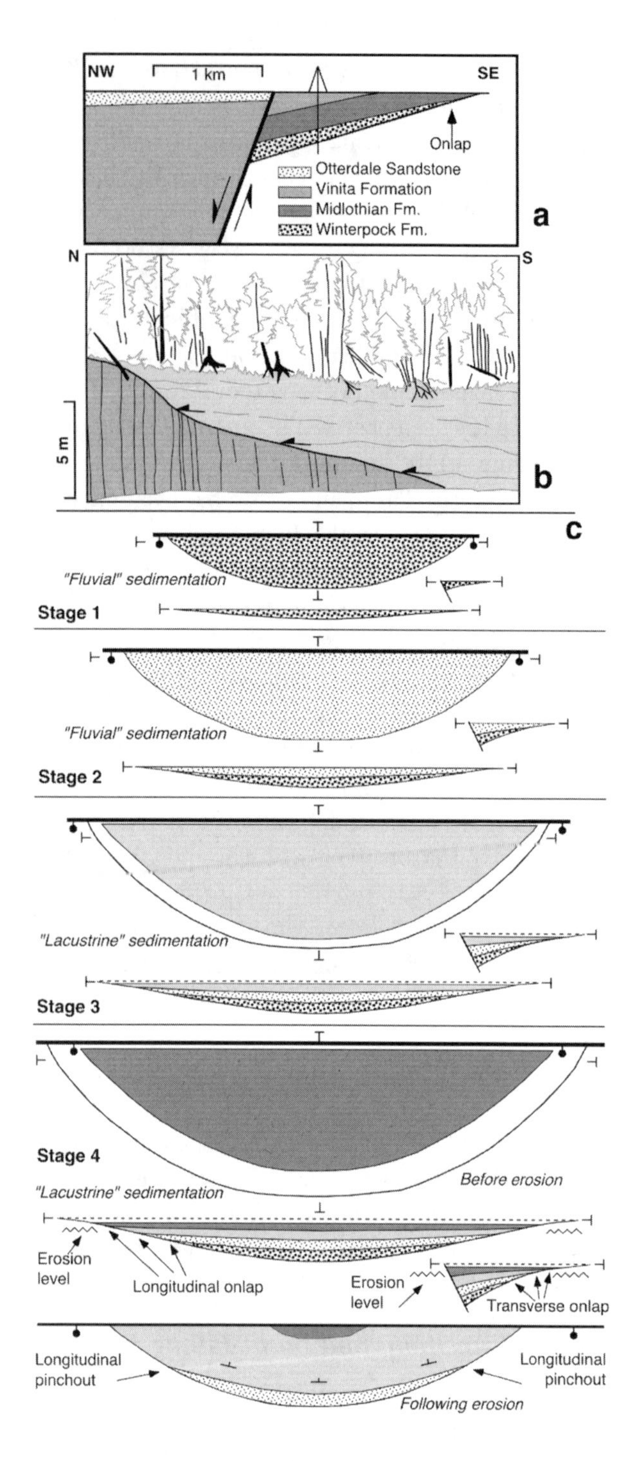

FIGURE 4.11 (*a* and *b*) Transverse hanging-wall onlap: (*a*) enlargement of the eastern end of cross section K-K′ in figure 4.2, in which onlap is inferred from outcrop and drill-hole data in the Richmond basin (nomenclature from LeTourneau, chapter 3 in volume 2 of *The Great Rift Valleys of Pangea*); (*b*) sketch of the seacliff at Tennycape, Nova Scotia, in which the Wolfville Formation onlaps subvertical strata of the Mississippian Horton Group (modified from Schlische and Anders 1996). (*c*) Growth and filling model for a basin bounded by a single fault. Basin growth results in larger depositional surface area, onlap geometries, and a transition from predominantly fluvial sedimentation to predominantly lacustrine sedimentation as the enlarging basin becomes sediment starved. Stage 4 shows a map view of the basin following erosional truncation. The oldest stratigraphic wedge has no surface expression; the second-oldest unit pinches out along strike (longitudinal pinchout) (modified from Schlische and Anders 1996).

monly deep-water) strata overlain in turn by shallow-water lacustrine strata and, in some cases, by fluvial strata (figure 4.12).

Basin geometry, onlap geometry, and major stratigraphic transitions are consistent with the filling of a basin growing wider, longer, and deeper through time under conditions of constant sediment-supply rate and a finite supply of water (Schlische and Olsen 1990; Schlische 1991) (figure 4.11c). These basin-growth and basin-filling models predict (1) initial fluvial sedimentation and associated hanging-wall onlap (sediment supply exceeds basin capacity), followed by a transition to lacustrine sedimentation (basin capacity exceeds sediment supply); (2) increasing lake depths following the fluvial–lacustrine transition (excess basin capacity is increasing), followed by a lake-depth maximum (available volume of water is exactly equal to excess basin capacity); and then (3) declining lake depths (the finite volume of water is spread out over an increasingly larger excess capacity) and a return to fluvial sedimentation (which requires decreasing subsidence rates toward the end of the extensional episode). Although the basin-filling models are in qualitative agreement with the observed stratigraphy, there are a number of important limitations and exceptions, which are described in detail by Schlische and Olsen (1990), Schlische (1991), and Schlische and Anders (1996).

According to these basin-filling models, the fluvial–lacustrine transition is a consequence of the gradual growth of the basin in length and width. Other workers have interpreted the fluvial–lacustrine transition to reflect a "tectonic" event that deepened the basin and allowed a lake to form (e.g., Manspeizer 1988; Ressetar and Taylor 1988; Manspeizer et al. 1989; Lambiase 1990). This explanation is somewhat unsatisfactory because the fluvial–lacustrine transition occurred at a different times in different basins, implying no re-

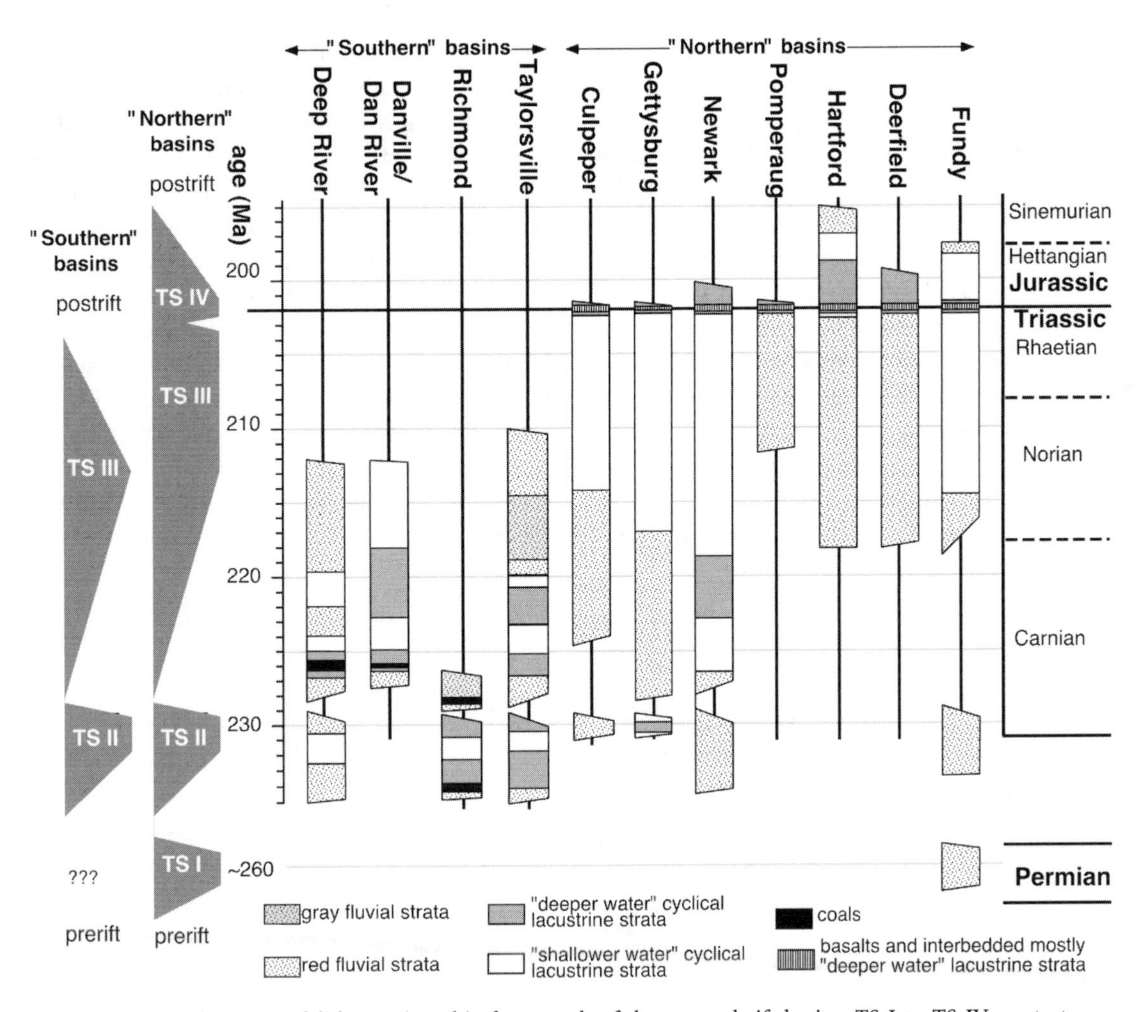

FIGURE 4.12 Chrono- and lithostratigraphic framework of the exposed rift basins. TS I to TS IV are tectonostratigraphic sequences. (Simplified from Olsen 1997 and revised based on Olsen et al. 2000 and LeTourneau, chapter 3 in volume 2)

gional tectonic control on sedimentation. Smoot (1991) hypothesized that reductions in fluvial gradients and clogging of the outlets may contribute to the fluvial–lacustrine transition. This hypothesis predicts fluvial–lacustrine transitions of variable age and successfully accounts for a change from initial braided-stream deposits to meandering-stream deposits observed in the basal fluvial sequences of many exposed rift basins. Lambiase (1990) and Lambiase and Bosworth (1995) related the fluvial–lacustrine transition to a reduction in the number of intrabasinal fault blocks through time and to the development of a closed basin bounded by the uplifted footwall blocks, the ramping margin of the basin, and high-relief accommodation zones (HRAZs) that are associated with antithetic overlapping normal faults. These models, however, have limited applicability to the eastern North America rift system, which lacks the HRAZs. Gupta et al. (1998) described a model of fault-system evolution and rift-basin subsidence that predicts an increase in local fault-displacement rates as a result of strain localization and stress enhancement on optimally located faults. This model, applied to terrestrial rift basins, is appealing because the fluvial–lacustrine transition need not occur at the same time throughout an extending region and the local increase in displacement rate on a given fault is not associated with a change in the regional extension rate. This model has yet to be applied formally to the eastern North American rift system.

Basin-growth and basin-filling models also make specific predictions about how accumulation rates should change through a vertical succession and with position in the basin; these predictions then can be compared with accumulation rate data derived from cyclostratigraphy (thickness of the cycle divided by its period). Using a full-graben basin growing wider and deeper through time under conditions of constant sediment-supply rate, Schlische and Olsen (1990) determined that accumulation rates are equal to subsidence rates during fluvial sedimentation (figure 4.13a); during lacustrine sedimentation, accumulation rates decrease exponentially through time (figure 4.13a). The major shortcomings of this model—the full-graben geometry and lack of longitudinal basin growth—were addressed by Schlische (1991), who used a geometric half-graben basin-growth model (similar to that in figure 4.11c) under conditions of constant sediment-supply rate. Accumulation rates (determined for a vertical drill hole through the synrift section) generally increase rapidly and then decrease somewhat more slowly (figure 4.13a). In addition, accumulation rate curves vary with position in the basin. Contreras, Scholz, and King (1997) used a more sophisticated numerical model of basin growth and sedimentation incorporating self-similar faulting ($n = 1$), flexure, isostasy, and diffusion of erosion and sedimentation; they confirmed that accumulation rate curves are sensitive to location within the basin as well as to the temporal evolution of the BFS (constant strain rate versus constant fault-lengthening rate) (figure 4.13b and c).

At the time of the publication of Schlische and Olsen (1990), accumulation rate data for the Newark

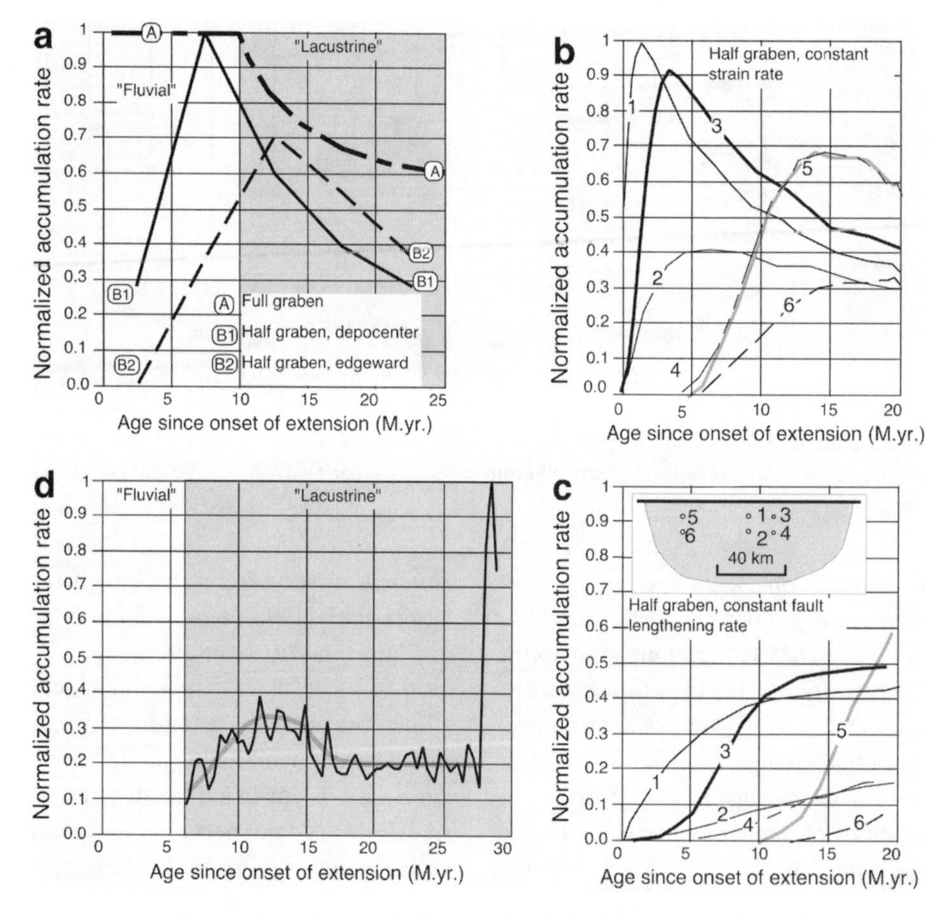

FIGURE 4.13 Comparison of accumulation rate data derived from various basin-filling models (*a–c*) and of cyclostratigraphically calibrated accumulation rate data from the Newark basin (*d*). The various accumulation rate data sets were normalized by the maximum accumulation rate in each data set to facilitate comparisons. The numbered curves in *b* and *c* were derived from vertical drill holes through the basin shown in *c* (*a*, data for full graben from Schlische and Olsen 1990 and for half-graben from Schlische 1991; *b*, constant strain rate and, *c*, constant fault-lengthening rate modified from Contreras, Scholz, and King 1997). (*d*) NBCP data based on a composite reference section scaled to the Rutgers core (Olsen et al. 1996; Olsen and Kent 1996).

basin were based solely on outcrops and appeared to confirm the model's predictions. However, the data were derived from outcrops scattered throughout the basin and were not corrected for spatial variations in subsidence rates. The NBCP has provided detailed accumulation rate data, based on cyclostratigraphy, for the individual cores containing lacustrine strata as well as for a composite section, which was constructed by scaling individual cored sections to the Rutgers locality using the thickening trends observed in the overlap sections (Olsen et al. 1996; Olsen and Kent 1996) (figure 4.10a). Although there is considerable scatter in the Triassic age section, which may reflect a climatically driven oscillation in sediment-supply rate (Silvestri and Schlische 1992), accumulation rates generally increase following the fluvial–lacustrine transition, reach a maximum in the lower Passaic Formation, decrease toward the middle Passaic Formation, and then remain essentially constant through the uppermost Passaic Formation (figure 4.13d). The first three trends are in broad agreement with the predictions of Schlische (1991) and with the constant strain rate models of Contreras, Scholz, and King (1997). However, neither model accounts for the fourth trend, which perhaps indicates that increases in sediment-supply rates offset increases in basin capacity as a result of basin growth.

A major deviation from the predictions of the basin-filling models occurs in the Early Jurassic strata, which are characterized by markedly higher accumulation rates and deeper lake facies than the immediately preceding stratigraphic section. Schlische and Olsen (1990) proposed that accelerated extension resulted in asymmetric deepening, which caused sediments and water to "pond" in the basin depocenter, leading to higher accumulation rates and lake depths. This scenario predicts that the contact between the two sequences should be in part unconformable and be marked by internal onlap. Such an unconformity has been documented only in the Connecticut Valley basin (Cornet 1977; J. P. Smoot, personal communication, cited in Olsen 1997). The Early Jurassic sequence is also associated with the emplacement of tholeiitic basalt flows. No doubt some of the marked increases in accumulation rate and water depth can be attributed to the isostatic consequences of the loading of up to several hundred meters of basalt. However, increases in lake depths (or the switch from fluvial to lacustrine sedimentation) preceded the igneous event in the Fundy, Connecticut Valley, and Newark basins (Olsen 1997) (figure 4.12). Thus the emplacement of the lava flows was not the cause of the stratigraphic changes but may have contributed to their magnitude. Perhaps the increase in extension rate was partially responsible for triggering the igneous activity (Schlische 1990). A case also can be made that a wetter climate resulted in more water (and more sediment) entering the basins. At least in the Newark basin, sedimentary facies become somewhat wetter in the upper 3 km of the Passaic Formation (Smoot and Olsen 1994), but these changes are not nearly as extreme or abrupt as those occurring between the latest Triassic and earliest Jurassic sequences. Tectonics was therefore responsible for the higher accumulation rates and deeper-water facies present in Early Jurassic age strata, although it is likely that accelerated faulting led to higher footwall uplift, resulting perhaps in increased orographic precipitation (e.g., Manspeizer 1982).

The foregoing discussion highlights the utility of the basin-filling models. These models can explain many of the elements of the stratigraphic record in terms of the growth and filling of half-graben basins, but they also highlight notable departures from the baseline conditions, which are the result of tectonic or climatic events. The Early Jurassic event discussed here is perhaps the most significant "anomaly." Others include the exclusively fluvial deposits in the Triassic sections of the Connecticut Valley, which are perhaps related to a larger than normal axial supply of sediment and/or to a lower basin subsidence rate, and the two deep-water lacustrine intervals in the Triassic section of the Danville basin (figure 4.12), which are as yet unexplained.

Lambiase (1990) described the Newark basin as a "dual-cycle" basin in that it contains a Late Triassic tripartite sequence overlain by an Early Jurassic sequence produced as a result of tectonic rejuvenation. Olsen (1997) refined this concept and defined tectonostratigraphic sequences as packages of sedimentary rock that compose all or part of a tripartite sequence. These tectonostratigraphic packages mostly likely are bounded by unconformities, which may be subtle in the depocenters of the basins. The majority of the Late Triassic section in the exposed basins falls into Olsen's tectonostratigraphic sequence (TS) III; the Early Jurassic belongs to TS IV. TS I is a Permian age sequence known from only the Fundy and Argana (Morocco) basins (Olsen et al. 2000). In the Fundy basin, the TS I package may not be a synrift deposit (Olsen et al. 2000). TS II deposits were deposited under considerably more humid conditions than TS I. The climatic milieu changed from more humid to more arid going from TS II to TS III. Well-developed angular unconformities between TS II and TS III are present in the Richmond-Taylorsville (LeTourneau, chapter 3 in volume 2) and Fundy (Olsen et al. 2000) basins. TS I and TS II are poorly exposed because they are onlapped and overlapped by TS III. Interestingly, TS I and TS II preferentially outcrop in the Richmond-Taylorsville and Fundy basins, both of which appear to be more strongly inverted than other Mesozoic rifts (see later in chapter).

Extrabasinal Drainages

Thus far, our attention has focused on the first-order effect of basin geometry on thickness and facies patterns in the basin fill. The geometry of the uplifted footwall block also influences sedimentation patterns. Based on the footwall displacement field and studies of uplifted footwall blocks from areas of active extensional tectonics, the geometry of footwall uplift consists of narrow escarpment sloping toward the sedimentary basin and a wider flank sloping away from the basin (figure 4.9b). Consequently, drainages entering the basin from the footwall are quite small and deliver only limited quantities of sediment (Frostick and Reid 1987; Cohen 1990); most footwall rivers flow away from the basin. Conversely, the hanging-wall block slopes toward the basin, and the number of streams and their sediment loads entering the basin from the hanging-wall margin are substantially larger than the direct footwall component (figure 4.9b). The magnitude of the footwall uplift commonly reaches a maximum near the center of the fault system and tapers off to zero near the ends (e.g., Zandt and Owens 1980; Anders and Schlische 1994). Therefore, streams flowing down the flank of the uplift may swing around and enter the basin at its longitudinal ends (Cohen 1990), contributing to an axial component of sediment transport (Lambiase 1990) (figure 4.9b). BFS segment boundaries are commonly associated with footwall elevation lows, especially during the early stages of fault linkage (but see also Anders and Schlische 1994). These lows may be exploited by streams flowing off the footwall block (Leeder and Gawthorpe 1987; Gawthorpe and Hurst 1993), and thus relay ramps are sites of significant flux of coarse clastics into the basin (figure 4.9b).

Sedimentologic data from the Newark basin generally support these relationships. Conglomeratic deposits along the BFS are relatively small and discontinuous in map view. However, some of the largest deposits are associated with relay ramps (figure 4.3a). Paleocurrent and provenance data indicate that most sediment was sourced from the hanging-wall block or from the axial ends of the basin (Glaeser 1966; Allen 1978; Oschudlak and Hubert 1988; Smoot 1991). The situation in the Hartford subbasin is considerably more complex. The footwall block was the predominant source of sediment during Triassic time (Hubert et al. 1978), perhaps because erosion could keep pace with uplift during a period of low extension rate. This hypothesis is to some extent supported by the exclusively fluvial deposits in the Hartford basin, in contrast to the Triassic seqences of all the other exposed basins. During the extrusive interval, when extension rates and footwall uplift rates may have been higher, the hanging wall was the predominant source (Hubert et al. 1978), whereas the footwall served as the predominant source in the postextrusive interval (McInerny 1993).

Regional Trends

Two regional trends have been noted for the eastern North American rift system. The most obvious is the apparent increase in aridity of contemporary sequences going from the southern basins to the northern basins (Olsen 1997). These trends reflect the paleogeography of the basins with respect to a zonal climatic regime; that is, in Late Triassic time, the southern basins were located closer to the paleoequator than the northern basins were. The second regional trend concerns the absence of TS IV deposits from the exposed southern basins. The absence of these deposits can be attributed to either nondeposition or erosion of Jurassic deposits. Thermal maturation indices and modeling studies in the Taylorsville basin suggest that the former is more likely (Malinconico, chapter 6 in this volume). As documented more fully in the next section, border faulting, basin subsidence, and synrift sedimentation ended prior to the Early Jurassic igneous episode in the southern basins (Withjack, Schlische, and Olsen 1998). This interpretation is bolstered by widespread subsurface basalt flows in the southeastern United States; these approximately 200 Ma lava flows (Ragland 1991) are interpreted as postrift because they are untilted and extend beyond the limits of synrift basins (McBride, Nelson, and Brown 1989).

Postrift Deformation and Basin Inversion

A large suite of structures formed during NW–SE extension; these include NE-striking normal faults, diabase dikes, joints, and veins; NW-trending fault-displacement folds; some NE-trending reverse-drag and normal-drag folds; and E-striking left-oblique faults. Many of these structures are associated with syndeformational sedimentation. However, many structures (especially reverse faults, associated folds, and faults exhibiting multiple slickenline orientations [e.g., de Boer and Clifton 1988]) cannot be attributed to NW–SE extension. Sanders (1963) was among the first to suggest that the stress regime changed after rifting and that reverse faults, strike-slip faults, and folds formed during postrift (generally postdepositonal) deformation. Swanson (1982), de Boer and Clifton (1988), and de Boer (1992) proposed a period of margin-parallel "shift" between the rifting and drifting stages. Although some of the structures that Sanders and others attributed to postrift shortening have subsequently been shown to have formed during NW–SE syndepositional extension, the general concept of postrift deformation is viable. In recent years, a growing body of evidence indicates that these structures are more widespread and involve more shortening than previously thought and that in some cases the basins themselves have undergone inversion (Withjack, Olsen, and Schlische 1995; Withjack, Schlische, and Olsen 1998).

Although there is some disagreement about the definition of the word "inversion" (Williams, Powell, and Cooper 1989), it is generally agreed that inversion involves a reversal in deformation style: positive inversion involves extension (or transtension) followed by compression (or transpression), whereas negative inversion involves compression (or transpression) followed by extension (or transtension). Basin inversion is positive inversion associated with the transformation of an area of subsidence to one of uplift. Classic basin-inversion structures consist of anticlinal folds, reverse faults, and strike-slip faults associated with the reactivation of normal faults (Harding 1983; Bally 1984; Lowell 1995) (figure 4.5). These structures have a harpoon or an arrowhead geometry and commonly are referred to as Sunda folds, based on the type locality in Indonesia (e.g., White and Wing 1978). Excellent reviews of basin inversion are given by Cooper and Williams (1989), Coward (1994, 1996), and Buchanan and Buchanan (1995). Basin inversion is one aspect of *postrift deformation,* which I define as any deformation postdating the synrift stage. This section reviews the features associated with postrift deformation, attempts to constrain the age of formation of these features, and presents a model for the tectonic evolution of eastern North America, which consists of a rifting stage and a diachronous transition to a postrift stage involving postrift shortening and basin inversion as well as the initiation of seafloor spreading.

Features Associated with Postrift Shortening and Basin Inversion

The largest-scale structures associated with postrift shortening and basin inversion have been observed in the Fundy basin (Withjack, Olsen, and Schlische 1995). Seismic-reflection data indicate that the relatively shallow-dipping border faults of the Fundy and Chignecto subbasins, which experienced greater than 10 km of normal slip, were reactivated as thrust faults (figure 4.14c) with as much as 4 km of reverse slip, and the Minas fault zone experienced less than 8 km of right-oblique reverse slip. Anticlines and synclines formed during reverse motion along the reactivated faults (figure 4.14a–c). Synrift strata uniformly thicken toward the associated faults and show no evidence of thinning on the crests of anticlines or of thickening in the troughs of synclines (figure 4.14b). The folds are therefore postdepositional structures. The folds generally trend in a SW or WSW direction, parallel to the large axial syncline of the Fundy basin (figure 4.1a), which is also thought to be related to basin inversion (Withjack, Olsen, and Schlische 1995). Basin inversion may account for enigmatic structural relations at the junction of the Fundy, Minas, and Chignecto subbasins. Schlische's (1990) kinematic model predicts less extension in the Chignecto subbasin than in the Fundy subbasin, yet seismic data show the depths of the two subbasins to be comparable. The BFS of the Fundy

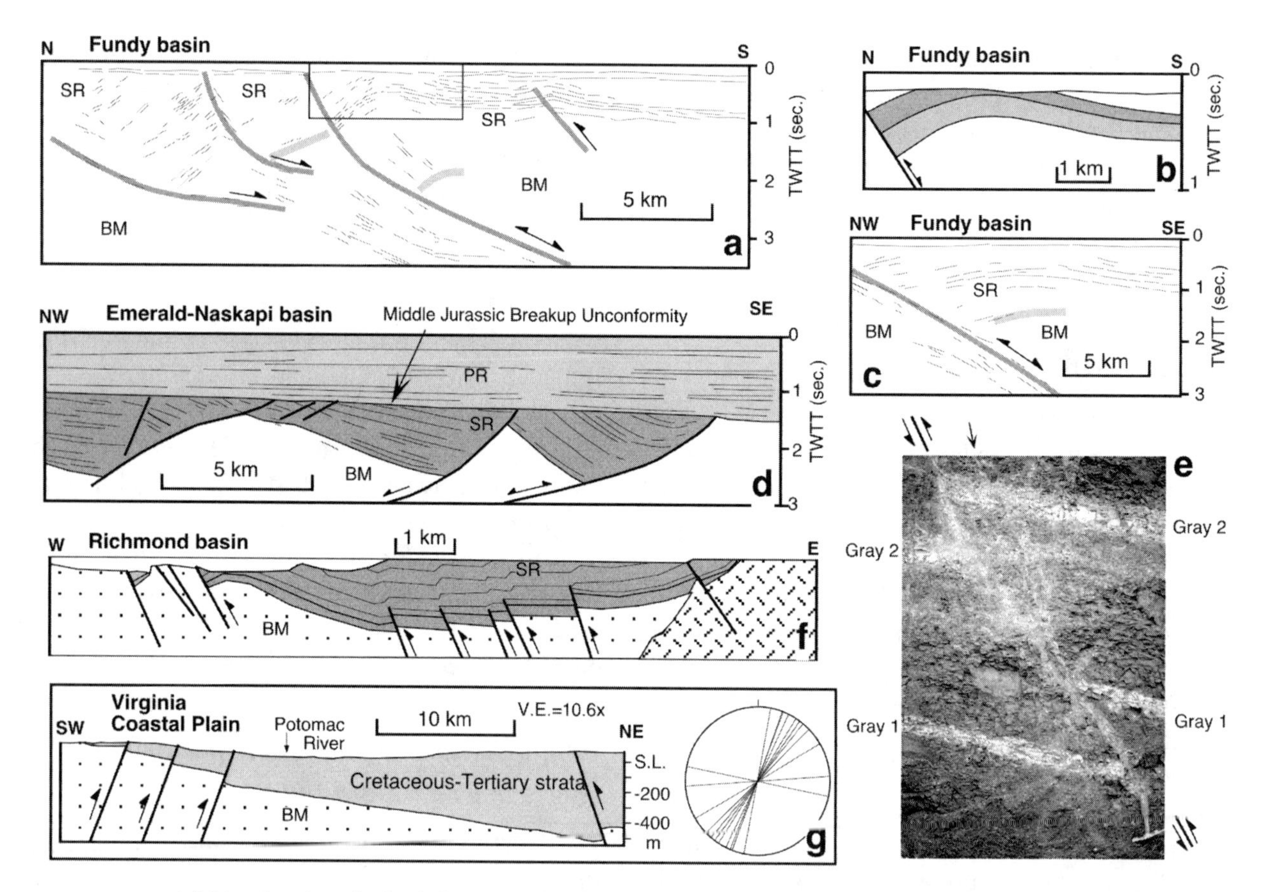

FIGURE 4.14 (*a*) Line drawing of seismic line 81-47 from the Fundy basin. (*b*) Enlargement of the boxed area in *a*, showing the thickening of units across the hinge of inversion-related anticline. (*c*) Line drawing of seismic line 81-79 from the Fundy basin, showing anticline related to reverse reactivation of border fault. (*d*) Line drawing and interpretation of seismic line across the Emerald-Naskapi basin on the Scotian Shelf; *BM*, basement; *PR*, postrift strata; *SR*, synrift strata. For locations, see figure 4.1a and b (*a–d* adapted from Withjack, Olsen, and Schlische 1995). (*e*) Reverse fault in the Newark basin, Edison, New Jersey. The main fault and subsidiary faults (e.g., *arrow*) and fractures exhibit extensive bleaching. (*f*) Cross section across the northern Richmond basin (adapted from Shaler and Woodworth 1899). (*g*) Cross section through the Coastal Plain of Virginia (adapted from Prowell 1988). The equal-area net shows strikes of all mapped faults offsetting the Coastal Plain units.

basin apparently underwent larger reverse reactivation than the BFS of the Chignecto subbasin (Withjack, Olsen, and Schlische 1995). Outcrop examples of inversion from the Minas subbasin include reverse-reactivated normal faults within a larger NE-striking normal-fault zone (figure 4.6b); normal faults exhibiting synrift growth that were rotated into a reverse-fault geometry (figure 4.6c); and en echelon folds associated with right-oblique reverse reactivation of a fault zone that underwent left-oblique normal movement during synrift faulting. All postrift structures in outcrop are consistent with NW–SE shortening, but this direction is biased by reactivated extensional structures formed by NW–SE extension (figure 4.6d and e).

Reverse faults, conjugate strike-slip faults, and minor folds within the Connecticut Valley basin have been attributed to postrift compression, which is thought to have rotated from NE–SW to NW–SE through time (de Boer and Clifton 1988; Wise 1993). The Newark basin contains NE-striking reverse faults associated with extensive fluid flow (figure 4.14e) as well as reverse-reactivated normal faults (Schlische 1992). Shaler and Woodworth (1899) documented basement-involved reverse faults and associated folds within the synrift strata of the Richmond basin (figure 4.14f). NE-striking reverse faults, NE-trending folds, and NNW-striking dikes mapped by Venkatakrishnan and Lutz (1988) are indicative of NNW–SSE shortening. In the Danville basin, NE-striking reverse faults are parallel to and occasionally deform small normal faults (Ackermann et al., chapter 8 in this volume); the exceptionally narrow width of this basin with steeply dipping strata may also be a manifestation of postrift shortening (figure 4.2).

Compressional structures are not limited to the rift basins. Generally, NE-striking reverse faults offset Cretaceous and younger rocks of the Atlantic coastal plain (Prowell 1988) (figure 4.14g). Unlike the reverse and strike-slip faults of the Fundy basin, these faults are not mineralized (Prowell 1988). Maximum separation is less than 80 m, much less than the minimum of 4 km of reverse displacement associated with some faults in the Fundy basin.

Age Relations

The Early Jurassic age eastern North American magmatic event provides a temporal benchmark that indicates that rifting, basin subsidence, and initiation of postrift shortening and basin inversion began earlier in the south than in the north (Withjack, Schlische, and Olsen 1998). This conclusion is based on the following lines of evidence:

1. Seismic data indicate that the NE-striking Cooke fault in South Carolina underwent at least 140 m of reverse displacement before the emplacement of Early Jurassic basalt (Behrendt et al. 1981; Hamilton, Behrendt, and Ackermann 1983). NE-trending compressional structures in the Richmond basin are cut by Early Jurassic age NNW-trending diabase dikes (Venkatakrishnan and Lutz 1988), indicating that shortening began prior to Early Jurassic time. In the Fundy basin, the shortening postdates the earliest Jurassic age basalts and postextrusive deposits (Withjack, Olsen, and Schlische 1995); structural relations in the nearby Orpheus graben and Emerald-Naskapi basin (figures 4.1a and 4.14d) indicate that most of the inversion ended before or during Early Cretaceous time and probably before or during early Middle Jurassic time (Withjack, Olsen, and Schlische 1995).

2. NW-striking diabase dikes are prevalent from Virginia southward and are indistinguishable in age from the earliest Jurassic NE-striking dikes in northeastern North America (McHone 1988; Ragland, Cummins, and Arthur 1992) (figure 4.4). Thus, two different stress fields existed in eastern North America during Early Jurassic time (McHone 1988): the southern one was characterized by NE–SW-oriented S_{Hmin}, with S_{Hmax} (maximum horizontal stress) oriented NW–SE; the northern one was characterized by NW–SE-oriented S_{Hmin}. The southern stress field is incompatible with the NW–SE extension responsible for faulting and basin formation. Indeed, the NW-striking dikes are subperpendicular to and cut the NE-striking BFSs of the Deep River and Danville basins (figure 4.4). Thus the southern basins, subjected to S_{Hmax} oriented NW–SE, were no longer extending in earliest Jurassic time. At the same time, the northern basins, subjected to S_{Hmax} oriented NE–SW, experienced accelerated extension and subsidence (Olsen, Schlische, and Gore 1989).

3. Cessation of extension prior to Early Jurassic time in the exposed southern basins is further suggested by the absence of Jurassic sedimentary rocks (figure 4.12). In addition, untilted postrift, Early Jurassic age basalt (e.g., McBride, Nelson, and Brown 1989) overlies tilted synrift strata in the subsurface of South Carolina and Georgia.

Tectonic Synthesis

Withjack, Schlische, and Olsen (1998) proposed the following tectonic history for the evolution of the continental margins of the central North Atlantic (figures 4.15 and 4.16):

1. Supercontinental assembly was complete by Permian time (figure 4.15a). In Middle to Late Triassic time (figure 4.15b), all of eastern North America underwent NW–SE extension, manifested primarily in the formation and filling of half-graben basins. Olsen's (1997) Triassic age tectonostratigraphic sequences suggest that there may have been two Triassic extensional pulses (figure 4.16).
2. Prior to Early Jurassic time, the southern basins stopped subsiding.
3. In earliest Jurassic time, the southern region experienced NW–SE shortening, resulting in the development of small-scale reverse faults, folds, and possible basin inversion as well as the intrusion of NW-striking diabase dikes (figure 4.15c). Coevally, the northern basins were actively extending in a NW–SE direction, resulting in the intrusion of NE-striking dikes and accelerated subsidence, which was responsible for Olsen's (1997) TS IV.

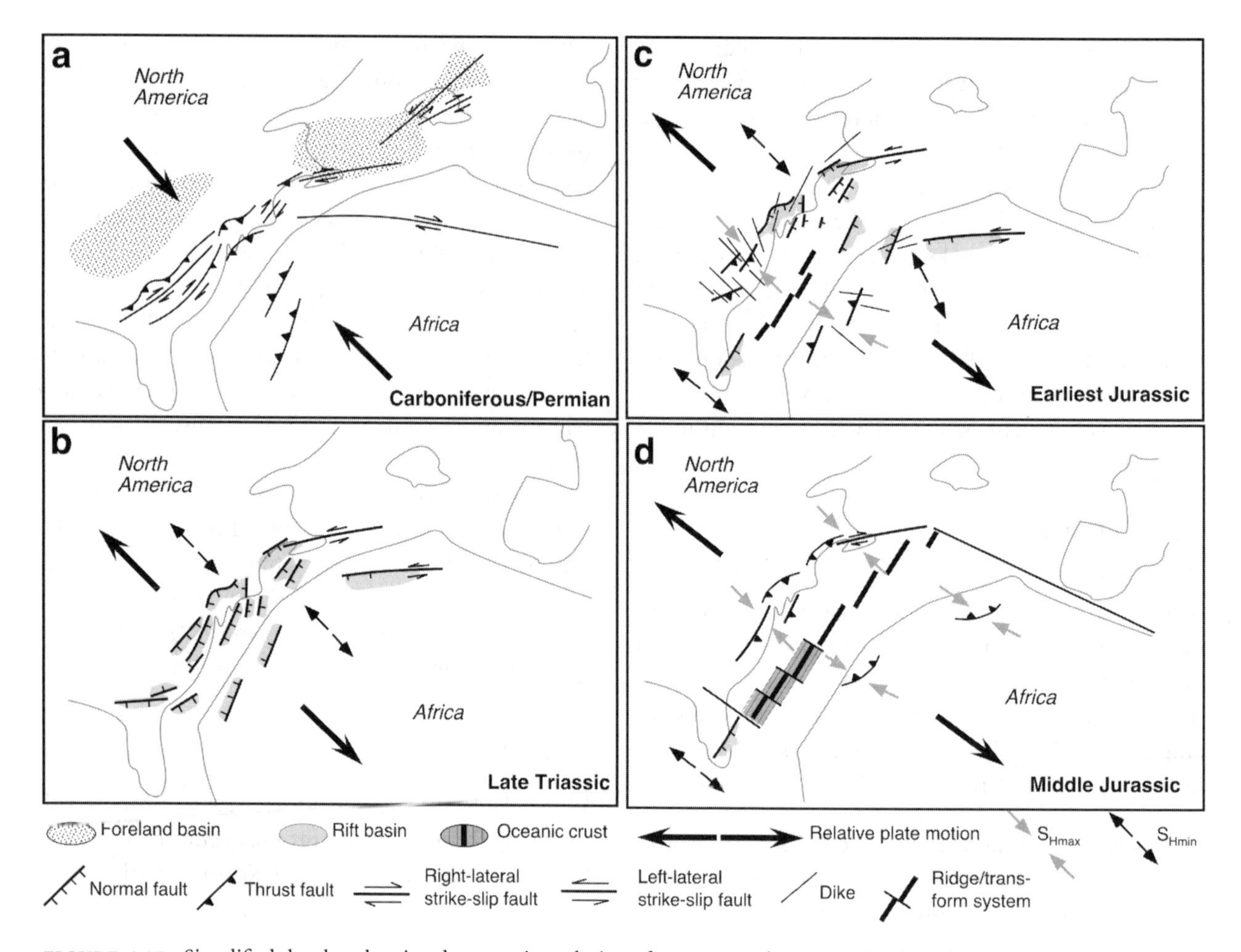

FIGURE 4.15 Simplified sketches showing the tectonic evolution of eastern North America (and northwestern Africa). For further discussion, see the text. (Simplified from Withjack, Schlische, and Olsen 1998)

4. By Middle Jurassic time, all of eastern North America was experiencing shortening, generally oriented NW–SE, which resulted in the development of small-scale reverse faults, folds, and basin inversion (figure 4.15d). This deformational regime persisted locally into Cenozoic time (based on NE-striking reverse faults affecting Cretaceous and early Cenozoic age coastal plain units). The present-day deformational regime is associated with S_{Hmax} oriented NE–SW (Zoback 1992).

The cause of the inversion is not immediately obvious. On the northwestern European continental margin, where basin inversion is quite common and has been studied for considerably longer than in eastern North America, tectonic inversion (in contrast to regional exhumation, which may be caused by isostatic unloading or igneous underplating [Brodie and White 1994; Coward 1994]) has been attributed to (1) transpression along strike-slip faults, resulting from a change in stress regime (e.g., Glennie and Boegner 1982); (2) the effects of the Alpine orogeny (e.g., Ziegler 1987, 1989; Roberts 1989); (3) the transition from rifting to drifting in the North Atlantic (Chapman 1989); (4) the interaction between the convergent southeastern plate boundary and the divergent northwestern margin (Cartwright 1989); and (5) the effects of ridge push and continental resistance to plate motion (Boldreel and Andersen 1993; see also Dewey 1988 and Bott 1992). Only mechanisms (3) and (5) are viable on a regional basis in eastern North America (Withjack, Olsen, and Schlische 1995; Withjack, Schlische, and Olsen 1998). The inferred NW–SE shortening direction is also compatible with the inferred early NW–SE spreading direction (Klitgord and Schouten 1986). However, the spreading direction in the central North Atlantic has not remained constant (Klitgord and Schouten 1986) (figure 4.16), which may explain some of the variability in the orientations of

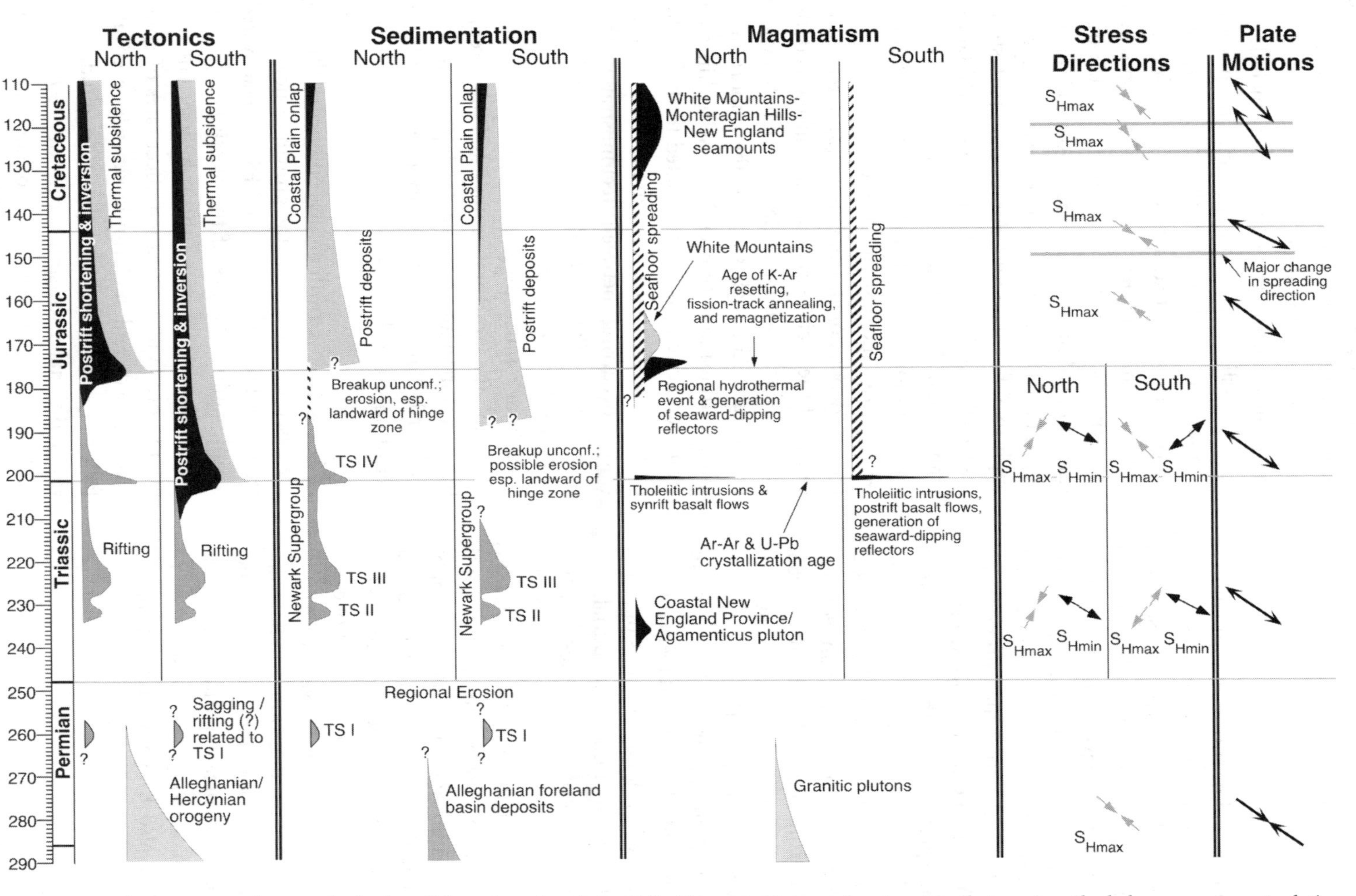

FIGURE 4.16 Summary of events in the late Paleozoic and early–middle Mesozoic history of eastern North America. Shaded areas represent relative and qualitative amounts of tectonic, sedimentary, and magmatic activity, increasing in magnitude to the right; no absolute scale is implied. (Numerical timescale based on Olsen, Schlische, and Gore 1989, as modified in Olsen 1997. Some data from Klitgord and Schouten 1986; de Boer et al. 1988; and McHone 1988. Updated from Olsen, Schlische, and Gore 1989, based principally on Olsen 1997; Withjack, Schlische, and Olsen 1998; and Olsen et al. 2000)

postrift structures reported in the Connecticut Valley basin (de Boer and Clifton 1988). In addition, the NE–SW opening of the Labrador Sea may have resulted in NE–SW compression.

The tectonic model outlined here has the following implications:

1. The diachronous end of rifting and initiation of postrift shortening imply that the initiation of seafloor spreading itself was diachronous (Withjack, Schlische, and Olsen 1998) (figures 4.15 and 4.16). Seafloor spreading is hypothesized to have begun by Early Jurassic time in the area adjacent to the Carolina trough; seafloor spreading began by late Early Jurassic or Middle Jurassic time in the region from the Baltimore Canyon trough to southeastern Canada. This two-step initiation of seafloor spreading for the central North Atlantic previously was unrecognized—not surprising, given that the magnetic anomaly record of the oldest oceanic crust is obscured by passive-margin sediments (e.g., Klitgord, Hutchinson, and Schouten 1988), which themselves have been dated directly only in the Georges Bank basin (e.g., Poag 1991).

2. Postrift deformation appears to have occurred closely following the end of synrift sedimentation. In basins that experienced widespread inversion, uplift shortly followed the period of maximum burial.

3. The initiation of postrift deformation coincides with the rift–drift transition. Basin inversion may thus be in part responsible for the breakup unconformity (terminology from Falvey 1974). The nascent continental margin may have been particularly susceptible to postrift shortening during the initial stages of seafloor spreading because lithospheric strength was lowest during the rift–drift transition as a result of extensional thinning and heating (e.g., Coward 1994).

4. Because the inferred direction of shortening (NW–SE) is parallel to the synrift extension direction and the inferred seafloor spreading direction, there is no need for a shift or shearing stage between the rift and drift stages (see also Klitgord, Hutchinson, and Schouten 1988).

Discussion

Let us now revisit the major points of contention outlined at the beginning of this chapter. Contention (1) concerns whether the large suite of structures contained in the eastern North American rift system is attributable either to a uniform stress field reactivating variably oriented Paleozoic structures (Ratcliffe–Burton model) or to a history of changing stress regimes. Both hypotheses are correct. Although it is true that more and more structures are recognized as having formed during the synrift stage (including folds and strike-slip faults), an important suite of structures also formed during postrift shortening and basin inversion, a process that appears to be an integral part of rifting and continental separation. Contention (2) concerns the timing of formation of structures with respect to sedimentation (syndepositional versus postdepositional). Again, both hypotheses are correct but apply to different periods. For the most part, structures (especially border faults, some intrabasinal faults, and many folds) that formed during the rifting phase are syndepositional; all structures associated with postrift deformation and basin inversion in the exposed basins are postdepositional. Contention (3) is related to the causes of the major stratigraphic transitions in the exposed basins. Once again, elements of both hypotheses are correct. Although some of the important stratigraphic transitions (especially those within tectonostratigraphic sequences) are related to the infilling of a basin growing in size through time, other transitions (those occurring between tectonostratigraphic sequences) are likely related to tectonic and, in some cases, to climatic changes.

Although much progress has been made in characterizing the structural geology and interpreting the basin evolution and tectonic development of the eastern North American rift system, there are still a number of unresolved issues:

1. The basin-growth models outlined in this chapter are based on models for the growth of newly formed faults, yet the majority of the BFSs are reactivated Paleozoic faults. The size of the reactivated part of the fault undoubtedly grows through time, but to what extent does fault reactivation affect rates of fault and basin growth and subbasin linkage? Is reactivation a factor in surpressing the degree of longitudinal onlap (for additional discussion, see Morley 1999)?

2. Why are transverse folds so poorly developed (or even absent) in the footwall blocks of normal faults? In the case of border faults in which the footwall block consists of "basement," the rocks may lack the subhorizontal layering required to recognize the trans-

verse folds. Another possibility (applying to both BFSs and intrabasinal faults) is that the magnitude of footwall uplift may be up to 5.5 times less than hangingwall displacement (Stein and Barrientos 1985), resulting in along-strike displacement variations too subtle to detect.

3. What are the minor structures associated with folds, and what controls their spatial distribution? Of particular interest is the pressure-solution cleavage present in parts of the Jacksonwald and Sassamannsville synclines in the southeastern Newark basin (Lucas, Hull, and Manspeizer 1988; Schlische 1992). Is the cleavage related to postrift deformation or to localized shortening in the inner arcs of units that buckled independently during warping? In the latter case, extensional structures should form in the outer arcs.

4. Why do the transverse fault-displacement folds related to small-scale BFS segmentation have no basinwide expression? One possibility is that the transverse folds are reflecting the near-fault displacement variations on individual fault segments, whereas the geometry of the basin reflects the overall displacement geometry. (Anders and Schlische [1994] showed that footwall blocks of basin-bounding faults in the Basin and Range are not affected by fault segmentation, which may also explain why there are no transverse folds in the footwall blocks.) Another possibility is that intrabasinal faults accommodate extension in areas of displacement deficit on the BFS, which requires a certain amount of "communication" among faults within the system (Cowie, Vanneste, and Sornette 1993).

5. How has postrift shortening modified fold geometry? The Ferndale dome-and-basin folds (figure 4.10a) and the Watchung syncline in the Newark basin (figure 4.10a; see also figure 4.2, section F–F′) are interference folds that combine elements of transverse and longitudinal folds. The transverse aspects are likely related to along-strike variations in fault displacement. The longitudinal aspects may be related to fault-propagation (drag) folding or to basin inversion. Folding should be syndepositional in the former case and postdepositional in the latter.

6. What accounts for the anomalous distribution of strata in the Newark basin? Erosional truncation of a structurally simple basin results in a partial bull's-eye pattern of stratigraphic units (figure 4.11c), with the youngest units preserved adjacent to the BFS at the center of the basin. In the Newark basin, the youngest units are preferentially preserved in the northeastern third of the basin (figure 4.1b). Schlische and Olsen (1988) hypothesized that basin subsidence was greatest in the northeast because the BFS has the steepest dip in this region. However, the facies and thickness data for Triassic strata do not support this contention. One viable possibility is that inversion preferentially affected the central and southwestern parts of the basin because the BFS dips less steeply and would be preferentially reactivated during compression. Another possibility is that the depocenter shifted to the northeast in Early Jurassic time. Regional thickness and facies trends for the Jurassic strata are much less well known than for the Triassic strata and are in need of closer examination.

7. What is the relationship between basin inversion/postrift shortening and fluid flow or igneous activity? The inversion in the southern basins is associated with the Early Jurassic age magmatic event and the emplacement of seaward-dipping reflectors at the continent–ocean boundary (Withjack, Schlische, and Olsen 1998). The inversion in the northern basins is broadly contemporaneous with an inferred hydrothermal event that reset many isotopic clocks (Sutter 1988), that remagnetized many rocks (Witte, Kent, and Olsen 1991), and that may be related to the emplacement of seaward-dipping reflectors during the initiation of seafloor spreading along the northern segment of the central North Atlantic margin (figure 4.16). Olsen, Schlische, and Gore (1989) and Schlische (1990) attributed the fluid flow to a period of accelerated normal faulting immediately prior to breakup, but faulting during postrift deformation and inversion is also possible. In fact, the reverse fault shown in figure 4.14e exhibits extensive hydrothermal alteration along the fault plane and subsidiary fractures. Interestingly, Sibson (1995) argued that high fluid pressures are required to reactivate normal faults that are not optimally oriented (i.e., those dipping steeper than 50°) and that fluid pressures increase during inversion as open subvertical fractures close during regional subhorizontal compression. The implications of these relationships for thermal maturity studies are as yet largely unexplored.

8. What is responsible for the N-trending set of dikes that intrude the southern basins and basement? Based on cross-cutting relations, these dikes are younger than the NW-striking set (Ragland, Cummins, and Arthur 1992) and may be related to the

emplacement of offshore plutons (de Boer et al. 1988) (figure 4.4). They indicate that the shortening direction shifted from NW–SE to N–S. However, the tectonic implications of this shift and the absolute age relationships of the two dike sets remain to be fully explored (M. O. Withjack, personal communication 1997).

9. What caused the extensional pulses thought to be responsible for the tectonostratigraphic sequences? Was TS I deposited in a rift basin or a sag basin (Olsen et al. 2000)? Are any sequences related to extensional collapse following the cessation of continental convergence rather than extension associated with continental fragmentation? Is the apparent marked increase in extension rate associated with the start of TS IV in any way related to the cessation of rifting and initiation of seafloor spreading in the south?

10. Although there is crude agreement between the predictions of half-graben basin-filling models and accumulation rate data from the Newark basin, more sophisticated models (that incorporate syndepositional intrabasinal faults, variable sediment supply rates, and sediment compaction) and more data on accumulation rates from across the basin are required before we can make specific predictions about fault-displacement and basin-growth rates.

Acknowledgments

My views on the eastern North American rift system and extensional tectonics benefited from discussions and collaborations with the following colleagues: Rolf Ackermann, Mark Anders, Amy Clifton, Juan Contreras, Patience Cowie, Nancye Dawers, Anu Gupta, Gregory Herman, Brian Jones, Marjorie Levy, Douglass Musser, Paul Olsen, David Reynolds, Christopher Scholz, Robert Sheridan, Neil Shubin, ShayMaria Silvestri, Martha Withjack, and Scott Young. I thank Don Monteverde and Richard Volkert for drawing my attention to the reverse fault shown in figure 4.14e. My research has been supported generously by the National Science Foundation, the Rutgers University Research Council, Mobil Technology Company, the Geological Society of America, and Sigma Xi. I thank Rolf Ackermann, Roger Buck, Amy Clifton, Peter LeTourneau, Diane Luken, Timothy Maguire, and especially Martha Withjack for reviewing various versions of the manuscript for this chapter.

Literature Cited

Allen, J. F., Jr. 1978. Paleocurrent and facies analysis of the Triassic Stockton Formation in western New Jersey. M.S. thesis, Rutgers University.

Anders, M. H., and R. W. Schlische. 1994. Overlapping faults, intrabasin highs, and the growth of normal faults. *Journal of Geology* 102:165–179.

Anderson, E. M. 1942. *The Dynamics of Faulting and Dyke Formation, with Applications to Britain*. Edinburgh: Oliver and Boyd.

Angelier, J. 1994. Fault slip analysis and paleostress reconstruction. In P. L. Hancock, ed., *Continental Deformation*, pp. 53–100. New York: Pergamon.

Arthaud, F., and P. Matte. 1977. Late Paleozoic strike-slip faulting in southern Europe and northern Africa: Result of right-lateral shear zone between the Appalachians and the Urals. *Geological Society of America Bulletin* 88:1305–1320.

Bain, G. L., and B. W. Harvey. 1977. *Field Guide to the Geology of the Durham Triassic Basin*. Carolina Geological Society Guidebook, no. 40. Durham, N.C.: Carolina Geological Society.

Bally, A. W. 1984. Tectogenese et sismique reflexion. *Bulletin de la Société Géologique de France* 24:279–285.

Barnett, J. A. M., J. Mortimer, J. H. Rippon, J. J. Walsh, and J. Watterson. 1987. Displacement geometry in the volume containing a single normal fault. *American Association of Petroleum Geologists Bulletin* 71:925–937.

Barrell, J. 1915. *Central Connecticut in the Geologic Past*. State Geological and Natural History Survey of Connecticut Bulletin, no. 23. Hartford: State of Connecticut.

Barrientos, S. E., R. S. Stein, and S. N. Ward. 1987. Comparison of the 1959 Hebgen Lake, Montana, and the 1983 Borah Peak, Idaho, earthquakes from geodetic observations. *Seismological Society of America Bulletin* 77:784–808.

Behrendt, J. C., R. M. Hamilton, H. D. Ackermann, and V. J. Henry. 1981. Cenozoic faulting in the vicinity of the Charleston, South Carolina, 1886 earthquake. *Geology* 9:117–122.

Bell, R. E., G. D. Karner, and M. S. Steckler. 1988. Detachments during extension: Application to the Newark series basins. *Tectonics* 7:447–462.

Benson, R. N., and R. G. Doyle. 1988. Early Mesozoic rift basins and the development of the United States middle Atlantic continental margin. In W. Manspeizer, ed., *Triassic–Jurassic Rifting: Continental Breakup and the Origin of the Atlantic Ocean and the Passive Mar-*

gins, pp. 99–127. Developments in Geotectonics, no. 22. Amsterdam: Elsevier.

Berg, T. M., comp. 1980. *Geologic Map of Pennsylvania* [scale 1:250,000]. Harrisburg: Commonwealth of Pennsylvania, Topographic and Geological Survey.

Boldreel, L. O., and M. S. Andersen. 1993. Late Paleocene to Miocene compression in the Faero-Rockall area. In J. R. Parker, ed., *Petroleum Geology of Northwest Europe,* pp. 1025–1034. London: Geological Society.

Bott, M. H. P. 1992. The stress regime associated with continental breakup. In B. C. Storey, T. Alabaster, and R. J. Pankhurst, eds., *Magmatism and the Causes of Continental Breakup,* pp. 125–136. Geological Society Special Publication, no. 68. London: Geological Society.

Brodie, J., and N. J. White. 1994. Sedimentary basin inversion caused by igneous underplating: Northwest European continental shelf. *Geology* 22:147–150.

Brown, D. E. 1986. The Bay of Fundy: Thin-skinned tectonics and resultant early Mesozoic sedimentation [abstract]. In *Basins of Eastern Canada and Worldwide Analogues,* p. 29. Atlantic Geoscience Society Symposium, August 13–15, 1986, Halifax, Nova Scotia.

Brown, P. M., and J. M. Parker III, comps. 1985. *Geologic Map of North Carolina* [scale 1:500,000]. Raleigh: North Carolina Department of Natural Resources and Community Development, Geological Survey Section.

Buchanan, J. G., and P. G. Buchanan, eds. 1995. *Basin Inversion.* Geological Society Special Publication, no. 88. London: Geological Society.

Buck, W. R. 1988. Flexural rotation of normal faults. *Tectonics* 7:959–973.

Burgess, C. F., B. R. Rosendahl, S. Sander, C. A. Burgess, J. Lambiase, S. Derksen, and N. Meader. 1988. The structural and stratigraphic evolution of Lake Tanganyika: A case study of continental rifting. In W. Manspeizer, ed., *Triassic–Jurassic Rifting: Continental Breakup and the Origin of the Atlantic Ocean and the Passive Margins,* pp. 859–881. Developments in Geotectonics, no. 22. Amsterdam: Elsevier.

Cartwright, J. A. 1989. The kinematics of inversion in the Danish Central Graben. In M. A. Cooper and G. D. Williams, eds., *Inversion Tectonics,* pp. 153–175. Geological Society Special Publication, no. 44. London: Geological Society.

Chapman, G. R., S. J. Lippard, and J. E. Martyn. 1978. The stratigraphy and structure of the Kamasia Range, Kenya Rift Valley. *Journal of the Geological Society of London* 135:265–281.

Chapman, T. J. 1989. The Permian to Cretaceous structural evolution of the Western Approaches Basin (Melville sub-basin), UK. In M. A. Cooper and G. D. Williams, eds., *Inversion Tectonics,* pp. 177–200. Geological Society Special Publication, no. 44. London: Geological Society.

Childs, C., J. Watterson, and J. J. Walsh. 1995. Fault overlap zones within developing normal fault systems. *Journal of the Geological Society of London* 152:535–549.

Cohen, A. S. 1990. Tectono-stratigraphic model for sedimentation in Lake Tanganyika, Africa. In B. J. Katz, ed., *Lacustrine Basin Exploration: Case Studies and Modern Analogues,* pp. 137–150. American Association of Petroleum Geologists Memoir, vol. 50. Tulsa, Okla.: American Association of Petroleum Geologists.

Contreras, J., C. H. Scholz, and G. C. P. King. 1997. A model of rift basin evolution constrained by first order stratigraphic observations. *Journal of Geophysical Research* 102:7673–7690.

Cooper, M. A., and G. D. Williams, eds. 1989. *Inversion Tectonics.* Geological Society Special Publication, no. 44. London: Geological Society.

Cornet, B. 1977. The palynostratigraphy and age of the Newark Supergroup. Ph.D. diss., Pennsylvania State University.

Cornet, B., and P. E. Olsen. 1985. A summary of the biostratigraphy of the Newark Supergroup of eastern North America, with comments on early Mesozoic provinciality. In R. Weber, ed., *Simposio sobre floras del Triásico tardío, su fitogeografía y paleoecología: III Congreso Latinoamericano de Paleontología, México,* pp. 67–81. Mexico City: Instituto de Geología, Universidad Nacional Autonoma de México.

Costain, J. K., and C. Coruh. 1989. Tectonic setting of Triassic half-grabens in the Appalachians: Seismic data, acquisition, processing, and results. In A. J. Tankard and H. R. Balkwill, eds., *Extensional Tectonics and Stratigraphy of the North Atlantic Margins,* pp. 155–174. American Association of Petroleum Geologists Memoir, vol. 46. Tulsa, Okla.: American Association of Petroleum Geologists.

Coward, M. P. 1994. Inversion tectonics. In P. L. Hancock, ed., *Continental Deformation,* pp. 289–304. New York: Pergamon.

Coward, M. P. 1996. Balancing sections through inverted basins. In P. G. Buchanan and D. A. Nieuwland, eds., *Modern Developments in Structural Interpretation, Validation, and Modelling,* pp. 51–77. Geological Society Special Publication, no. 99. London: Geological Society.

Cowie, P. A., R. Knipe, and I. G. Main. 1996. Introduction to *Scaling Laws for Fault and Fracture Populations: Analyses and Applications.* Special issue of *Journal of Structural Geology* 18:v–xi.

Cowie, P. A., and C. H. Scholz. 1992. Physical explanation for displacement-length relationship of faults using a post-yield fracture mechanics model. *Journal of Structural Geology* 14:1133–1148.

Cowie, P. A., C. Vanneste, and D. Sornette. 1993. Statistical physics model for the spatiotemporal evolution of faults. *Journal of Geophysical Research* 98:21809–21821.

Crespi, J. M. 1988. Using balanced cross sections to understand early Mesozoic extensional faulting. In A. J. Froelich and G. R. Robinson Jr., eds., *Studies of the Early Mesozoic Basins of the Eastern United States,* pp. 220–229. U.S. Geological Survey Bulletin, no. 1776. Washington, D.C.: Government Printing Office.

Davison, I. 1994. Linked fault systems: Extensional, strike-slip, contractional. In P. L. Hancock, ed., *Continental Deformation,* pp. 121–142. New York: Pergamon.

Dawers, N. H., and M. H. Anders. 1995. Displacement-length scaling and fault linkage. *Journal of Structural Geology* 17:607–614.

Dawers, N. H., M. H. Anders, and C. H. Scholz. 1993. Growth of normal faults: Displacement-length scaling. *Geology* 21:1107–1110.

de Boer, J. Z. 1992. Stress configurations during and following emplacement of ENA basalts in the northern Appalachians. In J. H. Puffer and P. C. Ragland, eds., *Eastern North American Mesozoic Magmatism,* pp. 361–378. Geological Society of America Special Paper, no. 268. Boulder, Colo.: Geological Society of America.

de Boer, J. Z., and A. E. Clifton [incorrectly printed as "Clifford"]. 1988. Mesozoic tectogenesis: Development and deformation of "Newark" rift zones in the Appalachians (with special emphasis on the Hartford basin, Connecticut). In W. Manspeizer, ed., *Triassic–Jurassic Rifting: Continental Breakup and the Origin of the Atlantic Ocean and the Passive Margins,* pp. 275–306. Developments in Geotectonics, no. 22. Amsterdam: Elsevier.

de Boer, J. Z., G. H. McHone, J. H. Puffer, P. C. Ragland, and D. Whittington. 1988. Mesozoic and Cenozoic magmatism. In R. E. Sheridan and J. A. Grow, eds., *The Atlantic Continental Margin,* pp. 217–241. Vol. I-2 of *The Geology of North America.* Boulder, Colo.: Geological Society of America.

Dewey, J. F. 1988. Lithospheric stress, deformation, and tectonic cycles: The disruption of Pangaea and the closure of the Tethys. In M. G. Audley-Charles and A. Hallam, eds., *Gondwana and Tethys,* pp. 23–40. Geological Society Special Publication, no. 37. London: Geological Society.

Donohoe, H. V., Jr., and P. I. Wallace. 1982. *Geological Map of the Cobequid Highlands, Colchester, Cumberland, and Pictou Counties, Nova Scotia* [scale 1:50,000]. Halifax: Nova Scotia Department of Mines and Energy.

Dula, W. F., Jr. 1991. Geometric models of listric normal faults and rollover folds. *American Association of Petroleum Geologists Bulletin* 75:1609–1625.

Dunning, G. R., and J. D. Hodych. 1990. U-Pb zircon and baddeleyite age for the Palisade and Gettysburg sills of northeast United States: Implications for the age of the Triassic–Jurassic boundary. *Geology* 18: 795–798.

Faill, R. T. 1973. Tectonic development of the Triassic Newark-Gettysburg basin in Pennsylvania. *Geological Society of America Bulletin* 84:725–740.

Faill, R. T. 1988. Mesozoic tectonics of the Newark basin, as viewed from the Delaware River. In J. M. Husch and M. J. Hozik, eds., *Geology of the Central Newark Basin, Field Guide and Proceedings,* pp. 19–41. Fifth Meeting of the Geological Association of New Jersey. Lawrenceville: Geological Association of New Jersey.

Falvey, D. A. 1974. The development of continental margins in plate tectonic theory. *APEA Journal* 14:95–106.

Faulds, J. E., and R. J. Varga. 1998. The role of accommodation zones and transfer zones in the regional segmentation of extended terranes. In J. E. Faulds and J. H. Stewart, eds., *Accommodation Zones and Transfer Zones: The Regional Segmentation of the Basin and Range Province,* pp. 1–45. Geological Society of America Special Paper, no. 323. Boulder, Colo.: Geological Society of America.

Froelich, A. J., and P. E. Olsen. 1984. Newark Supergroup: A revision of the Newark Group in eastern North America. *U.S. Geological Survey Bulletin* 1537-A:55–58.

Frostick, L. E., and I. Reid. 1987. Tectonic control of desert sediments in rift basins ancient and modern. In L. Frostick and I. Reid, eds., *Desert Sediments: Ancient and Modern,* pp. 53–68. Geological Society Special Publication, no. 35. London: Geological Society.

Gawthorpe, R. L., and J. M. Hurst. 1993. Transfer zones in extensional basins: Their structural style and influence on drainage development and stratigraphy. *Journal of the Geological Society of London* 150:1137–1152.

Gibbs, A. D. 1983. Balanced cross-section construction from seismic sections in areas of extensional tectonics. *Journal of Structural Geology* 5:153–160.

Gibbs, A. D. 1984. Structural evolution of extensional basin margins. *Journal of the Geological Society of London* 141:609–620.

Gibson, J. R., J. J. Walsh, and J. Watterson. 1989. Modelling of bed contours and cross-sections adjacent to planar normal faults. *Journal of Structural Geology* 11:317–328.

Gillespie, P. A., J. J. Walsh, and J. Watterson. 1992. Limitations of dimension and displacement data from single faults and the consequences for data analysis and interpretation. *Journal of Structural Geology* 14:1157–1172.

Glaeser, J. D. 1966. *Provenance, Dispersal, and Depositional Environments of Triassic Sediments in the Newark-Gettysburg Basin.* Pennsylvania Geological Survey Bulletin, 4th ser., no. G43. Harrisburg: Pennsylvania Geological Survey.

Glennie, K. W., and P. L. E. Boegner. 1982. Sole Pit inversion tectonics. In L. V. Illing and G. D. Hobson, eds., *Petroleum Geology of the Continental Shelf of Northwest Europe,* pp. 110–120. London: Institute of Petroleum Geology.

Gupta, S., P. A. Cowie, N. H. Dawers, and J. R. Underhill. 1998. A mechanism to explain rift-basin subsidence and stratigraphic patterns through fault-array evolution. *Geology* 26:595–598.

Gupta, A., and C. H. Scholz. 1998. Utility of elastic models in predicting fault displacement fields. *Journal of Geophysical Research* 103:823–834.

Hamblin, W. K. 1965. Origin of "reverse drag" on the downthrown sides of normal faults. *Geological Society of America Bulletin* 76:1145–1164.

Hamilton, R. M., J. C. Behrendt, and H. D. Ackermann. 1983. Land multichannel seismic-reflection evidence for tectonic features near Charleston, S.C. In G. S. Gohn, ed., *Studies Related to the Charleston, S.C., Earthquake of 1886: Tectonics and Seismicity,* pp. I1–I18. U.S. Geological Survey Professional Paper, no. 1313. Washington, D.C.: Government Printing Office.

Harding, T. P. 1983. Structural inversion at Rambutan oil field, south Sumatra basin. In A. W. Bally, ed., *Seismic Expression of Structural Styles,* pp. 13–18. American Association of Petroleum Geologists Studies in Geology, no. 15, vol. 3. Tulsa, Okla.: American Association of Petroleum Geologists.

Harding, T. P. 1984. Graben hydrocarbon occurrences and structural style. *Amercan Association of Petroleum Geologists Bulletin* 68:333–362.

Hodych, J. P., and G. R. Dunning. 1992. Did the Manicouagan impact trigger end-of-Triassic mass extinction? *Geology* 20:51–54.

Hubert, J. F., A. A. Reed, W. L. Dowdall, and J. M. Gilchrist. 1978. *Guide to the Mesozoic Redbeds of Central Connecticut.* State Geological and Natural History Survey of Connecticut Guidebook, no. 4. Hartford: Connecticut Department of Environmental Protection.

Hutchinson, D. R., and K. D. Klitgord. 1988a. Deep structure of rift basins from the continental margin around New England. In A. J. Froelich and G. R. Robinson Jr., eds., *Studies of the Early Mesozoic Basins of the Eastern United States,* pp. 211–219. U.S. Geological Survey Bulletin, no. 1776. Washington, D.C.: Government Printing Office.

Hutchinson, D. R., and K. D. Klitgord. 1988b. Evolution of rift basins on the continental margin off southern New England. In W. Manspeizer, ed., *Triassic–Jurassic Rifting: Continental Breakup and the Origin of the Atlantic Ocean and the Passive Margins,* pp. 81–98. Developments in Geotectonics, no. 22. Amsterdam: Elsevier.

Hutchinson, D. R., K. D. Klitgord, and R. S. Detrick. 1986. Rift basins of the Long Island platform. *Geological Society of America Bulletin* 97:688–702.

Hutchinson, D. R., K. D. Klitgord, M. W. Lee, and A. M. Trehu. 1988. U.S. Geological Survey deep seismic reflection profile across the Gulf of Maine. *Geological Society of America Bulletin* 100:172–184.

Jackson, J. A. 1987. Active normal faulting and crustal extension. In M. P. Coward, J. F. Dewey, and P. L. Hancock, eds., *Continental Extensional Tectonics,* pp. 3–17. Geological Society Special Publication, no. 28. London: Geological Society.

Jackson, J. A., and D. McKenzie. 1983. The geometrical evolution of normal fault systems. *Journal of Structural Geology* 5:471–482.

Jackson, J. A., and N. J. White. 1989. Normal faulting in the upper continental crust: Observations from regions of active extension. *Journal of Structural Geology* 11:15–36.

Jones, B. D. 1994. Structure and stratigraphy of the Hopewell fault block, New Jersey and Pennsylvania. M.S. thesis, Rutgers University.

Kent, D. V., and P. E. Olsen. 1997. Paleomagnetism of Upper Triassic continental sedimentary rocks from the Dan River–Danville rift basin (eastern North America). *Geological Society of America Bulletin* 109:366–377.

Kent, D. V., and P. E. Olsen. 2000. Paleomagnetism of the Triassic–Jurassic Blomidon Formation in the Fundy basin: Implications for early Mesozoic tropical climate gradients. *Earth and Planetary Science Letters* 179:311–324.

Kent, D. V., P. E. Olsen, and W. K. Witte. 1995. Late Triassic–earliest Jurassic geomagnetic polarity sequence and paleolatitudes from drill cores in the Newark rift basin, eastern North America. *Journal of Geophysical Research* 100:14965–14998.

Keppie, J. D., comp. 1979. *Geological Map of Nova Scotia* [scale 1:500,000]. Halifax: Nova Scotia Department of Mines and Energy.

Keppie, J. D. 1982. The Minas geofracture. In P. St. Julien and J. Beland, eds., *Major Structural Zones and Faults of the Northern Appalachians,* pp. 1–34. Geological Association of Canada Special Paper, no. 24. Ottawa: Geological Association of Canada.

King, G., and M. Ellis. 1990. The origin of large local uplift in extensional regions. *Nature* 348:689–693.

King, G. C. P., R. S. Stein, and J. B. Rundle. 1988. The growth of geological structures by repeated earthquakes. 1. Conceptual framework. *Journal of Geophysical Research* 93:13307–13318.

King, P. B. 1971. Systematic pattern of Triassic dikes in the Appalachian region, second report. *U.S. Geological Survey Professional Papers* 750-D: 84–88.

Klitgord, K. D., D. R. Hutchinson, and H. Schouten. 1988. U.S. Atlantic continental margin: Structural and tectonic framework. In R. E. Sheridan and J. A. Grow, eds., *The Atlantic Continental Margin,* pp. 19–55. Vol. I-2 of *The Geology of North America.* Boulder, Colo.: Geological Society of America.

Klitgord, K. D., and H. Schouten. 1986. Plate kinematics of the central Atlantic. In P. R. Vogt and B. E. Tucholke, eds., *The Western North Atlantic Region,* pp. 351–378. Vol. M of *The Geology of North America.* Boulder, Colo.: Geological Society of America.

Lambiase, J. J. 1990. A model for tectonic control of lacustrine stratigraphic sequences in continental rift basins. In B. J. Katz, ed., *Lacustrine Exploration: Case Studies and Modern Analogues,* pp. 265–276. American Association of Petroleum Geologists Memoir, vol. 50. Tulsa, Okla.: American Association of Petroleum Geologists.

Lambiase, J. J., and W. Bosworth. 1995. Structural controls on sedimentation in continental rifts. In J. J. Lambiase, ed., *Hydrocarbon Habitat in Rift Basins,* pp. 117–144. Geological Society Special Publication, no. 80. London: Geological Society.

Larsen, P-H. 1988. Relay structures in a Lower Permian basement–involved extension system, East Greenland. *Journal of Structural Geology* 10:3–8.

Leavy, B. D., A. J. Froelich, and E. C. Abram. 1983. *Bedrock Map and Geotechnical Properties of Rocks of the Culpeper Basin and Vicinity, Virginia and Maryland* [scale 1:125,000]. U.S. Geological Survey Map I-1313-C. Washington, D.C.: Government Printing Office.

Leeder, M. R., and R. L. Gawthorpe. 1987. Sedimentary models for extensional tilt-block/half-graben basins. In M. P. Coward, J. F. Dewey, and P. L. Hancock, eds., *Continental Extensional Tectonics,* pp. 139–152. Geological Society Special Publication, no. 28. London: Geological Society.

LeTourneau, P. M. 1985. Alluvial fan development in the lower Portland Formation, central Connecticut: Implications for tectonics and climate. In G. R. Robinson Jr. and A. J. Froelich, eds., *Proceedings of the Second U.S. Geological Survey Workshop on the Early Mesozoic Basins of the Eastern United States,* pp. 17–26. U.S. Geological Survey Circular, no. 946. Washington, D.C.: Government Printing Office.

LeTourneau, P. M. 1999. Depositional history and tectonic evolution of Late Triassic age rifts of the U.S. central Atlantic margin: Results of an integrated stratigraphic, structural, and paleomagnetic analysis of the Taylorsville and Richmond basins. Ph.D. diss., Columbia University.

LeTourneau, P. M., and N. G. McDonald. 1988. Facies analysis of Early Jurassic lacustrine sequences, Hartford basin, Connecticut and Massachusetts. *Geological Society of America, Abstracts with Programs* 20:32.

Lindholm, R. C. 1978. Triassic–Jurassic faulting in eastern North America: A model based on pre-Triassic structures. *Geology* 6:365–368.

Lorenz, J. C. 1988. *Triassic–Jurassic Rift Basin Sedimentology.* New York: Van Nostrand Reinhold.

Lowell, J. D. 1995. Mechanics of basin inversion from worldwide examples. In J. G. Buchanan and P. G. Buchanan, eds., *Basin Inversion,* pp. 39–57. Geological Society Special Publication, no. 88. London: Geological Society.

Lucas, M., J. Hull, and W. Manspeizer. 1988. A foreland-type fold and related structures of the Newark rift basin. In W. Manspeizer, ed., *Triassic–Jurassic Rifting: Continental Breakup and the Origin of the Atlantic*

Ocean and the Passive Margins, pp. 307–332. Developments in Geotectonics, no. 22. Amsterdam: Elsevier.

Lyttle, P. T., and J. B. Epstein. 1987. *Geologic Map of the Newark 1° × 2° Quadrangle, New Jersey, Pennsylvania, and New York* [scale 1:250,000]. U.S. Geological Survey Miscellaneous Investigation Series Map I-1715. Washington, D.C.: Government Printing Office.

Manning, A. H., and J. Z. de Boer. 1989. Deformation of Mesozoic dikes in New England. *Geology* 17:1016–1019.

Manspeizer, W. 1980. Rift tectonics inferred from volcanic and clastic structures. In W. Manspeizer, ed., *Field Studies of New Jersey Geology and Guide to Field Trips: 52nd Annual Meeting, New York State Geological Association,* pp. 314–350. Newark, N.J.: Newark College of Arts and Sciences, Rutgers University.

Manspeizer, W. 1981. Early Mesozoic basins of the central Atlantic passive margin. In A. W. Bally, ed., *Geology of Passive Continental Margins: History, Structure, and Sedimentologic Record (with Special Emphasis on the Atlantic Margin),* pp. 1–60. American Association of Petroleum Geologists Education Course Note Series, no. 19, vol. 4. Tulsa, Okla.: American Association of Petroleum Geologists.

Manspeizer, W. 1982. Triassic–Liassic basins and climate of the Atlantic passive margins. *Geologische Rundschau* 73:895–917.

Manspeizer, W. 1988. Triassic–Jurassic rifting and opening of the Atlantic: An overview. In W. Manspeizer, ed., *Triassic–Jurassic Rifting: Continental Breakup and the Origin of the Atlantic Ocean and the Passive Margins,* pp. 41–79. Developments in Geotectonics, no. 22. Amsterdam: Elsevier.

Manspeizer, W., and H. L. Cousminer. 1988. Late Triassic–Early Jurassic synrift basins of the U.S. Atlantic margin. In R. E. Sheridan and J. A. Grow, eds., *The Atlantic Continental Margin,* pp. 197–216. Vol. I-2 of *The Geology of North America.* Boulder, Colo.: Geological Society of America.

Manspeizer, W., J. Z. de Boer, J. K. Costain, A. J. Froelich, C. Coruh, P. E. Olsen, J. G. McHone, J. H. Puffer, and D. C. Prowell. 1989. Post-Paleozoic activity. In R. D. Hatcher Jr., W. A. Thomas, and G. W. Viele, eds., *The Appalachian-Oachita Orogen in the United States,* pp. 319–374. Vol. F-2 of *The Geology of North America.* Boulder, Colo.: Geological Society of America.

Marrett, R., and R. W. Allmendinger. 1991. Estimates of strain due to brittle faulting: Sampling of fault populations. *Journal of Structural Geology* 13:735–738.

Mawer, C. K., and J. C. White. 1987. Sense of displacement on the Cobequid-Chedabucto fault system, Nova Scotia, Canada. *Canadian Journal of Earth Sciences* 24:217–223.

McBride, J. H., K. D. Nelson, and L. D. Brown. 1989. Evidence and implications of an extensive early Mesozoic rift basin and basalt/diabase sequence beneath the southeast Coastal Plain. *Geological Society of America Bulletin* 101:512–520.

McHone, J. G. 1988. Tectonic and paleostress patterns of Mesozoic intrusions in eastern North America. In W. Manspeizer, ed., *Triassic–Jurassic Rifting: Continental Breakup and the Origin of the Atlantic Ocean and the Passive Margins,* pp. 607–620. Developments in Geotectonics, no. 22. Amsterdam: Elsevier.

McInerny, D. P. 1993. Fluvial architecture and contrasting fluvial styles of the lower New Haven Arkose and mid–upper Portland Formation, early Mesozoic Hartford basin, central Connecticut. M.S. thesis, University of Massachusetts.

McLaughlin, D. B. 1963. Newly recognized folding in the Triassic of Pennsylvania. *Proceedings of the Pennsylvania Academy of Sciences* 37:156–159.

Meyertons, C. T. 1963. *Triassic Formations of the Danville Basin.* Virginia Division of Mineral Resources Reports of Investigations, vol. 6. Charlottesville: Virginia Division of Mineral Resources.

Morley, C. K. 1995. Developments in the structural geology of rifts over the last decade and their impact on hydrocarbon exploration. In J. J. Lambiase, ed., *Hydrocarbon Habitat in Rift Basins,* pp. 1–32. Geological Society Special Publication, no. 80. London: Geological Society.

Morley, C. K. 1999. Patterns of displacement along large normal faults: Implications for basin evolution and fault propagation, based on examples from East Africa. *American Association of Petroleum Geologists Bulletin* 83:613–634.

Muraoka, H., and H. Kamata. 1983. Displacement distribution along minor fault trace. *Journal of Structural Geologists* 5:483–495.

Musser, D. L. 1993. Structure, stratigraphy, and evolution of the Norfolk rift basin. M.S. thesis, Rutgers University.

Nadon, G. C., and G. V. Middleton. 1985. The stratigraphy and sedimentology of the Fundy Group (Triassic) of the St. Martins area, New Brunswick. *Canadian Journal of Earth Sciences* 22:1183–1203.

Oh, J., J. A. Austin Jr., J. D. Phillips, M. F. Coffin, and

P. L. Stoffa. 1995. Seaward-dipping reflectors offshore the southeastern United States: Seismic evidence for extensive volcanism accompanying sequential formation of the Carolina trough and Blake Plateau basin. *Geology* 23:9–12.

Olsen, P. E. 1980. Fossil great lakes of the Newark Supergroup in New Jersey. In W. Manspeizer, ed., *Field Studies of New Jersey Geology and Guide to Field Trips: 52nd Annual Meeting, New York State Geological Association,* pp. 352–398. Newark, N.J.: Newark College of Arts and Sciences, Rutgers University.

Olsen, P. E. 1997. Stratigraphic record of the early Mesozoic breakup of Pangea in the Laurasia–Gondwana rift system. *Annual Review of Earth and Planetary Science* 25:337–401.

Olsen, P. E., A. J. Froelich, D. L. Daniels, J. P. Smoot, and P. J. W. Gore. 1990. Rift basins of early Mesozoic age. In J. W. Horton Jr. and V. A. Zullo, eds., *The Geology of the Carolinas: Carolina Geological Society 50th Anniversary Volume,* pp. 142–170. Knoxville: University of Tennessee Press.

Olsen, P. E., and D. V. Kent. 1996. Milankovitch climate forcing in the tropics of Pangaea during the Late Triassic. *Palaeogeography, Palaeoclimatology, Palaeoecology* 122:1–26.

Olsen, P. E., D. V. Kent, B. Cornet, W. K. Witte, and R. W. Schlische. 1996. High-resolution stratigraphy of the Newark rift basin (early Mesozoic, eastern North America). *Geological Society of America Bulletin* 108: 40–77.

Olsen, P. E., D. V. Kent, S. J. Fowell, R. W. Schlische, M. O. Withjack, and P. M. LeTourneau. 2000. Implications of a comparison of the stratigraphy and depositional environments of the Argana (Morocco) and Fundy (Nova Scotia, Canada) Permian–Jurassic basins. In M. Oujidi and M. Et-Touhami, eds., *Le Permien et le Trias du Maroc: Actes de la Première Réunion du Groupe Marocain du Permien et du Trias,* pp. 165–183. Oujda, Morocco: Hilal Impression.

Olsen, P. E., and R. W. Schlische. 1990. Transtensional arm of the early Mesozoic Fundy rift basin: Penecontemporaneous faulting and sedimentation. *Geology* 18:695–698.

Olsen, P. E., R. W. Schlische, and M. S. Fedosh. 1996. 580 ky duration of the Early Jurassic flood basalt event in eastern North America estimated using Milankovitch cyclostratigraphy. In M. Morales, ed., *The Continental Jurassic,* pp. 11–22. Museum of Northern Arizona Bulletin, no. 60. Flagstaff: Museum of Northern Arizona.

Olsen, P. E., R. W. Schlische, and P. J. W. Gore, eds. 1989. *Tectonic, Depositional, and Paleoecological History of Early Mesozoic Rift Basins, Eastern North America.* International Geological Congress Field Trip T-351. Washington, D.C.: American Geophysical Union.

Oschchudlak, M. E., and J. F. Hubert. 1988. Petrology of Mesozoic sandstones in the Newark basin, central New Jersey and adjacent New York. In W. Manspeizer, ed., *Triassic–Jurassic Rifting: Continental Breakup and the Origin of the Atlantic Ocean and the Passive Margins,* pp. 333–352. Developments in Geotectonics, no. 22. Amsterdam: Elsevier.

Parker, R. A., H. F. Houghton, and R. C. McDowell. 1988. Stratigraphic framework and distribution of early Mesozoic rocks of the northern Newark basin, New Jersey and New York. In A. J. Froelich and G. R. Robinson Jr., eds., *Studies of the Early Mesozoic Basins of the Eastern United States,* pp. 31–39. U.S. Geological Survey Bulletin, no. 1776. Washington, D.C.: Government Printing Office.

Peacock, D. J. P., and D. J. Sanderson. 1991. Displacements, segment linkage, and relay ramps in normal fault zones. *Journal of Structural Geology* 13:721–733.

Pienkowski, A., and R. P. Steinen. 1995. Perennial lake cycles in the lower Portland Formation, Hartford, CT. *Geological Society of America, Abstracts with Programs* 27:74.

Plint, A. G., and H. W. van de Poll. 1984. Structural and sedimentary history of the Quaco Head area, southern New Brunswick. *Canadian Journal of Earth Sciences* 21:753–761.

Poag, C. W. 1991. Rise and demise of the Bahama–Grand Banks gigaplatform, northern margin of the Jurassic proto-Atlantic seaway. *Marine Geology* 102:63–130.

Prowell, D. C. 1988. Cretaceous and Cenozoic tectonism on the Atlantic coastal margin. In R. E. Sheridan and J. A. Grow, eds., *The Atlantic Continental Margin,* pp. 557–564. Vol. I-2 of *The Geology of North America.* Boulder, Colo.: Geological Society of America.

Ragland, P. C. 1991. Mesozoic igneous rocks. In J. W. Horton Jr. and V. A. Zullo, eds., *The Geology of the Carolinas: Carolina Geological Society 50th Anniversary Volume,* pp. 171–190. Knoxville: University of Tennessee Press.

Ragland, P. C., L. E. Cummins, and J. D. Arthur. 1992. Compositional patterns for early Mesozoic diabases from South Carolina to central Virginia. In J. H. Puffer and P. C. Ragland, eds., *Eastern North American Mesozoic Magmatism,* pp. 301–331. Geological Society of

America Special Paper, no. 268. Boulder, Colo.: Geological Society of America.

Ratcliffe, N. M. 1971. The Ramapo fault system in New York and adjacent New Jersey: A case of tectonic heredity. *Geological Society of America Bulletin* 82:127–141.

Ratcliffe, N. M. 1980. Brittle faults (Ramapo fault) and phyllonitic ductile shear zones in the basement rocks of the Ramapo seismic zones, New York and New Jersey, and their relationship to current seismicity. In W. Manspeizer, ed., *Field Studies of New Jersey Geology and Guide to Field Trips: 52nd Annual Meeting, New York State Geological Association,* pp. 278–311. Newark, N.J.: Newark College of Arts and Sciences, Rutgers University.

Ratcliffe, N. M., and W. C. Burton. 1985. Fault reactivation models for the origin of the Newark basin and studies related to U.S. eastern seismicity. In G. R. Robinson Jr. and A. J. Froelich, eds., *Proceedings of the Second U.S. Geological Survey Workshop on the Early Mesozoic Basins of the Eastern United States,* pp. 36–45. U.S. Geological Survey Circular, no. 946. Washington, D.C.: Government Printing Office.

Ratcliffe, N. M., and W. C. Burton. 1988. Structural analysis of the Furlong fault and the relationship of mineralization to faulting and diabase intrusion, Newark basin, Pennsylvania. In A. J. Froelich and G. R. Robinson Jr., eds., *Studies of the Early Mesozoic Basins of the Eastern United States,* pp. 176–193. U.S. Geological Survey Bulletin, no. 1776. Washington, D.C.: Government Printing Office.

Ratcliffe, N. M., W. C. Burton, R. M. D'Angelo, and J. K. Costain. 1986. Low-angle extensional faulting, reactivated mylonites, and seismic reflection geometry of the Newark basin margin in eastern Pennsylvania. *Geology* 14:766–770.

Ressetar, R., and G. K. Taylor. 1988. Late Triassic depositional history of the Richmond and Taylorsville basins, eastern Virginia. In W. Manspeizer, ed., *Triassic–Jurassic Rifting: Continental Breakup and the Origin of the Atlantic Ocean and the Passive Margins,* pp. 423–443. Developments in Geotectonics, no. 22. Amsterdam: Elsevier.

Reynolds, D. J. 1994. Sedimentary basin evolution: Tectonic and climatic interaction. Ph.D. diss., Columbia University.

Reynolds, D. J., and R. W. Schlische. 1989. Comparative studies of continental rift systems. *Geological Society of America, Abstracts with Programs* 21:61.

Roberts, D. G. 1989. Basin inversion in and around the British Isles. In M. A. Cooper and G. D. Williams, eds., *Inversion Tectonics,* pp. 131–150. Geological Society Special Publication, no. 44. London: Geological Society.

Rodgers, J. 1985. *Bedrock Geological Map of Connecticut* [scale 1:125,000]. Hartford: State Geological and Natural History Survey of Connecticut, Connecticut Department of Environmental Protection.

Root, S. I. 1988. Structure and hydrocarbon potential of the Gettysburg basin, Pennsylvania and Maryland. In W. Manspeizer, ed., *Triassic–Jurassic Rifting: Continental Breakup and the Origin of the Atlantic Ocean and the Passive Margins,* pp. 353–367. Developments in Geotectonics, no. 22. Amsterdam: Elsevier.

Root, S. I. 1989. Basement control of structure in the Gettysburg rift basin, Pennsylvania and Maryland. *Tectonophysics* 166:281–292.

Rosendahl, B. R. 1987. Architecture of continental rifts with special reference to East Africa. *Annual Review of Earth and Planetary Science* 15:445–503.

Sanders, J. E. 1963. Late Triassic tectonic history of northeastern United States. *American Journal of Science* 261:501–524.

Schlische, R. W. 1990. Aspects of the structural and stratigraphic evolution of early Mesozoic rift basins of eastern North America. Ph.D. diss., Columbia University.

Schlische, R. W. 1991. Half-graben basin filling models: New constraints on continental extensional basin evolution. *Basin Research* 3:123–141.

Schlische, R. W. 1992. Structural and stratigraphic development of the Newark extensional basin, eastern North America: Implications for the growth of the basin and its bounding structures. *Geological Society of America Bulletin* 104:1246–1263.

Schlische, R. W. 1993. Anatomy and evolution of the Triassic–Jurassic continental rift system, eastern North America. *Tectonics* 12:1026–1042.

Schlische, R. W. 1995. Geometry and origin of fault-related folds in extensional settings. *American Association of Petroleum Geologists Bulletin* 79:1661–1678.

Schlische, R. W., and R. V. Ackermann. 1995. Kinematic significance of sediment-filled fissures in the North Mountain Basalt, Fundy rift basin, Nova Scotia, Canada. *Journal of Structural Geology* 17:987–996.

Schlische, R. W., and M. H. Anders. 1996. Stratigraphic effects and tectonic implications of the growth of normal faults and extensional basins. In K. K. Beratan, ed., *Reconstructing the Structural History of Basin and*

Range Extension Using Sedimentology and Stratigraphy, pp. 183–203. Geological Society of America Special Paper, no. 303. Boulder, Colo.: Geological Society of America.

Schlische, R. W., and P. E. Olsen. 1988. Structural evolution of the Newark basin. In J. M. Husch and M. J. Hozik, eds., *Geology of the Central Newark Basin, Field Guide and Proceedings,* pp. 43–65. Fifth Meeting of the Geological Association of New Jersey. Lawrenceville: Geological Association of New Jersey.

Schlische, R. W., and P. E. Olsen. 1990. Quantitative filling model for continental extensional basins with applications to early Mesozoic rifts of eastern North America. *Journal of Geology* 98:135–155.

Schlische, R. W., M. O. Withjack, and G. Eisenstadt. 2002. An experimental study of the secondary deformation produced by oblique-slip normal faulting. *American Association of Petroleum Geologists Bulletin* 86:885–906.

Schlische, R. W., S. S. Young, R. V. Ackermann, and A. Gupta. 1996. Geometry and scaling relations of a population of very small rift-related normal faults. *Geology* 24:683–686.

Shaler, N. S., and J. B. Woodworth. 1899. Geology of the Richmond basin, Virginia. In *U.S. Geological Survey 19th Annual Report,* pp. 1246–1263. Washington, D.C.: Government Printing Office.

Shelton, J. W. 1984. Listric normal faults: An illustrated summary. *American Association of Petroleum Geologists Bulletin* 68:801–815.

Sheridan, R. E., and J. A. Grow, eds. 1988. *The Atlantic Continental Margin.* Vol. I-2 of *The Geology of North America.* Boulder, Colo.: Geological Society of America.

Sheridan, R. E., D. L. Musser, L. Glover III, M. Talwani, J. I. Ewing, W. S. Holbrook, G. M. Purdy, R. Hawman, and S. Smithson. 1993. Deep seismic reflection data of EDGE U.S. mid-Atlantic continental-margin experiment: Implications for Appalachian sutures and Mesozoic rifting and magmatic underplating. *Geology* 21:563–567.

Sibson, R. H. 1995. Selective fault reactivation during basin inversion: Potential for fluid redistribution through fault-valve action. In J. G. Buchanan and P. G. Buchanan, eds., *Basin Inversion,* pp. 3–19. Geological Society Special Publication, no. 88. London: Geological Society.

Silvestri, S. M. 1994. Facies analysis of Newark basin cores and outcrops. *Geological Society of America, Abstracts with Programs* 26:A-402.

Silvestri, S. M. 1997. Cycle correlation, thickening trends, and facies changes of individual paleolake highstands across the Newark basin, New Jersey and Pennsylvania. *Geological Society of America, Abstracts with Programs* 29:80.

Silvestri, S. M., and R. W. Schlische. 1992. Analysis of Newark basin drill cores and outcrop section: Tectonic and climatic controls on lacustrine facies architecture. *Geological Society of America, Abstracts with Programs* 24:76.

Smoot, J. P. 1991. Sedimentary facies and depositional environments of early Mesozoic Newark Supergroup basins, eastern North America. *Palaeogeography, Palaeoclimatology, Palaeoecology* 84:369–423.

Smoot, J. P., and P. E. Olsen. 1994. Climatic cycles as sedimentary controls of rift basin lacustrine deposits in the early Mesozoic Newark basin based on continuous cores. In A. J. Lomando, B. C. Schreiber, and P. M. Harris, eds., *Lacustrine Reservoirs and Depositional Systems,* pp. 201–237. Society of Economic Paleontologists and Mineralogists Core Workshop, no. 19. Tulsa, Okla: Society of Economic Paleontologists and Mineralogists.

Stein, R. S., and S. E. Barrientos. 1985. Planar high-angle faulting in the Basin and Range: Geodetic analysis of the 1983 Borah Peak, Idaho, earthquake. *Journal of Geophysical Research* 90:11355–11366.

Sutter, J. F. 1988. Innovative approaches to dating igneous events in the early Mesozoic basins, eastern North America. In A. J. Froelich and G. R. Robinson Jr., eds., *Studies of the Early Mesozoic Basins of the Eastern United States,* pp. 194–200. U.S. Geological Survey Bulletin, no. 1776. Washington, D.C.: Government Printing Office.

Swanson, M. T. 1982. Preliminary model for early transform history in central Atlantic rifting. *Geology* 10:317–320.

Swanson, M. T. 1986. Preexisting fault control for Mesozoic basin formation in eastern North America. *Geology* 14:419–422.

Thayer, P. A. 1970. Stratigraphy and geology of Dan River Triassic basin, North Carolina. *Southeastern Geology* 12:1–31.

Thayer, P. A., and N. S. Summer. 1996. Petrology and diagenesis of Upper Triassic fluvial sandstones, subsurface Dunbarton basin, S.C. In P. M. LeTourneau and P. E. Olsen, eds., *Aspects of Triassic–Jurassic Rift Basin Geoscience: Abstracts,* p. 53. State Geological and Natural History Survey of Connecticut Miscellaneous Reports, no. 1. Hartford: Connecticut Department of Environmental Protection.

Trudgill, B., and J. Cartwright. 1994. Relay ramp morphologies and normal fault linkages, Canyonlands National Park, Utah. *Geological Society of America Bulletin* 106:1143–1157.

Unger, J. D. 1988. A simple technique for analysis and migration of seismic reflection profiles from the Mesozoic basins of eastern North America. In A. J. Froelich and G. R. Robinson Jr., eds., *Studies of the Early Mesozoic Basins of the Eastern United States,* pp. 229–235. U.S. Geological Survey Bulletin, no. 1776. Washington, D.C.: Government Printing Office.

Venkatakrishnan, R., and R. Lutz. 1988. A kinematic model for the evolution of the Richmond Triassic basin. In W. Manspeizer, ed., *Triassic–Jurassic Rifting: Continental Breakup and the Origin of the Atlantic Ocean and the Passive Margins,* pp. 445–462. Developments in Geotectonics, no. 22. Amsterdam: Elsevier.

Volckmann, R. P., and W. L. Newell. 1978. New information on the tectonic framework of the Culpeper Triassic–Jurassic basin. *U.S. Geological Survey Professional Papers* 1100:51–52.

Walsh, J. J., and J. Watterson. 1987. Distributions of cumulative displacement and seismic slip on a single normal fault surface. *Journal of Structural Geology* 9:1039–1046.

Walsh, J. J., and J. Watterson. 1988. Analysis of the relationship between displacements and dimensions of faults. *Journal of Structural Geology* 10:239–247.

Walsh, J. J., and J. Watterson. 1991. Geometric and kinematic coherence and scale effects in normal fault systems. In A. M. Roberts, G. Yielding, and B. Freeman, eds., *The Geometry of Normal Faults,* pp. 193–203. Geological Society Special Publication, no. 56. London: Geological Society.

Watterson, J. 1986. Fault dimensions, displacements, and growth. *Pure and Applied Geophysics* 124:365–373.

Wheeler, G. 1939. Triassic fault-line deflections and associated warping. *Journal of Geology* 47:337–370.

White, J. M., and R. S. Wing. 1978. Structural development of the South China Sea with particular reference to Indonesia. In *Proceedings of the Seventh Annual Convention of the Indonesian Petroleum Association,* pp. 159–177. Jakarta: Indonesian Petroleum Association.

Williams, G. D., C. M. Powell, and M. A. Cooper. 1989. Geometry and kinematics of inversion tectonics. In M. A. Cooper and G. D. Williams, eds., *Inversion Tectonics,* pp. 3–15. Geological Society Special Publication, no. 44. London: Geological Society.

Williams, H. 1978. *Tectonic Lithofacies Map of the Appalachian Orogen* [scale 1:1,000,000]. St. John's: Memorial University of Newfoundland.

Wise, D. U. 1982. New fault and fracture domains of southwestern New England: Hints on localization of the Mesozoic basins. In O. C. Farquhar, ed., *Geotechnology in Massachusetts,* pp. 447–453. Amherst: University of Massachusetts Graduate School.

Wise, D. U. 1988. Mesozoic stress history of the upper Connecticut Valley at Turners Falls, Massachusetts. In W. A. Bothner, ed., *Guidebook for Field Trips in Southwestern New Hampshire, Southeastern Vermont, and North-Central Massachusetts,* pp. 351–372. New England Intercollegiate Geological Conference, 80th Annual Meeting. Durham: University of New Hampshire.

Wise, D. U. 1993. Dip domain method applied to the Mesozoic Connecticut Valley rift basins. *Tectonics* 11:1357–1368.

Withjack, M. O., and D. J. Drickman-Pollock. 1984. Synthetic seismic-reflection profiles and rift-related structures: *American Association of Petroleum Geologists Bulletin* 68:1160–1178.

Withjack, M. O., P. E. Olsen, and R. W. Schlische. 1995. Tectonic evolution of the Fundy basin, Canada: Evidence of extension and shortening during passive-margin development. *Tectonics* 14:390–405.

Withjack, M. O., R. W. Schlische, and P. E. Olsen. 1998. Diachronous rifting, drifting, and inversion on the passive margin of eastern North America: An analog for other passive margins. *American Association of Petroleum Geologists Bulletin* 82:817–835.

Witte, W. K., D. V. Kent, and P. E. Olsen. 1991. Magnetostratigraphy and paleomagnetic poles from Late Triassic–earliest Jurassic strata of the Newark basin. *Geological Society of America Bulletin* 103:1648–1662.

Xiao, H., and J. Suppe. 1992. Origin of rollover. *American Association of Petroleum Geologists Bulletin* 76:509–529.

Zandt, G., and T. J. Owens. 1980. Crustal flexure associated with normal faulting and implications for seismicity along the Wasatch front, Utah. *Seismological Society of America Bulletin* 70:1501–1520.

Zen, E., R. Goldsmith, N. M. Ratcliffe, P. Robinson, and R. S. Stanley. 1983. *Bedrock Geologic Map of Massachusetts* [scale 1:250,000]. Washington, D.C.: U.S. Geological Survey, Government Printing Office.

Zhang, P., D. B. Slemmons, and F. Mao. 1991. Geometric pattern, rupture termination, and fault segmentation of the Dixie Valley–Pleasant Valley active normal fault system, Nevada, U.S.A. *Journal of Structural Geology* 13:165–176.

Ziegler, P. A. 1987. Compressional intraplate tectonics in the Alpine foreland. *Tectonophysics* 137:389–420.

Ziegler, P. A. 1989. Geodynamic model for Alpine intraplate compressional deformation in western and central Europe. In M. A. Cooper and G. D. Williams, eds., *Inversion Tectonics*, pp. 63–85. Geological Society Special Publication, no. 44. London: Geological Society.

Zoback, M. L. 1992. Stress field constraints on intraplate seismicity in eastern North America. *Journal of Geophysical Research* 97:11761–11782.

5

Tectonics of the Lantern Hill Fault, Southeastern Connecticut: Embryonic Rifting of Pangea Along the Central Atlantic Margin

Robert J. Altamura

The Lantern Hill fault is a 16 km long, N–S-trending, hypersilicified, breccia zone that transgresses meta-igneous and metasedimentary units of the Proterozoic Z Avalonian terrane in southeastern Connecticut. The fault zone is massively silicified along its length, especially at its northern end, where it intersects the Honey Hill fault—the terrane boundary between Avalonian and Gander terranes.

An analysis of stress and sense-of-slip indicators as recorded by faults and quartz-filled fractures in the Lantern Hill fault zone indicates that it formed in a region of crustal extension with σ_1 subvertical and σ_3 subhorizontal along an azimuth of 105°. Analysis of borehole data indicates that the main silicified zone strikes N–NE and dips approximately 50° west. The silicified zone consists of steeply dipping, 0.6 m thick, parallel quartz veins. Quartz-vein arrays, which intersect the trend of the Lantern Hill fault at an angle of approximately 15°, represent evidence of oblique-slip normal faulting. Based on radiometric age dating, main-stage fracturing and mineralization in the Lantern Hill fault occurred during the Middle Triassic. Deposition of redbeds in the Hartford basin, located approximately 45 km to the west of the Lantern Hill fault, began during the Late Triassic. The tectonic setting and regional stress field in which this rift basin formed is the same as that responsible for the Lantern Hill fault. Hypersilicified breccia zones similar to the Lantern Hill fault occur in the footwall of the eastern border fault of the Hartford basin and in the Narragansett basin, Rhode Island, and may be time correlative with the Lantern Hill fault silicified zones.

In the proposed model, regional WNW–ESE extension was accompanied by silicification and brecciation during the embryonic rifting of the central Atlantic margin during the Middle Triassic (238 Ma)—approximately 23 Ma prior to deposition of graben-fill sediments and emplacement of tholeiitic basalts in the Hartford basin. Whereas establishment of the Hartford rift marked the end of significant extensional tectonics along the Lantern Hill fault, the establishment of the mid-Atlantic Ocean spreading center during the Jurassic probably marked an end to extensional deformation in the region of the Hartford basin. At this time, the regional stress configuration changed to approximately E–W compression because of ridge push.

An unusual system of quartz veins found in the region of Lantern Hill in southeastern Connecticut is related to the early Mesozoic tectonic evolution of southern New England and the incipient rifting of Pangea during the Middle Triassic.

Geologic Setting

The end of the Paleozoic and the beginning of the Mesozoic was marked by a transition from regional crustal compression to regional crustal extension. In southern New England, evidence for late Paleozoic compressional tectonics is present within the Carboniferous deposits of the Narragansett basin in Rhode Island (Mosher 1983; Skehan, Rast, and Mosher 1986) and of the Worcester basin in eastern Massachusetts (Skehan, Rast, and Mosher 1986; Goldstein 1994). A late Paleozoic thermal event is indicated by metamorphic cooling ages for southern Rhode Island (Dallmeyer 1982), southeastern Connecticut (this study), south-central Connecticut (Wintsch et al. 1992), and Permian granitic intrusions in southern New England (Quinn 1971; Rodgers 1985; Hermes, Gromet, and Murray 1994). Felsic pegmatites associated with the Narragansett Pier Granite were emplaced in the Lantern Hill study area around 272 Ma or later (O. D. Hermes, personal commmunication 1990).

O'Hara and Gromet (1983) identified Permian to Triassic Rb-Sr apparent ages for micas and plagioclase in mylonites of the Honey Hill fault zone in southeastern Connecticut. Based on field study and analysis of microstructures, Goldstein (1984, 1989) proposed that low-angle normal faulting was the last operative phase of deformation along the Honey Hill–Lake Char fault zone. Getty (1988) proposed a late Paleozoic extension in the Willimantic fault zone (a correlative of the Honey Hill–Lake Char fault zone) based on radiometric ages.

Wintsch et al. (1992) concluded that in southern New England the Avalonian plate was still underthrusting the Gander plate (their central Maine terrane) during the Pennsylvanian at approximately 300 Ma. Major differences in the cooling histories of these two terranes led these researchers to conclude that collision between the Avalonian and Gander terranes ended in Late Permian time (< 250 Ma). In contrast, however, Goldstein (1989, 1994) concluded that low-angle normal faulting associated with the Alleghenian orogeny occurred approximately from 260 Ma to 235 Ma. This latter tectonic scenario was a result of either syncompressional normal faulting, similar to Himalayan examples, or extension caused by a releasing bend in a major Alleghenian fault system. In Goldstein's view (1989, 1994, personal communication 1996), the final assembly of Pangean terranes occurred early during the Acadian. In contrast, Wintsch et al. (1992) found evidence for the complete assembly of Pangea during the Late Permian.

Early Mesozoic tectonics that followed are referred to as the Newark phase of tectonogenesis (de Boer and Clifton 1988) and are believed by many geologists working in the Newark Supergroup to be related in time to the basin-fill deposits. Paleontological data (Cornet 1977, personal communication 1995), cyclostratigraphy (Olsen 1997), and regional magnetostratigraphy (Olsen et al. 1996) suggest that the Hartford rift basin in southern New England started to fill at approximately 215 Ma and continued until at least 195 Ma. There is substantial evidence that an early phase of rifting occurred in the Taylorsville basin, Virginia (LeTourneau 1999), and in the Fundy basin, Canada (Olsen and Kent 2000), approximately from 230 Ma to 215 Ma. Tectonic and radiogenic age data (this study) indicate that the initiation of crustal extension associated with the Newark tectonogenesis is significantly earlier than previously believed.

Location of Study Area

The Lantern Hill fault zone is located in southeastern Connecticut, near North Stonington. It is approximately 45 km from the eastern edge of the Triassic–Jurassic Hartford basin and approximately 40 km from the western edge of the Carboniferous Narragansett basin in Rhode Island (figures 5.1 and 5.2). It is remarkable for its predominance of regional-scale tectonic features, including the Honey Hill–Lake Char fault, which is a likely suture of the Gander and Avalonian terranes (figure 5.2).

The Lantern Hill Silicified Zone

The Lantern Hill and Snake Meadow Brook Fault System

The Lantern Hill fault is a 16 km long, N–S-trending, hypersilicified, breccia zone that transgresses meta-

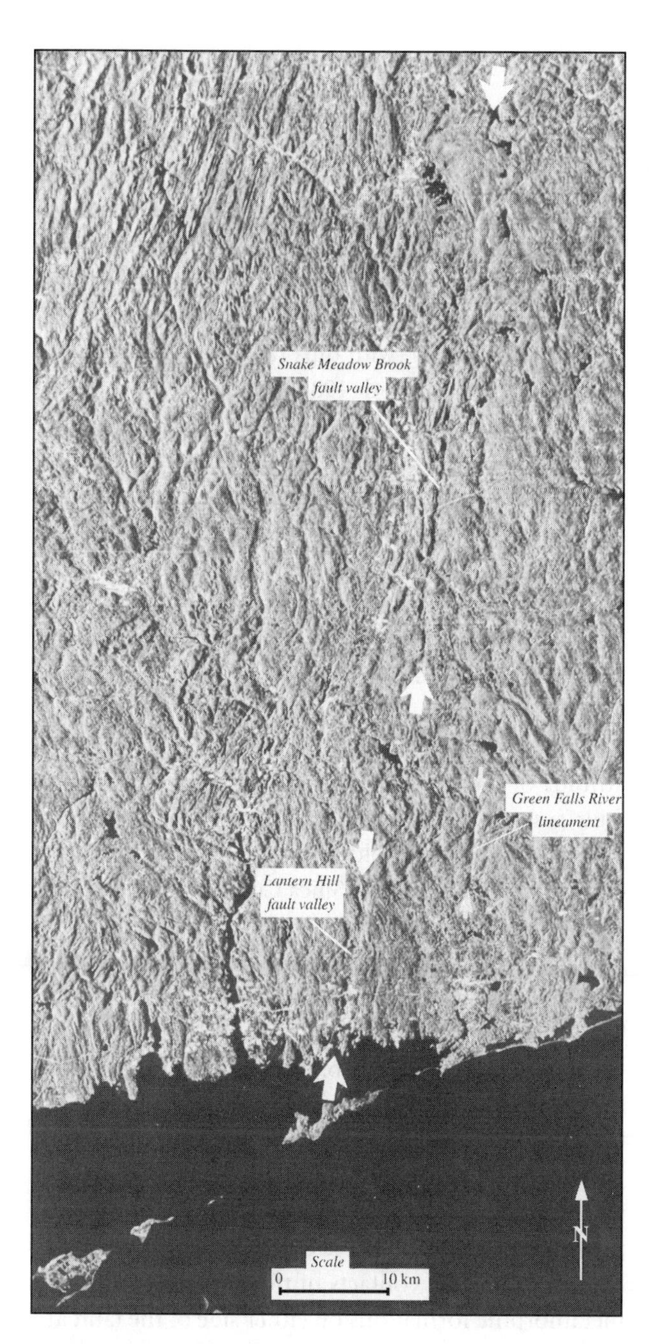

FIGURE 5.1 Synthetic aperture radar (SAR) image of central and eastern Connecticut, showing the fault valleys of the Snake Meadow Brook and the Lantern Hill faults. Parallel lineament to the east, here identified as the Green Falls River lineament, may represent a related fault zone (With permission of the State Geological and Natural History Survey of Connecticut)

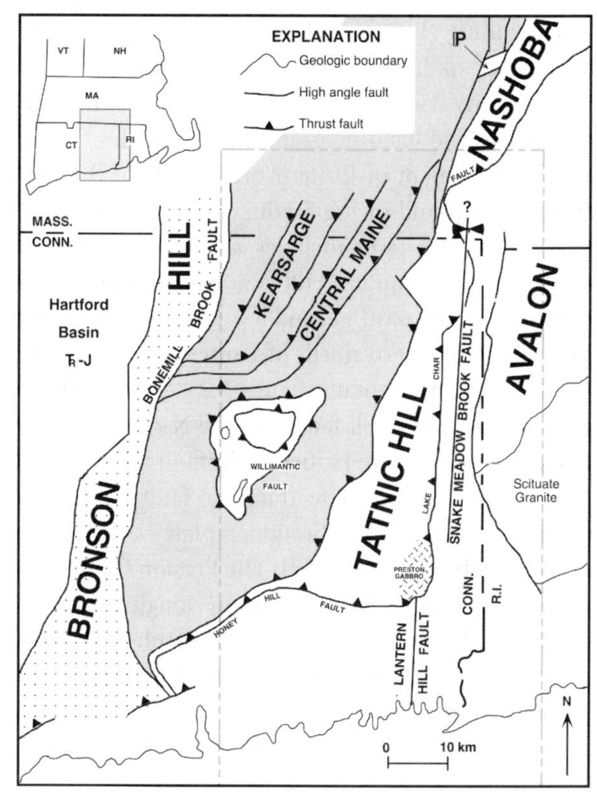

FIGURE 5.2 Terrane map of eastern Connecticut and adjacent areas in Massachusetts and Rhode Island. Note the location of the Lantern Hill fault with respect to the terrane boundary between Avalon and the Tatnic Hill–Nashoba belt (Gander terrane). The Lantern Hill vein complex occurs at the intersection of the Lantern Hill fault with this boundary, the Honey Hill fault.

igneous and metasedimentary units of the Proterozoic Z Avalonian terrane in southeastern Connecticut. The Snake Meadow Brook fault is a remarkably linear, 50 km long, N- to NE-trending silicified fault in eastern Connecticut that was identified through analysis and field checking of synthetic aperture radar imagery (Altamura 1985, 1987; Altamura and Quarrier 1986) (figure 5.1). The Snake Meadow Brook fault cuts the Avalonian terrane and is interpreted as the en echelon continuation of the Lantern Hill fault (Altamura 1987; State Geological and Natural History Survey of Connecticut 1990). This composite system, the Lantern Hill–Snake Meadow Brook fault system, traverses the entire state of Connecticut in a N–S direction over a distance of more than 70 km. The silicified Wekepeke fault in Massachusetts identified by Peck (1975) may represent a farther northward continuation of the fault system. To the south, the system may be extended to the Mesozoic basins interpreted from geophysical data (Hutchinson, Klitgord, and Detrick 1986) below the continental shelf of the Atlantic Ocean.

The surface traces of both the Lantern Hill fault and the Snake Meadow Brook fault are well expressed as

NNE-trending linear valleys. Whereas the Snake Meadow Brook fault parallels the strike of the tectonic grain in its Avalonian host rocks, the Lantern Hill fault transgresses the tectonic grain at right angles (figure 5.2). Displacement of Proterozoic Z units of the Waterford Group and of the Sterling Plutonic Group by the Lantern Hill fault includes a 0.4 km left-lateral component. The Lantern Hill fault is mapped south to the Connecticut coastline, and it is traceable to a point approximately 400 m north of Lantern Hill in North Stonington, Connecticut (figure 5.2). The northern end of the Lantern Hill fault cuts the N-dipping, low-angle, late Paleozoic Honey Hill fault (Goldsmith 1985), but it is difficult to trace the fault northward into the main body of the upper plate rocks of the Preston Gabbro (Dixon 1982). The Preston Gabbro, in combination with the mechanically tough mylonites and ultramylonites of the Honey Hill fault zone, may have acted as a resistant boundary, forcing the eastward en echelon side step of the Lantern Hill fault toward its northward continuation—the Snake Meadow Brook fault.

Geology of the Lantern Hill Silicified Zone

The Lantern Hill silicified zone is located within the moderate- to high-angle Lantern Hill fault zone, in the footwall of the Honey Hill fault. The map outline of this quartz-vein complex delineates a planar body approximately 1.6 km long and 70 m wide that appears to truncate and flare at its northern end near the contact with mylonites of the Honey Hill fault (Altamura 1993) (figure 5.3). The Lantern Hill silicified zone may continue beneath the Honey Hill fault. Minor quartz veins, similar in composition and attributed to the Lantern Hill silicified zone, occur locally along the Honey Hill–Lake Char fault and within the upper plate (Sclar 1958; Dixon 1982; Goldsmith 1985). The silicified zone consists of parallel arrays of subvertical quartz veins that strike slightly oblique to the trend of the regional fault zone (Altamura 1987). These veins are oriented such that they contain the σ_2 axis and are at 30° to the Lantern Hill fault zone. This relationship suggests that faulting and mineralization were coeval.

Rock units of the Avalonian terrane that are transgressed by the Lantern Hill fault in the study area include Hope Valley Alaskite Gneiss, Plainfield Formation (interbedded quartzite and mica schist and gneiss), and Waterford Group units, including the Mamacoke Formation (gneiss). Gander terrane rock units in the study area are Preston Gabbro, Preston Gabbro dioritic phase, Quinebaug Formation (gneiss), and mylonite along Paleozoic faults. Outcrops of mylonite and ultramylonites of gabbroic composition occur approximately 35 m due north of the Lantern Hill silicified zone. Foliation in Avalon units is parallel to geologic contacts, and in the vicinity of the Lantern Hill fault it strikes E–W, with dips ranging from 30 to 45° to the north (Goldsmith 1985; this study).

Within the fault zone, foliation (S_1) in metasomatized Plainfield Formation units shows evidence of mesoscale folding and shearing within the ductile regime (Altamura 1995a). Fold axes plunge to the west and southwest from 45 to 90°, and foliation defining the folds is truncated along shear planes (S_2). These features represent evidence of an earlier deformation and are restricted to wall rock caught up in the fault zone. Quartz veins cut these ductile folds and shears.

Lack of exposures make it difficult to determine the attitude of the Lantern Hill and Snake Meadow Brook fault systems. However, detailed mapping (Altamura 1993; Altamura and Gold 1994) and analysis of drilling data obtained by the U.S. Silica Company indicate that the dip of the silicified zone—and, by inference, of the fault zone—is between 30 and 60° to the west. The dip is steepest in the south and shallowest in the north. The mean dip obtained for 10 cross sections is 51° west (R. Blodgett, U.S. Silica Co., personal communication 1986). Based on field mapping, Goldsmith (1985) estimated the cumulative throw on the fault at Lantern Hill to be approximately 420 m (west side down), whereas this study indicates a maximum throw of approximately 1,200 m. These estimates are approximate because lithologic contacts of the moderately dipping metamorphic formations on either side of the fault are difficult to place accurately.

Vein arrays of the silicified zone vary as much as 15° clockwise from the NNE trend of the fault zone. Vein textures throughout the complex range from massive (homogeneous and without crystal outline) to vuggy. Field observations identified multiple vein sets that comprise three principal stages of emplacement: early stage, main stage, and late stage. Early- and main-stage quartz is brecciated and lithified to varying degrees by main-stage quartz. It is possible to distinguish a sequence of ages among the vein sets from crosscutting relationships. The vein sets are denoted QV_1 for the oldest through QV_9 for the youngest. Early-stage veins (QV_1 to QV_5) are found only in the metasomatized rock of the hanging wall and footwall; main-stage veins (QV_6) occur only in the core silicified zone, whereas later veins (QV_7 to QV_9) can be found in both wall rock and the silicified core zone. Friable and silicified quartz breccias occur locally in zones, the largest occurring at the southern end of the silicified zone. Breccia zones do not appreciably offset the silicified zone. Mapping has shown that the friable breccia grades into a well-lithified breccia in the north.

Net horizontal displacement of the Lantern Hill fault zone is 700 m and is sinistral based on offset of contacts in the hanging wall and footwall. There is also a rotation of the true dip of metamorphic units from 30 to 50° between the hanging wall and the footwall (Goldsmith 1985; Rodgers 1985) (figure 5.3).

The topographic high of the Lantern Hill silicified zone and quarry exposures provide a fortuitous opportunity to collect mesoscale structural data in the heart of a fault zone. The Lantern Hill fault occurs as a valley elsewhere, and glacial drift covers all but the margins of the fault zone. Differential weathering between the quartz silicified zone and the surrounding schists and gneisses has resulted in approximately 128 m of relief, with a maximum elevation of 148 m, the highest point along Connecticut's coastal slope.

Veins of the silicified zone range from 1 to 60 cm in thickness. Most of these veins and minor fault planes that bound and cut the veins trend north to northeast. Although veins of the silicified zone generally are oriented at a slight oblique angle to the trend of the regional fault valley, a few are parallel to it (Altamura 1987) (figure 5.3). Other minor veins are present in stockwork patterns within the metasomatized wall rock.

More than 550 quartz veins (figure 5.4) throughout the Lantern Hill complex were identified and characterized on the basis of orientation, thickness, texture, accessory minerals, and crosscutting relationships (Altamura 1995a). The orientation of a stress field may vary locally within a fault zone at a given time, and it is possible that orientation of the resulting quartz veins may also vary. For this reason, grouping of quartz veins into vein sets was based on orientation and on as many of the criteria given here as possible. The nine vein sets for the Lantern Hill complex (QV_1 to QV_9) are classified on the basis of physical similarity within a group,

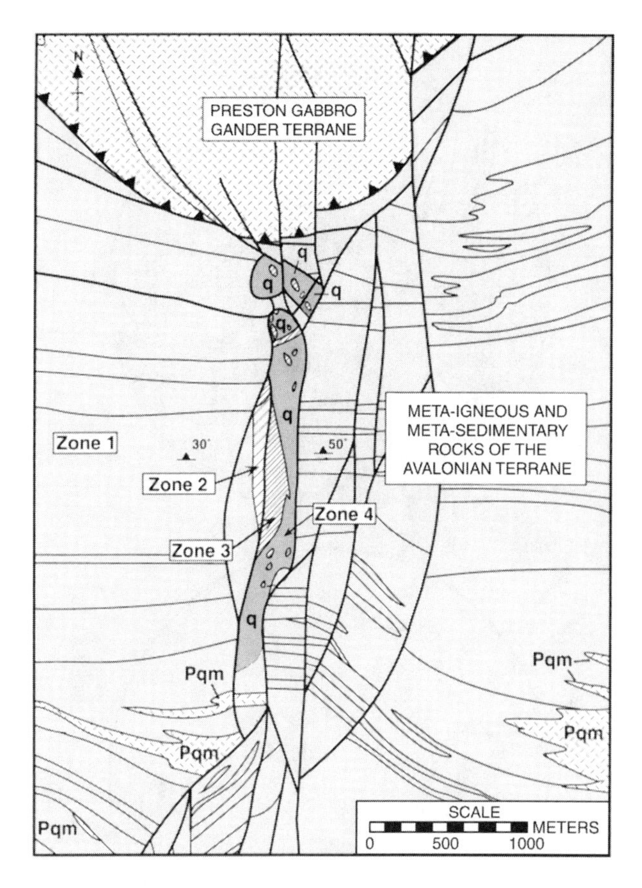

FIGURE 5.3 Generalized bedrock geologic map of the study area, showing fundamental geologic units. Four petrologic zones pertinent to the Lantern Hill fault zone are indicated: Zone l, unaltered country rock; Zone 2, distal alteration aureole; Zone 3, proximal alteration aureole; and Zone 4, the silicified zone; *Pqm,* Permian quartz monzonite of Goldsmith (1985); *q,* quartzose rock of the silicified zone.

with an emphasis on crosscutting relationships (figure 5.5A). With the exception of QV_2 and QV_9, vein sets are pervasive throughout the Lantern Hill vein complex. QV_2 and QV_9 are so few in number that it is assumed that they represent distinct deformation events; therefore, any interpretation of their significance should be viewed with caution.

Nevertheless, the Lantern Hill fault zone shows a clear pattern of strain variation through time (figure 5.5). The regional extension pattern (QV_6) is distinguishable from other minor vein sets that resulted from strain along localized faults as the Lantern Hill fault and extension zone evolved. The Lantern Hill fault zone is modeled from vein data as a NNE-striking zone resulting from perpendicular to subperpendicular extension oriented WNW–ESE across the zone.

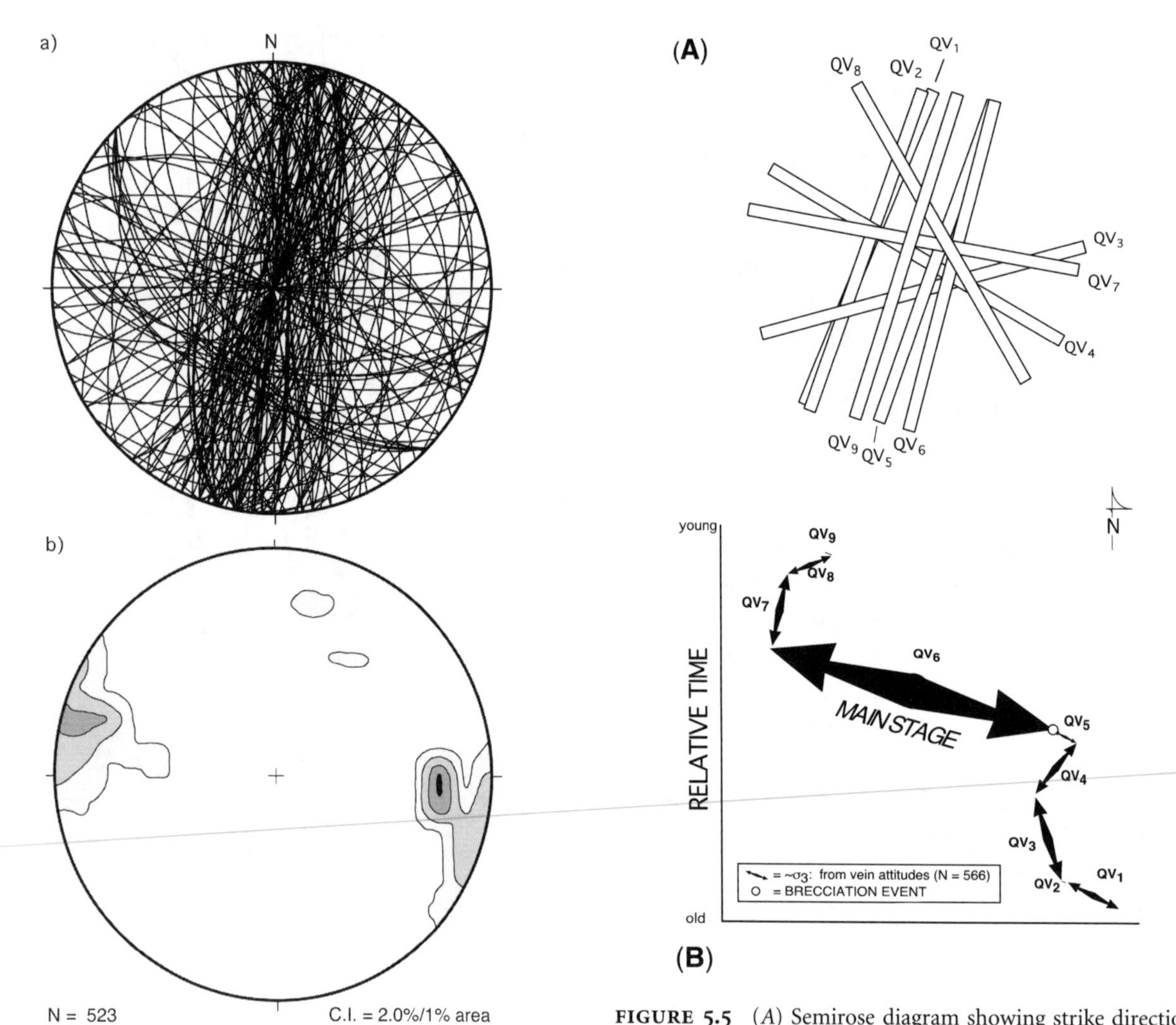

FIGURE 5.4 (*a*) Equal-area stereogram depicting veins as great circles on the lower hemisphere. Note that the vast majority of veins trend NNE and are steeply dipping. Veins with other orientations are common in the stockwork veins that make up the wall rock. (*b*) Equal-area–contoured stereogram representing the normal to vein attitudes.

FIGURE 5.5 (*A*) Semirose diagram showing strike directions for the major quartz veins in the Lantern Hill vein complex, North Stonington, Connecticut. From oldest to youngest, these vein sets are labeled QV_1 through QV_9. Most veins are steeply dipping to vertical, and mean strike and dip data are provided in the text. Note that the dominant strike direction is NNE and at a slight angle to the main trend of the Lantern Hill fault zone. (*B*) Chronological extensional stress summary for quartz veins of the Lantern Hill quartz-vein complex. The doubly pointing arrow represents the interpreted extension direction during vein growth, and its size is scaled to reflect abundance within the complex. QV_1 through QV_5 occur as stockwork veins in wall rock (Zone 3) adjacent to the silicified core zone (Zone 4). QV_6 is the dominant vein set that composes the silicified zone and represents main-stage silicification. QV_7 through QV_9 are late-stage and cut both the silicified zone and wall rocks.

Evidence of right-lateral shear along the fault zone indicates that from time to time the buildup of shear stress may have caused minor strike-slip movement within the dominant WNW extension field (figure 5.5B). The vein data likely reflects the changing dominance of the two stress orientations.

An alternative model that accounts for multiple vein sets at Lantern Hill recognizes that in extensional regimes where σ_1 is vertical it is not uncommon for σ_2 and σ_3 to be of similar intensity or even to change position (J. Z. de Boer, personal communication 1997). If this alternative model is applied to the Lantern Hill vein data, then the dominant NE-trending vein sets of main-stage silicification of the Lantern Hill fault formed during NW maximum extension (i.e., σ_3 orientation; see figure 5.5). Under this stress regime, right-lateral slip along the fault would have accompa-

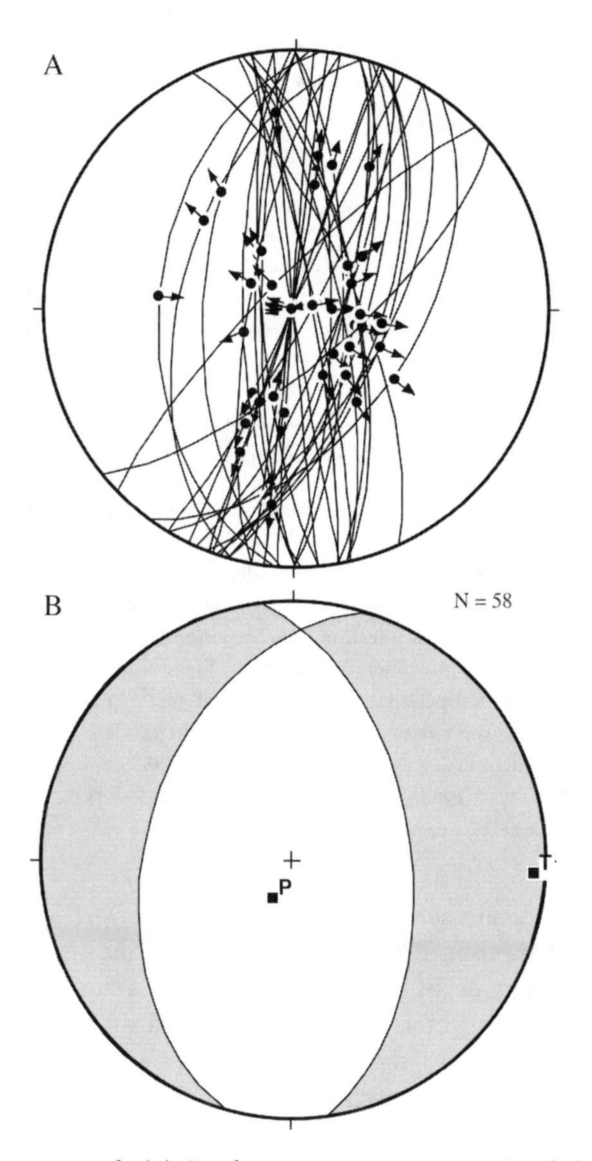

FIGURE 5.6 (*A*) Equal-area stereogram representing fault planes from the Lantern Hill fault zone as great circles on the lower hemisphere. Slip vectors for faults are indicated by the arrows. The population is dominated by normal faults, although a few reverse and strike-slip faults were found. The dominant trend is NNE, and dips are moderate to steep. (*B*) Equal-area stereogram showing the fault plane solution for Lantern Hill faults represented in *A*. The principal shortening direction (*P*) is steeply dipping to the SSW, and the orientation of principal extension (*T*) is horizontal and ESE.

nied the normal faulting. It is also possible that at times σ_2 and σ_3 changed position, such that maximum extension was NE, and the NW-trending veins formed (figure 5.5). Additional Lantern Hill veins sets with other orientations may have formed during the σ_2 to σ_3 transitions.

Minor faults occur in the silicified zone and adjacent metamorphic rocks on either side of the Lantern Hill fault. The majority of these trend N–NE and dip steeply east or west (figure 5.6A). Slickenlines indicate principally a subvertical dip-slip (down) motion. However, left-lateral faults with slickenlines that plunge 25° cut main-stage quartz veins of the silicified zone (QV_6). Some of these faults in turn are cut by a few late-stage quartz veins. Kinematic analysis of fault data (figure 5.6B) and vein data (figure 5.4) indicates that an overall paleostress field associated with brittle deformation and silica mineralization along the Lantern Hill fault had a subhorizontal σ_3 (azimuth of ESE–WNW; 105°/285°) and a subvertical σ_1. The direction of regional extension near the Lantern Hill fault zone is similar to that which formed basin-bounding normal faults of the Hartford basin located west of the Lantern Hill area (de Boer and Clifton 1988).

Age of Main-Stage Silicification of the Lantern Hill Fault

Several minor fault planes in the Lantern Hill and Snake Meadow Brook fault system contain unaltered muscovite crystals measuring 0.1 to 0.5 cm across. These crystals occur on slickensided quartz veins and as euhedra within crystal-lined vugs within quartz veins. This muscovite is inferred to be syntectonic; it was emplaced along with quartz during brittle deformation along steeply dipping extensional fractures. Vein muscovites in main-stage veins (QV_6) (figure 5.7) of the Lantern Hill silicified zone have been dated at 238 Ma by $^{40}Ar/^{39}Ar$ and K/Ar methods (Altamura 1987; Altamura and Lux 1994, in preparation). Muscovite in unaltered Avalonian host rock adjacent to the Lantern Hill silicified zone provided ages of 255 Ma (i.e., cooling ages [Altamura and Lux 1994]). These absolute ages demonstrate that main-stage silicification and coeval fracturing of the Lantern Hill fault zone continued into the early Middle Triassic (Anisian). Muscovite from quartz veins in the Snake Meadow Brook fault have provided a 234 Ma K/Ar age (Altamura 1987), and although cooling ages have not been obtained for metamorphic muscovite in its host rocks, it seems likely that this age represents a crystallization age as well. The transition from convergent Alleghenian tectonics to the divergent Triassic–Jurassic Newark tectonics in southern New England thus may have been relatively rapid.

FIGURE 5.7 Scanning electron microscope image of intergrown euhedral quartz and muscovite crystals lining a vug within a main-stage quartz vein (QV_6) of the Lantern Hill quartz-vein complex. Muscovite yielded a 238 Ma plateau age, and fluid inclusions in quartz yielded mean trapping temperatures of 265°C. For discussion, see the text. (Photograph by H. E. Belkin, U.S. Geological Survey)

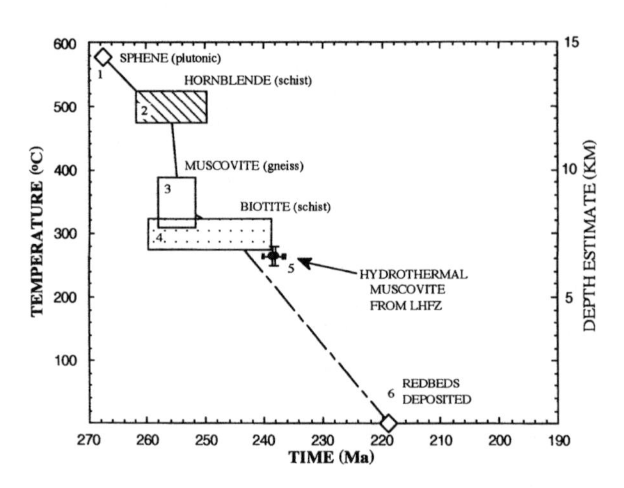

FIGURE 5.8 Cooling curve for host rocks of the Lantern Hill silicified fault, showing the position of the hydrothermal muscovites from the vein complex. Hydrothermal muscovites plot off the regional cooling curve, providing evidence that the absolute age is a crystallization age. The closure temperatures for the metamorphic minerals are from the literature, and the crystallization temperature of hydrothermal muscovites is from fluid inclusion data from cogenetic quartz. (Data from *1,* Wintsch and Aleinikoff 1987; *2,* Dallmeyer 1982; *3,* this study; *4,* Dallmeyer 1982; *5,* this study; and *6,* Cornett 1977; Olsen and Kent 2000)

A cooling/uplift curve for the late Paleozoic to Middle Triassic (i.e., Alleghenian through Newark tectonic periods) was constructed from both published radiometric ages of minerals from Avalonian country rock formations adjacent to the Lantern Hill fault and the $^{40}Ar/^{39}Ar$ and K/Ar age determinations for mineral separates from the Lantern Hill study area (figure 5.8). This curve provides a framework for understanding crustal conditions for southeastern Connecticut during the transition from convergent to extensional tectonic stresses and after faulting and silicification.

U-Pb ages of sphene from south-central Connecticut indicate that Avalonian rocks were at 580°C at 270 Ma (Wintsch and Aleinikoff 1987). Retention of argon in hornblendes in eastern Connecticut and western Rhode Island occurred at approximately 254 Ma (Dallmeyer 1982). Hornblende has a blocking temperature of approximately 500°C (McDougall and Harrison 1988). Retention of argon in muscovite in the Hope Valley Alaskite Gneiss host rocks of the Lantern Hill lode occurred at 255 Ma (Altamura and Lux 1994). The blocking temperature for the retention of argon in muscovite is approximately 350°C, assuming a moderate cooling rate (McDougall and Harrison 1988). Biotite from this area yielded age determinations from 244 to 254 Ma (Dallmeyer 1982) in an analysis using a blocking temperature of approximately 280°C and assuming a 1°C/Ma cooling rate (McDougall and Harrison 1988). By around 215 Ma, the first sedimentary deposits (i.e., the New Haven Arkose) were laid down in the neighboring Hartford rift basin on top of cold metamorphic rocks of the Gander terrane.

Analysis of syntectonic vein muscovites from the Lantern Hill fault zone has provided an age of 238 Ma. A relatively short period of time separates the Late Permian (255 Ma, Kazanian–Ufimian) cooling ages of country rock metamorphic muscovite and early Middle Triassic (238 Ma, Anisian) age determinations for emplacement of quartz veins and coeval fracturing. Quartz veins and fault breccia in the Lantern Hill silicified zone are cut by faults that are in turn cut by later quartz veins, providing evidence that silicification was coeval with fracturing.

Availability of the silica-bearing aqueous fluids would have had the effect of increasing pore pressure, causing a reduction in effective normal stress across potential slip planes, thus making brittle deformation more likely. A drop in fluid pressure caused by faulting was probably the mechanism for the deposition of quartz. To produce the 24 million m^3 of hydrothermal quartz known to be in reserve at Lantern Hill, enormous volumes of silica-bearing aqueous fluid must have migrated through the fault zone during silicification. Formation of the Lantern Hill silicified zone occurred during repeated episodes of fracturing, followed by quartz deposition (QV_1 to QV_9). Following a silicification event, permeability would have been decreased, perhaps allowing stress and fluid pressure to build until the next fracturing and mineralizing event.

Estimation of Pressure and Temperature Conditions During Main-Stage Silicification

Fluid inclusion data from main-stage vein quartz (QV_6) intergrown with dated hydrothermal muscovite (figure 5.7) suggest that the minimum solution temperature during main-stage silicification was approximately 232°C and that salinities were less than 0.35 weight percent NaCl equivalent.

Using the pressure-temperature-time (*P-T-t*) curve for Avalonian terrane rocks in south-central Connecticut (Wintsch et al. 1992), lithostatic pressure of Lantern Hill country rock was estimated to be 180 MPa at the time of main-stage quartz emplacement (i.e., 238 Ma). An independent pressure estimate for Lantern Hill silicification was established from three-phase CO_2–H_2O and two-phase H_2O fluid inclusions in vein quartz. With Kuehn and Rose's (1995) geobarometer, a fluid pressure estimate of 65 MPa was obtained from mean homogenization temperatures for these inclusions (Altamura 1995b). And with this estimated pressure (i.e., 65 MPa) of main-stage quartz, the minimum temperature of 232°C for crystallization was corrected to a trapping temperature of 265°C using Roedder's (1984) table. A depth of emplacement of 6.5 km was estimated by assuming a geothermal gradient of 41°C/km at the time of emplacement and dividing it into the corrected trapping temperature (T_t) for main-stage quartz (265°C). At 6.5 km depth, hydrostatic pressure is approximately 65 MPa. It is therefore inferred that largely hydrostatic conditions prevailed in the Lantern Hill fault zone during main-stage silicification at 238 Ma (Altamura 1995a).

Regional Implications

The oldest formation in the neighboring Hartford basin is the New Haven Arkose (figures 5.9 and 5.10). Paleontological data indicate that the age of the base of the New Haven Arkose is approximately 215 Ma (B. Cornet, personal communication 1995; Olsen and

FIGURE 5.9 Angular unconformity juxtaposing the Late Triassic New Haven Arkose against the schists and gneisses of the Ordovician Hoosac Formation near Southington, Connecticut. The exposure is near the western border fault of the Hartford basin. The arkose at other exposures contains pollen and fossils that have been used to date deposition at approximately 215 Ma.

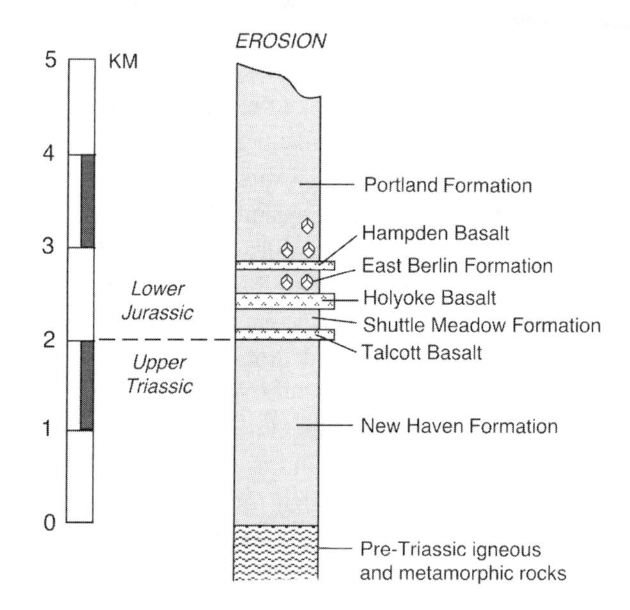

FIGURE 5.10 Stratigraphic column for Mesozoic rocks of central Connecticut, with generalized thicknesses of units. Hydrothermal quartz clasts from the North Branford lode are reported in sedimentary units above the Holyoke Basalt, indicating that the source of the quartz is that age or older. (Modified from Hubert et al. 1978)

Kent 2000). The New Haven Arkose is silicified near its western border fault (Clifton 1987) and in a small western outlier known as the Cherry Brook basin. Clasts of quartzose rock characterized by Russell as "whitish or bluish granular quartz traversed by a network of veins, and altered inclusions cemented by quartz—and some of them were angular and a foot in diameter" (1922:488) are found in sedimentary formations above the Early Jurassic Holyoke Basalt (i.e., East Berlin and Portland Formations). These clasts were derived from a silicified fault zone in the footwall of the eastern border fault of the Hartford basin near the town of North Branford, Connecticut. The North Branford silicified zone was emplaced along a fault, and because of the "law of inclusion" the age of the quartz veins and coeval fracturing must predate the East Berlin Formation. The North Branford silicified zone probably also predates significant motion along the eastern border fault of the Hartford basin because vein-quartz clasts are present throughout a stratigraphic section that is very thick (i.e., East Berlin Formation to Portland Formation).

North of the North Branford silicified zone, near the town of Middletown, Connecticut, late quartz veins cut a Permian pegmatite body in the footwall of a N–S-trending segment of the eastern border fault of the Hartford rift (Altamura 1995a). A 240.6 ± 2.0 Ma plateau age (Middle Triassic, Anisian) was obtained on vein muscovite, suggesting that the silicified zones represent a record of regional silicification.

Structures similar to the Lantern Hill fault occur throughout southern New England. Quartz-vein complexes similar to the Lantern Hill lode occur along and near the border fault of the Late Triassic–Early Jurassic Hartford rift basin, including near the towns of North Branford, Manchester, and Rockville, Connecticut (figure 5.11). Late silicified zones along the eastern and western border faults of the Hartford basin are also common (Rodgers 1985; Clifton 1987; Stopen 1988). A mass of quartz approximately the size of the Lantern Hill quartz-vein complex and several other smaller bodies occur along a NNE-trending continuation of the eastern border fault of the Hartford basin in the Paleozoic age metamorphic basement near Keene, New Hampshire (Moore 1979) (figure 5.11). In the Carboniferous Narragansett basin, silicified zones are especially well developed near Diamond Hill (western border fault) and Bristol, Rhode Island (near the center of the basin) (figure 5.11). The age of these latter silicified zones is unknown, but they appear to be post-orogenic, and Mesozoic emplacement ages are likely.

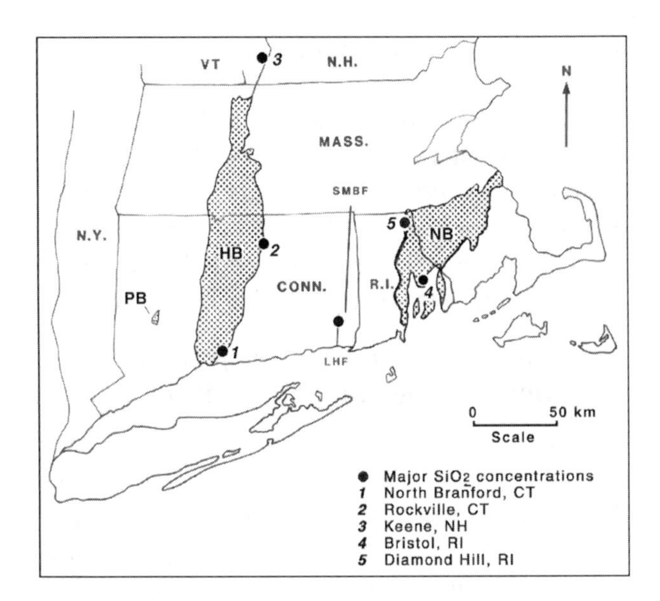

FIGURE 5.11 Major basins and major silica concentrations in southern New England; *HB* and *PB*, Mesozoic Hartford and Pomperaug basins; *NB*, Carboniferous Narragansett basin; *LHF*, Lantern Hill fault; *SMBF*, Snake Meadow Brook fault.

Examples of silicified faults with minor concentrations of quartz include the Snake Meadow Brook fault (Altamura 1987) and the Wekepeke fault near Worcester, Massachusetts (R. Goldsmith, personal communication 1992). A K/Ar age date for vein muscovite from the Snake Meadow Brook fault has yielded a 234 Ma age (Altamura 1987). Other candidates for this class of silicified faults may include the Diamond Ledge lineament near Stafford Springs, Connecticut, and the Green Falls River lineament near Voluntown, Connecticut. Another significant silicified fault zone occurs near Nottingham, New Hampshire (Billings 1956; Robinson 1988), and is approximately on strike with the Snake Meadow Brook and Wekepeke faults.

In general, metamorphic rocks between the Narragansett and Hartford basins are intersected by numerous subvertical quartz veins that trend NNE (average N11°E) (Eberly 1985). Many of these quartz veins are similar in trend and physical attributes to portions of the Lantern Hill fault. It is considered likely that they were emplaced at the time of silicification of the Lantern Hill fault.

From a study of the tectonic control of Triassic sedimentation in the New Brunswick failed-rift basin,

Nadon and Middleton (1984) concluded that sedimentation began during the Middle Triassic (Ladinian age). Olsen (1988, personal communication 1996) indicated that some sedimentary rocks in the Fundy basin are as old as Anisian. Emplacement of NNE-trending quartz veins and concomitant fracturing in the Lantern Hill fault also occurred during the Anisian.

Tectonic Setting of the Lantern Hill Silicified Zone

The Hartford rift basin formed during the Late Triassic and Early Jurassic by roughly E–W extension followed by WNW–ESE extension, a stress slightly oblique to the generally N–S tectonic grain of the pre-Mesozoic crust (Wise 1979; Clifton 1987). De Boer and Clifton (1988) proposed a slightly different model, according to which southern New England was subjected to NW–SE extension during the Late Triassic ("rifting phase") and WNW–ESE extension during the Early Jurassic ("shifting phase"). This extension was followed by NW–SE compression due to ridge push once a spreading center was developed in the Atlantic Ocean ("drifting phase"). Analysis of mesoscale faults and quartz veins from the Lantern Hill fault indicates WNW–ESE extension with subvertical σ_1, evidence that the stress conditions extended into the Middle Triassic.

The NNE-trending Higganum dolerite dike is a major tectonic structure in eastern Connecticut and central Massachusetts. This cross-grain intrusion has been linked geochemically to the Early Jurassic Talcott Basalt of the Hartford basin (Philpotts and Martello 1986), which rests conformably on the Late Triassic New Haven Arkose. The base of the Talcott Basalt represents the Triassic–Jurassic boundary (Cornet and Traverse 1975). The Higganum dike intersects and offsets several NNE-trending quartz veins (J. Z. de Boer, personal communication 1991), suggesting that silicification was restricted to an age prior to earliest Jurassic. NNE-trending monchiquite dikes have been found in southwestern Rhode Island, approximately 30 km east of the Lantern Hill fault. Radiometric dating has yielded a Middle Jurassic age (Hermes 1987).

The extensional stress that opened NNE-trending fractures along which the lamprophyres were emplaced is similar to that found for fractures in the Hartford basin, the Higganum dolerite dike, and the quartz veins of eastern Connecticut and central Massachusetts. All these fractures were emplaced under WNW–ESE extension. However, available age dates for quartz-vein emplacement (i.e., Lantern Hill fault and Snake Meadow Brook fault) are Middle Triassic, whereas all the age dates and correlations for the mafic dikes are Jurassic.

Association of the Lantern Hill fault to known Mesozoic age (Newarkian) structures and igneous rocks indicates that it is logical to assume that formation of NNE-trending extensional faults, emplacement of the quartz veins, and later emplacement of the dolerite and lamprophyre dikes occurred during the Newark tectonic phase.

The tectonic grain of Avalonian terrane units in the region of the Lantern Hill fault zone is E–W. However, to the east it bends to N–S subparallel to the map trace of the Honey Hill–Lake Char fault zone (Goldsmith 1985) (figure 5.2). In the Hartford basin, the initiating faults are N–S, dip-slip vertical to steeply west-dipping normal faults parallel to the principal tectonic grain (e.g., de Boer and Clifton 1988). Later NNE-trending transverse faults are oblique to the tectonic grain (Rodgers 1985) and imply a WNW–ESE extension.

Using seismic data, Hutchinson, Klitgord, and Detrick (1986) proposed that dip-slip motion along low-angle E- and W-dipping faults controlled formation of the Late Triassic–Early Jurassic rift zones beneath the continental shelf off Connecticut's shore. The Lantern Hill fault lies 100 km to the north of the Long Island rift basin envisioned by Hutchinson, Klitgord, and Detrick (1986) and is approximately parallel with the basin borders. The Lantern Hill fault may represent the site where an embryonic rift zone developed high in the hanging block of such a low-angle fault (figure 5.12). Extensional faulting along the Lantern Hill fault was taken up by other (E- or W-dipping) low-angle faults that may have controlled the formation of the Hartford rift basin (Clifton 1987; Altamura, Gold, and Lux 1996) and of the offshore rift basins.

Emplacement of the vein complex in the Lantern Hill fault during the Middle Triassic may be related to tectonic activity along preexisting (i.e., Alleghenian or older) low-angle faults. The latter were reactivated during the early phases of crustal extension associated with the embryonic splitting of Pangea in southern New England. The extensional Lantern Hill fault zone may have evolved smoothly from Alleghenian low-

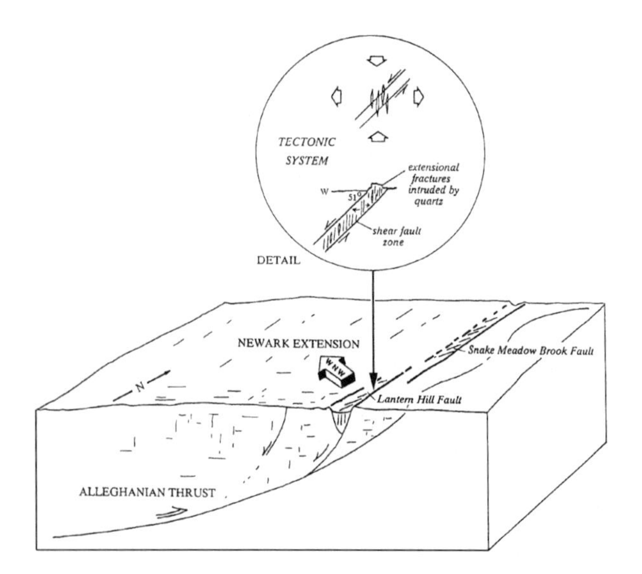

FIGURE 5.12 Simplified geologic block diagram, illustrating crustal extension at the beginning of the Triassic and the development of the Snake Meadow Brook and Lantern Hill fault system. This portrays synthetic or antithetic fractures to a master low-angle normal fault, which itself is a reactivated reverse fault associated with compressional Alleghenian deformation during the assembly of Pangea.

angle normal faulting (e.g., the Honey Hill–Lake Char fault in the Goldstein [1989] model) or may have developed more abruptly (see the model in Wintsch et al. 1992). In either case, deformation along the Lantern Hill fault zone clearly occurred later because it cuts the Honey Hill–Lake Char fault. This deformation was followed by the formation of the Hartford failed rift basin during the Late Triassic and ultimately by the opening of the Atlantic Ocean during the Jurassic.

Conclusions

The Lantern Hill fault zone developed as a response to extensional tectonic conditions (subvertical σ_1 and subhorizontal σ_3 with a 105° azimuth). Stress conditions responsible for the formation of the Lantern Hill fault zone are similar to those documented for Newark age structures (e.g., de Boer and Clifton 1988).

^{40}Ar/^{39}Ar plateau ages for muscovite separates from Lantern Hill host country rocks provide metamorphic cooling ages of 255 Ma and indicate that an Alleghenian thermal event affected the Avalonian terrane of southeastern Connecticut. These data provide a midpoint between the ages obtained for western Rhode Island (Dallmeyer 1982) and south-central Connecticut (Wintsch and Aleinikoff 1987), where similar cooling ages for Avalonian rocks have been obtained. These data support Wintsch and colleagues' (1992) suggestion that a regional Alleghenian thermal event affected the Avalonian terrane of southern New England.

Repeated fracturing and emplacement of main-stage quartz veins in the Lantern Hill silicified fault zone occurred during the early Middle Triassic (Anisian) at 238 Ma, approximately 20 Ma before the first sedimentary deposits (the New Haven Arkose) were laid down in the Hartford rift basin.

In the Lantern Hill fault zone, mesoscale faults cut quartz veins that compose the silicified zone, and later quartz veins in turn cut the faults. Emplacement of the majority of quartz veins and formation of coeval fractures in the Lantern Hill fault zone thus represent evidence for early rifting associated with the embryonic opening of the Atlantic Ocean. Increased activity along the eastern border fault of the Hartford basin and the development of this rift basin may have been responsible for the end of significant tectonic activity in the Lantern Hill fault zone.

Acknowledgments

I wish to thank H. E. Belkin, J. Z. de Boer, D. P. Gold, R. Goldsmith, D. R. Lux, R. P. Wintsch, and D. U. Wise for encouragement and assistance, in particular de Boer's and Wise's technical reviews of the manuscript for this chapter. H. E. Belkin and D. R. Lux graciously collaborated for the fluid inclusion and ^{40}Ar/^{39}Ar investigations, respectively. An early draft of the manuscript benefited from comments by W. R. Brice and A. G. Doden. I thank the State Geological and Natural History Survey of Connecticut, the P. D. Krynine Fund of the Department of Geosciences at Pennsylvania State University, Sigma Xi, and the U.S. Silica Company for financial support. I am also grateful to P. M. LeTourneau for considerable editorial commentary that improved the manuscript and to P. E. Olsen for improvements to the figures.

Literature Cited

Altamura, R. J. 1985. *Preliminary Radar (SLAR) Lineament Map of Connecticut* [scale 1:125,000]. Hartford: State Geological and Natural History Survey of Connecticut, Connecticut Department of Environmental Protection.

Altamura, R. I. 1987. The Snake Meadow Brook–Lantern Hill fault system, an en echelon(?) brittle fault zone, eastern Connecticut. *Geological Society of America, Abstracts with Programs* 19:2.

Altamura, R. J. 1993. A report on two lithological cross-sections across the Lantern Hill quartz ore deposit and wall rock, U.S. Silica Company quartz quarry, North Stonington, Connecticut [with map (scale 1:1,200)]. Report, U.S. Silica Company.

Altamura, R. J. 1995a. Tectonics, wall-rock alteration, and emplacement history of the Lantern Hill giant quartz lode, Avalonian terrane, southeastern Connecticut. In N. W. McHone, ed., *Guidebook for Fieldtrips in Eastern Connecticut and the Hartford Basin,* pp. E1–E36. State Geological and Natural History Survey of Connecticut Guidebook, no. 7. Hartford: Connecticut Department of Environmental Protection.

Altamura, R. J. 1995b. Pressure, temperature, and kinematic history of a fault zone: The Lantern Hill fault, Connecticut. *Geological Society of America, Abstracts with Programs* 27:283.

Altamura, R. J., and D. P Gold. 1994. Ore fluids, fluid flow, and geochemistry of a one-mile long sheeted quartz lode in southern New England. *Association Géologique du Canada et Association Mineralogique du Canada, Programme et Résumés* 19:A3.

Altamura, R J., D. P. Gold, and D. R. Lux. 1996. Tectonics of the Lantern Hill fault, southeastern Connecticut: Evidence for the embryonic opening of the Atlantic Ocean. *Geological Society of America, Abstracts with Programs* 28:3.

Altamura, R. J., and D. R. Lux. 1994. $^{40}Ar/^{39}Ar$ and K-Ar ages for muscovite from a giant quartz lode and alaskite host rocks, Avalonian terrane, southern New England. *Geological Society of America, Abstracts with Programs* 26:529.

Altamura, R. J., and D. R. Lux. In preparation. An $^{40}Ar/^{39}Ar$ age investigation of the Lantern Hill fault zone, Avalonian terrane, southern New England: Implications for rifting associated with the embryonic opening of the Atlantic Ocean.

Altamura, R. J., and S. S. Quarrier. 1986. Side-looking airborne radar lineament mapping in Connecticut. *Geological Society of America, Abstracts with Programs* 18:2.

Billings, M. 1956. Geologic map of New Hampshire [in pocket (scale 1:250,000)]. In M. P. Billings, ed., *The Geology of New Hampshire.* Part II, *Bedrock Geology.* Concord: New Hampshire State Planning and Development Commission.

Clifton, A. E. 1987. Tectonic analysis of the western border fault of the Mesozoic Hartford basin, Connecticut and Massachusetts. M.S. thesis, Wesleyan University.

Cornet, B. 1977. The palynostratigraphy and age of the Newark Supergroup. Ph.D. diss., Pennsylvania State University.

Cornet, B., and A. Traverse. 1975. Palynological contributions to the chronology and stratigraphy of the Hartford basin in Connecticut and Massachusetts. *Geoscience and Man* 11:1–33.

Dallmeyer, R. D. 1982. $^{40}Ar/^{39}Ar$ ages from the Narragansett basin and southern Rhode Island basement terrain: Their bearing on the extent and timing of Alleghanian tectonothermal events in New England. *Geological Society of America Bulletin* 93:1118–1130.

de Boer, J. Z., and A. Clifton [incorrectly printed as "Clifford"]. 1988. Mesozoic tectonogenesis: Development and deformation of Newark rift zones in the Appalachians. In W. Manspeizer, ed., *Triassic–Jurassic Rifting: Continental Breakup and the Origin of the Atlantic Ocean and the Passive Margins,* pp. 275–306. Developments in Geotectonics, no. 22. Amsterdam: Elsevier.

Dixon, H. R. 1982. Multi-stage deformation of the Preston Gabbro, eastern Connecticut. In R. Joeston and S. Quarrier, eds., *Guidebook for Field Trips in Connecticut and South Central Massachusetts,* pp. 453–464. New England Intercollegiate Geological Conference. Hartford: Connecticut Department of Environmental Protection.

Eberly, P. O. 1985. *Brittle Fracture Fabric Along a West–East Traverse from the Connecticut Valley to the Narragansett Basin.* Department of Geology and Geography, Contribution no. 57. Amherst: University of Massachusetts.

Getty, S. R. 1988. Basement-cover relations about the Willimantic dome, Connecticut: The significance of pinch-and-swell structures. *Geological Society of America, Abstracts with Programs* 20:21.

Goldsmith, R. 1985. *Bedrock Geologic Map of the Old Mystic and Part of the Mystic Quadrangles, Connecticut, New York, and Rhode Island* [scale 1:24,000]. U.S. Geological Survey Miscellaneous Investigations Series Map I-1524. Washington, D.C.: Government Printing Office.

Goldstein, A. G. 1984. Kinematic indicators in Lake Char fault mylonites. *Geological Society of America, Abstracts with Programs* 16:18.

Goldstein, A. G. 1989. Tectonic significance of multiple motions on terrane-bounding faults in the northern Appalachians. *Geological Society of America Bulletin* 101:927–938.

Goldstein, A. G. 1994. A shear zone origin for Alleghenian (Permian) multiple deformation in eastern Massachusetts. *Tectonics* 13:62–77.

Hermes, O. D. 1987. Geologic relationships of Permian Narragansett Pier and Westerly granites and Jurassic lamprophyric dike rocks, Westerly, Rhode Island. In D. C. Roy, ed., *Decade of North American Geology Centennial Field Guide,* vol. 5, pp. 181–186. Boulder, Colo.: Geological Society of America, Northeastern Section.

Hermes, O. D., L. P. Gromet, and D. P. Murray. 1994. *Bedrock Geologic Map of Rhode Island* [scale 1:100,000]. Rhode Island Map Series, no. 1. Kingston: University of Rhode Island.

Hubert, J. F., A. A. Reed, W. L. Dowdall, and J. M. Gilchrist. 1978. *Guide to the Mesozoic Redbeds of Central Connecticut.* State Geological and Natural History Survey of Connecticut Guidebook, no. 4. Hartford: Connecticut Department of Environmental Protection.

Hutchinson, D. R., K. D. Klitgord, and R. S. Detrick. 1986. Rift basins of the Long Island platform. *Geological Society of America Bulletin* 97:688–702.

Kuehn, C. A., and A. W. Rose. 1995. Carlin gold deposit, Nevada: Origin in a deep zone of mixing between normally pressured and overpressured fluids. *Economic Geology* 90:17–36.

LeTourneau, P. M. 1999. Depositional history and tectonic evolution of Late Triassic age rifts of the U.S. central Atlantic margin: Results of an integrated stratigraphic, structural, and paleomagnetic analysis of the Taylorsville and Richmond basins. Ph.D. diss., Columbia University.

McDougall, I., and T. M. Harrison. 1988. *Geochronology and Thermochronology by the* $^{40}Ar/^{39}Ar$ Method. New York: Oxford University Press.

Moore, G. E. 1979. *The Geology of the Keene-Brattleboro Quadrangle, New Hampshire and Vermont* [with map (scale 1:62,500)]. Concord, N.H.: Division of Forest and Lands, Department of Resources and Economic Development.

Mosher, S. 1983. Kinematic history of the Narragansett basin, Massachusetts and Rhode Island: Constraints on late Paleozoic plate reconstructions. *Tectonics* 2:327–344.

Nadon, G. E., and G. V. Middleton. 1984. Tectonic control of Triassic sedimentation in southern New Brunswick: Local and regional implications. *Geology* 12:619–622.

O'Hara, K., and L. P. Gromet. 1983. Textural and Rb-Sr isotopic evidence for late Paleozoic mylonitization within the Honey Hill fault zone, southeastern Connecticut. *American Journal of Science* 283:762–779.

Olsen, P. E. 1988. Paleoecology and paleoenvironments of the eastern North American (early Mesozoic) Newark Supergroup. In E. Manspeizer, ed., *Triassic–Jurassic Rifting: Continental Breakup and the Origin of the Atlantic Ocean and the Passive Margins,* pp. 185–230. Developments in Geotectonics, no. 22. Amsterdam: Elsevier .

Olsen, P. E. 1997. Stratigraphic record of the early Mesozoic breakup of Pangea in the Laurasia–Gondwana rift system. *Annual Review of Earth and Planetary Science* 25:337–401.

Olsen, P. E., and D. V. Kent. 2000. High resolution early Mesozoic Pangean climatic transect in lacustrine environments. *Zentralblatt für Geologie und Paläontologie,* part I, 11–12:1475–1495.

Olsen, P. E., D. V. Kent, B. Cornet, W. K. Witte, and R. W. Schlische. 1996. High-resolution stratigraphy of the Newark rift basin (early Mesozoic, eastern North America). *Geological Society of America Bulletin* 108:40–77.

Peck, J. H. 1975. *Preliminary Bedrock Geologic Map of the Clinton Quadrangle, Worcester County, Massachusetts* [scale 1:24,000]. U.S. Geological Survey Open-File Report 75–658. Washington, D.C.: Government Printing Office.

Philpotts, A. R., and A. A Martello. 1986. Diabase feeder dikes for the Mesozoic basalts in southern New England. *American Journal of Science* 286:105–126.

Quinn, A. 1971. *Bedrock Geology of Rhode Island.* U.S. Geological Survey Bulletin, no. 1295. Washington, D.C.: Government Printing Office.

Robinson, J. S. 1988. A brittle fracture analysis of the Flint Hill fault zone, southeastern New Hampshire. M.S. thesis, University of Massachusetts.

Rodgers, J. 1970. *The Tectonics of the Appalachians.* New York: Wiley.

Rodgers, J. 1985. *Bedrock Geological Map of Connecticut* [scale 1:125,000]. Hartford: State Geological and Natural History Survey of Connecticut, Connecticut Department of Environmental Protection.

Roedder, E. 1984. *Fluid Inclusions.* Reviews in Mineralogy. Washington, D.C.: Mineralogical Society of America.

Russell, W. L. 1922. The structural and stratigraphic relations of the great Triassic fault of southern Connecticut. *American Journal of Science* 4:483–497.

Sclar, C. B. 1958. *The Preston Gabbro and the Associated Gneisses, New London County. Connecticut.* State Geological and Natural History Survey of Connecticut Bulletin, no. 88. Hartford: State of Connecticut.

Skehan, J. W., N. Rast, and S. Mosher. 1986. Paleoenvironmental and tectonic controls of sedimentation in coal-forming basins of southeastern New England. In P. C. Lyons and C. L. Rice, eds., *Paleoenvironemtnal and Tectonic Controls in Coal-Forming Basins of the United States,* pp. 9–30. Geological Society of America Special Paper, no. 210. Boulder, Colo.: Geological Society of America.

State Geological and Natural History Survey of Connecticut. 1990. *Generalized Bedrock Geologic Map of Connecticut* [scale 1:1,161,860]. Hartford: Connecticut Department of Environmental Protection.

Stopen, L. E. 1988. Geometry and deformation history of mylonitic rocks and silicified zones along the Mesozoic Connecticut Valley border fault, western Massachusetts. M.S thesis, University of Massachusetts.

Wintsch, R. P., and J. N. Aleinikoff. 1987. U-Pb isotopic and geologic evidence for late Paleozoic anatexis, deformation, and accretion of the Late Proterozoic Avalon terrane, south-central Connecticut. *American Journal of Science* 287:107–126.

Wintsch, R. P., J. F. Sutter, M. J. Kunk, J. N. Aleinikoff, and M. J. Dorais. 1992. Contrasting *P-T-t* paths: Thermochronologic evidence for a late Paleozoic final assembly of the Avalon composite terrane in the New England Appalachians. *Tectonics* 11:672–689.

Wise, D. U. 1979. *Fault, Fracture, and Lineament Data for Western Massachusetts and Western Connecticut.* Report no. NUREG/CR2292 RA. Washington, D.C.: Nuclear Regulatory Commission.

6

Estimates of Eroded Strata Using Borehole Vitrinite Reflectance Data, Triassic Taylorsville Rift Basin, Virginia: Implications for Duration of Synrift Sedimentation and Evidence of Structural Inversion

MaryAnn Love Malinconico

Linear regression of the base 10 logarithm ($\log_{10}$) of percent mean random vitrinite reflectance versus depth for boreholes in the early Mesozoic Taylorsville basin, Virginia, was used to estimate the amount of synrift strata eroded and, with estimates of sedimentation rate, the duration of synrift sedimentation. The basin is one of several Triassic–Jurassic continental rift basins containing Newark Supergroup lacustrine and fluvial sediments that formed during the breakup of Pangea. The amount of missing section varies across the basin from approximately 0.9 to 2.6 km, the maximum occurring over a basement antiform, which is attributed to postrift inversion. The estimates of the end of synrift sedimentation range from 210.1 to 191.5 million years ago. Based on recent stratigraphic evidence from other parts of the basin, the probable end of synrift sedimentation, and perhaps of rifting, however, can be narrowed to between 208 and 202 Ma, just at or before the Triassic–Jurassic boundary. These calculations suggest an earlier cessation of rifting in the Taylorsville basin than in the Newark basin, to the north, whose youngest extant strata is approximately 200 to 199 million years old and which has been estimated to be missing at least 2 to 3 km more of younger synrift strata. This study supports other workers' hypothesis, based on structural data and regional stress patterns, that the rift–drift transition was diachronous along the eastern North American passive margin with the cessation of rifting and the initiation of compressional inversion tectonics by the Triassic–Jurassic boundary in Newark Supergroup basins from Virginia south.

The exposed Triassic–Jurassic rift basins containing lacustrine and fluvial sediments of the Newark Supergroup were formed during the initial breakup of the supercontinent of Pangea, presaging the initiation of seafloor spreading and the opening of the modern Atlantic Ocean. The basins are located along the Atlantic margin of the United States and maritime Canada from South Carolina to Nova Scotia (figure 6.1). Sedimentary deposition probably began in the ?Middle Triassic (Schamel et al. 1986; Olsen 1997). The youngest extant strata are lowest Jurassic (Hettangian to

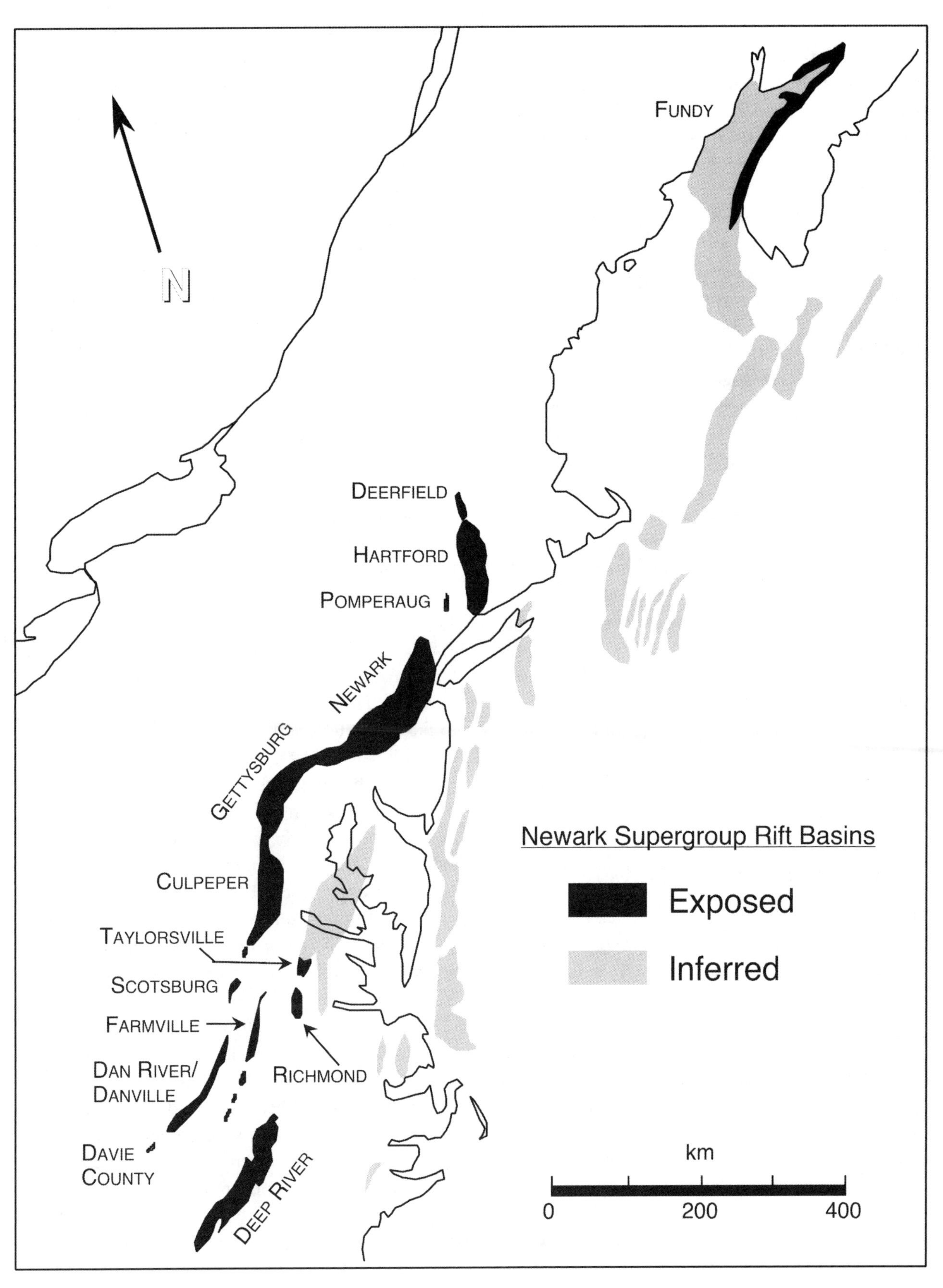

FIGURE 6.1 Location map of early Mesozoic Newark Supergroup rift basins, along the east coast of the United States and maritime Canada. (After Olsen 1990)

?Pliensbachian) but are found only in Newark Supergroup basins north of the state of Virginia (Wade et al. 1996; Olsen 1997). The absence of Jurassic strata in the southern basins has led to speculation that Jurassic formations there were not just removed by erosion but were not deposited in the first place (Olsen, Schlische, and Gore 1989; Schlische and Ackermann 1995; Withjack, Schlische, and Olsen 1998). Workers using crosscutting relationships of synrift sediments, dated Mesozoic basalts and postrift reverse faults, and regional stress patterns based on igneous dike orientation (Schlische and Ackermann 1995; Withjack, Schlische, and Olsen 1998) have suggested that rifting and extension, along with sedimentary deposition, ceased and that structural inversion was initiated in the southern basins before or in the earliest Jurassic. Using estimates of eroded strata from borehole vitrinite reflectance data from the Taylorsville basin of Virginia, stratigraphic correlations of Taylorsville strata with Newark basin strata, and estimates of sedimentation rates, this study tests the foregoing hypothesis and produces additional evidence that sedimentation ceased in the Taylorsville basin and probably in other basins to the south around the Triassic–Jurassic boundary, several million years before the end of rifting in the Newark basin and in other northern basins. In addition, the change in amount of estimated eroded strata illustrates the style of structural inversion across the basin.

Taylorsville Rift Basin

The Taylorsville basin is one of the early Mesozoic Newark Supergroup rift basins that lie in a N- to NE-striking belt along the eastern coast of the United States and maritime Canada (figure 6.1). The name Newark Supergroup properly applies to formations in the subaerially exposed basins, but several similarly formed premodern Atlantic rift basins are located under the Cretaceous and younger Coastal Plain sediments and on the continental shelf (Benson 1992).

The rift basins lie within the Paleozoic Appalachian orogen, and the border faults of the half-graben are normally reactivated thrust faults (Robinson 1979; Ratcliffe et al. 1986; Gates 1997; Withjack, Schlische, and Olsen 1998) that were active during rifting (Schlische 1993; Withjack, Olsen, and Schlische 1995). Rifting began as early as the Middle Triassic, a hypothesis based on paleontological dating of sediments (Olsen, Schlische, and Gore 1989) and on $^{40}Ar/^{39}Ar$ dating of fault breccia (Altamura 1996). Sedimentary fill of the Newark rift basins is entirely continental and consists of fluvial and lacustrine sedimentary rock plus basalt flows (Olsen et al. 1996; Olsen 1997).

The Taylorsville basin is a half-graben located along the metamorphic axis of the late Paleozoic Alleghanian orogen. The western border fault of the basin, the Hylas fault zone, is a normally reactivated late Paleozoic compressional/transpressive mylonitic ductile shear zone that separates the Proterozoic Goochland terrane of the footwall from the Pennsylvanian age Petersburg Granite (Lee and Williams 1993) of the hanging wall (Bobyarchick and Glover 1979) (figure 6.2). The basin itself is buried mostly under the Aptian age and younger (Milici et al. 1991) Coastal Plain sediments deposited during the thermal subsidence phase of the passive margin after the opening of the modern Atlantic Ocean.

In the late 1980s, Texaco drilled six Taylorsville basin coreholes (Campbell, Butler, Bowie Fogg, Ellis, Payne, and Roberts) and three drill holes (Wilkins, Gouldman, and Thornhill) for petroleum exploration purposes. The company donated the cores and cuttings to the Lamont-Doherty Earth Observatory of Columbia University in 1993. Figure 6.2 shows the map locations of the boreholes and two of the many seismic lines across the basin. The Butler, Payne, and Ellis coreholes lie on the southern line (TX85T11); Bowie Fogg and Roberts, on the northern line (TX85T12). A simplified cross section based on the structural and stratigraphic interpretations of LeTourneau (1999, chapter 3 in volume 2 of *The Great Rift Valleys of Pangea*), with actual and projected locations of the boreholes, is found in figure 6.3.

Weems (1980) originally mapped the geology of the exposed Taylorsville basin, and LeTourneau (chapter 3 of volume 2) describes the stratigraphy of the buried part of the basin. Weems assigned the surface exposures of the basin to three members of the Doswell Formation. In the subsurface, LeTourneau (1999) recognized two continental fluvial-lacustrine-fluvial basin-fill sequences. Such tripartite basin-fill sequences are typical of tectonically active nonmarine basins and have been described in detail by Olsen (1997) for the Mesozoic central Atlantic margin basins and by Katz and Liu (1998) for lacustrine sequences in general. The

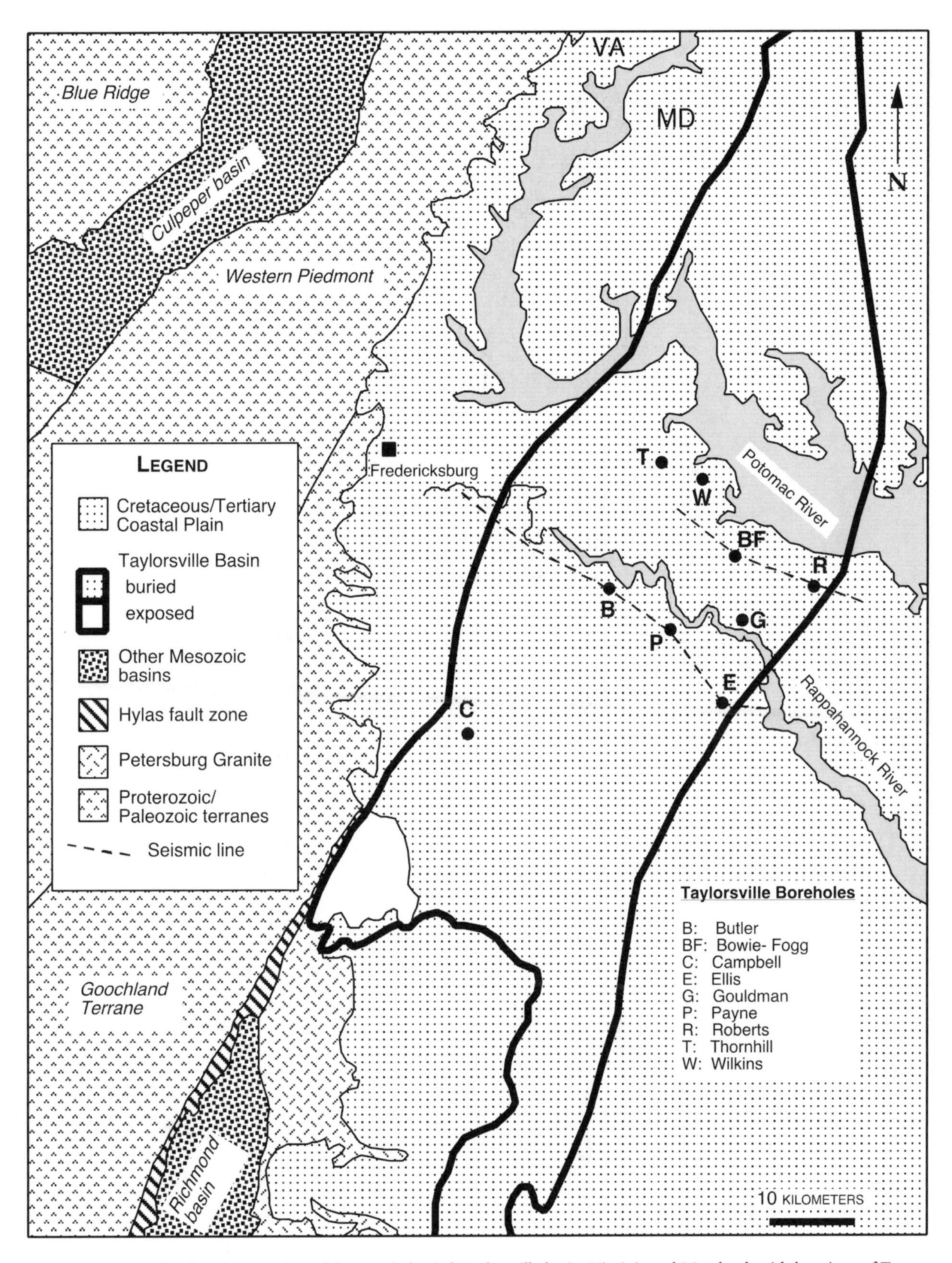

FIGURE 6.2 Regional geologic setting of the mostly buried Taylorsville basin, Virginia and Maryland, with locations of Texaco boreholes and two seismic lines: TX85T12 (northern) and TX85T11 (southern).

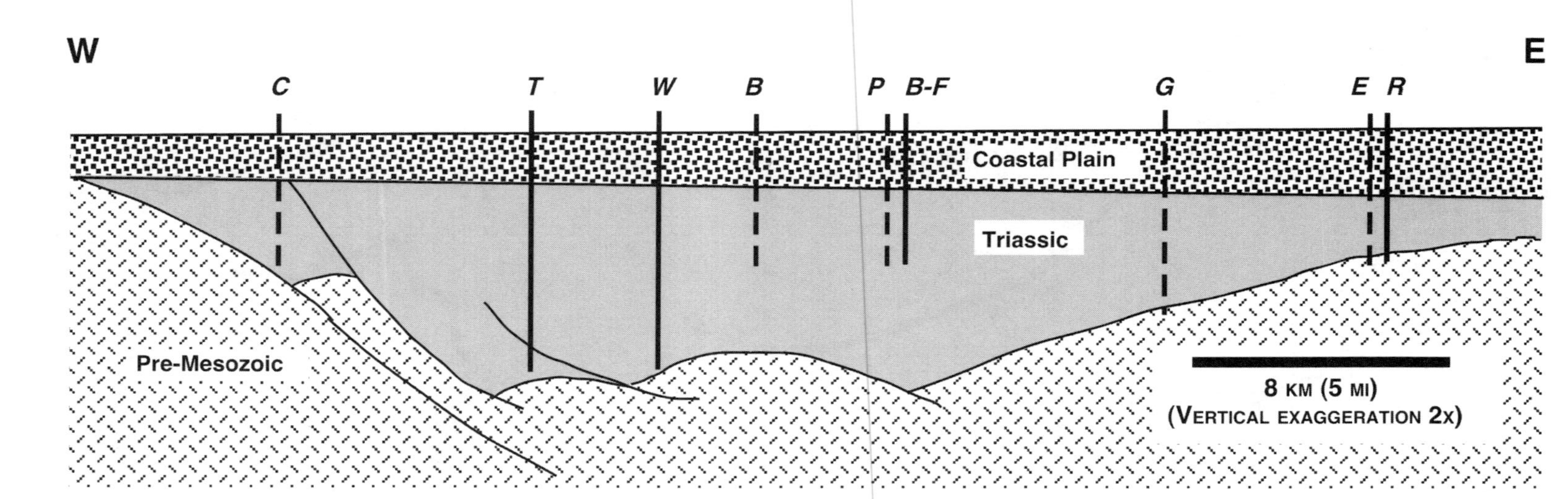

FIGURE 6.3 Generalized cross section of the Taylorsville basin along seismic line TX85T12 (figure 6.2) through the Roberts, Bowie Fogg, and Thornhill boreholes, and showing basement configuration, Coastal Plain, and projected positions of the other boreholes used in this study. The Gouldman, Roberts, and Wilkins boreholes reached basement. (Based on LeTourneau, chapter 3 in volume 2 of *The Great Rift Valleys of Pangea*)

oldest sequence in the Taylorsville basin, Basin-Fill Sequence I, is correlated with the lower two members of the Doswell Formation of Weems (1980) and has been renamed the Doswell Group (LeTourneau 1999). It is also correlated with Olsen's (1997) tectonostratigraphic sequence (TS) II. The King George Group is Basin-Fill Sequence II (TS III of Olsen 1997) and includes the Newfound, Port Royal, and Leedstown Formations. The Port Royal Formation, which is the lacustrine interval of that sequence, is gray to black sandstone, siltstone, and shale, including characteristic deep-lake-laminated black shale; the overlying Leedstown Formation includes brown, red, and gray fluvial channel and floodplain sandstone, siltstone, and shale. There is an obvious unconformity, visible also on seismic sections, between Basin-Fill Sequences I and II. Sequence II is basinwide, whereas Sequence I was deposited in small, early half-graben in the hanging wall of the basin floor. Based on magnetic stratigraphy, LeTourneau (1999) has correlated the top of the Port Royal Formation with the upper part of polarity zone E11r in the Newark basin, which occurs within the Skunk Hollow Member of the Lockatong Formation (Kent, Olsen, and Witte 1995) (figure 6.4).

The Port Royal Formation is found in the Wilkins, Thornhill, Butler, Gouldman, and Campbell holes (LeTourneau 1999). In the Campbell core, near the border fault, the lacustrine black shales of the Port Royal Formation are shallow, truncated by the Cretaceous unconformity at 138 m depth. In contrast, the Port Royal Formation in the Wilkins and Thornhill holes is thousands of meters deep (figure 6.4). The Campbell corehole, therefore, appears to be located in a rider block and has been projected to such a position on the southern seismic line (figure 6.3).

The youngest strata that LeTourneau (1999) identified in the Taylorsville basin is Upper Triassic (upper Norian), in contrast to the Early Jurassic (Hettangian) age of the youngest extant strata in the Newark basin. Estimates of the thickness of post–lowest Jurassic missing synrift strata in the Newark basin are 2 to 3 km, based on surficial vitrinite reflectance data (Katz et al. 1988; Pratt, Shaw, and Burruss 1988) and on fission track ages (Steckler et al. 1993); total thickness of eroded synrift strata in parts of the Newark basin where older stratigraphy is exposed may be substantially more than 3 km. In contrast, Weems (1980) suggested that there probably has been little erosion of Taylorsville synrift sediments because, in his opinion, the highest exposed beds of the Doswell Formation are nearly unconsolidated or uncompacted. Milici and colleagues (1991) also did not include any uplift and erosion of Taylorsville synrift strata prior to Cretaceous Coastal Plain deposition in their time-temperature model of the basin.

Notable among the Early Jurassic strata in the Newark basin are basaltic lava flows. Lava flows are found in the exposed northern Newark Supergroup basins and are dated fairly accurately, with accompanying intrusive dikes and sills, as having been emplaced during a 580 ± 100 ky interval at approximately 201 Ma (Olsen, Schlische, and Fedosh 1996). Although diabase dikes intrude the Taylorsville basin and surrounding basement, there are no basalt flows in the exposed or buried parts of the basin. Because of the lack of lava flows or other Jurassic strata in the exposed southern basins, it has been suggested that rifting and sedimentation in these basins ended before or soon after the beginning of the Jurassic (Olsen, Schlische, and Gore 1989; Schlische and Ackermann 1995; Withjack, Schlische, and Olsen 1998). There are subsurface basalt flows in South Carolina and Georgia (Carolina Crossroads Basalt); however, their affinity has been controversial. These flows have been correlated both with the earliest Jurassic Newark Supergroup flows (Olsen, Schlische, and Fedosh 1996) and with the East Coast Magmatic Province, an offshore volcanic wedge of mafic volcanic rocks at the continent–ocean transition (Kelemen and Holbrook 1995; Olsen, Schlische, and Fedosh 1996). Previous dates for the Clubhouse Crossroads Basalt range from Late Triassic to Late Cretaceous; however, Ragland, Cummins, and Arthur (1992) have concluded that the age is probably 200 ± 5 Ma, consistent with the 201 ± 2 Ma age for the Newark Supergroup basalts. From this reasoning, as well as from crosscutting and overlapping relationships, Withjack, Schlische, and Olsen (1998) suggested that the wedge off South Carolina formed at least by approximately 200 Ma.

Because the flood basalt event, including dike and sill intrusion, occurred in a relatively short time (Olsen, Schlische, and Fedosh 1996), dike orientation can be interpreted as representing the stress configuration of the region at approximately 201 Ma, with dike strike

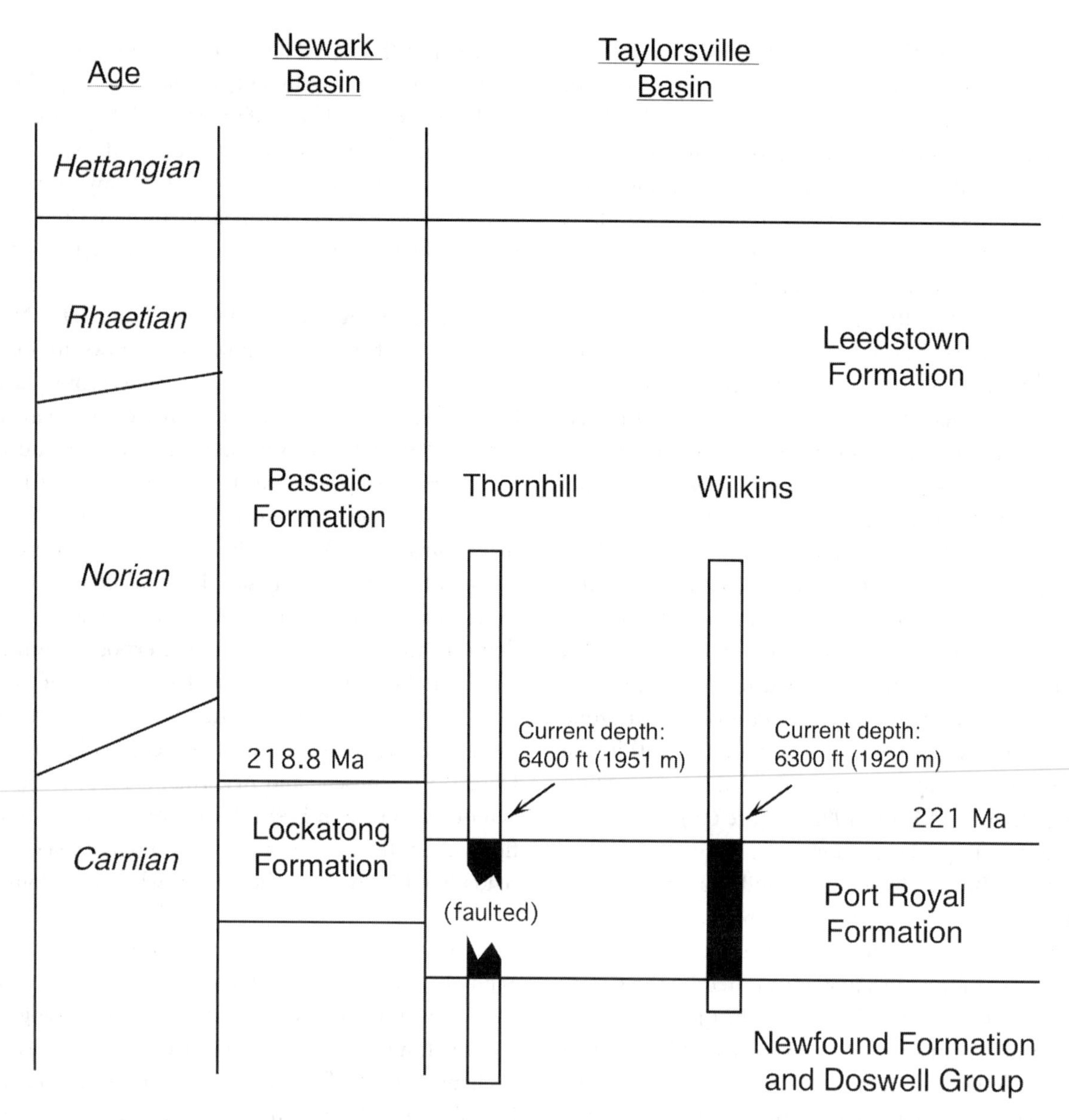

FIGURE 6.4 Stratigraphy of the Wilkins and Thornhill boreholes and correlations with the Newark basin. Data from Wilkins and Thornhill were used for calculations of the duration of synrift sedimentation for the basin. Stratigraphic thicknesses are not representative. (Based on LeTourneau 1999)

perpendicular to the minimum horizontal stress (Schlische and Ackermann 1995; Withjack, Schlische, and Olsen 1998). The strike of the Early Jurassic diabase dikes changes systematically from south to north: in Virginia N–NW, and in the northern United States and maritime Canada NE. Various reasons have been proposed for the regional deviation in minimum horizontal stress. Hill (1991) suggested that the fanning pattern of dikes resulted from regional doming caused by a hot spot located offshore of the modern southeastern United States. However, Kelemen and Holbrook (1995) and Holbrook and Kelemen (1993) argued against the existence of such a hot spot, basing their conclusion on geochemical and seismic studies of the East Coast Magmatic Province.

Alternatively, the stress pattern depicted by the dikes may have been created by a cessation of rifting and an initiation of shortening and structural inversion during the earliest stages of seafloor spreading, with a corresponding NW-striking *maximum* horizontal stress, from Virginia southward, while rifting continued to the north with a NW-striking *minimum* horizontal stress (Schlische and Ackermann 1995; Withjack, Schlische, and Olsen 1998). In addition to the change in dike orientation from south to north, evidence supporting this hypothesis includes basement-involved reverse faults with fault-propagation folds, basin boundary faults that do not deform crosscutting dikes, and constant thickness of synrift strata across reverse faults. Withjack, Schlische, and Olsen (1998) suggested, therefore, that seafloor spreading began along the current passive margin of the United States south of Virginia by approximately 200 Ma, just before the Triassic–Jurassic boundary, almost 15 million years before the rift–drift transition occurred to the north.

Vitrinite Reflectance Data and Estimation of Eroded Synrift Strata

Methods

Vitrinite reflectance for this study was measured on samples from three cores (Campbell, Bowie Fogg, and Ellis) and on well cuttings from three drill holes (Wilkins, Gouldman, and Thornhill). The cuttings or pieces of whole rock (not kerogen concentrates) were mounted in epoxy and polished. Reflectance was measured in oil immersion in plane polarized light at 547 nm on a Leitz MPV II microscope at the Coal Characterization Laboratory, Department of Geology, Southern Illinois University, Carbondale, and on a Leitz MPV I at the Department of Geological Sciences, University of Missouri, Columbia. Both microscopes are outfitted with rotational polarization as described in Bensley and Crelling (1991) and in Houseknecht et al. (1993). In reflectance measurement with rotational polarization, the maximum, minimum, and random reflectances of a particle are measured during computer-guided motorized rotation of the polarizer rather than during manual rotation of the microscope stage with a fixed polarizer. Appropriate corrections for polarizer and optical path anisotropy are done by the computer; these correction factors are calculated during routine standardization procedures. The random reflectance value reported using rotational polarization is that value measured when the polarizer is at 45°, the same orientation as traditional random reflectance measurement with stationary polarizer (Hevia and Virgos 1977; Stach et al. 1982). The rotational polarization method also calculates the "rotational" reflectance, which is the average of all reflectance data points collected during the rotation of the polarizer. The mean rotational reflectance for a sample is theoretically close to the mean random reflectance measured with no polarizer or close to the mean random reflectance with polarizer for randomly oriented grains, and it eliminates any skewness in data that may occur in highly anisotropic samples with preferential grain orientation. The mean random reflectance in oil immersion (%R_0), traditionally measured in studies using sedimentary or petroleum source rocks, is reported in this chapter.

Reflectance data was collected primarily on particles identified as indigenous nonoxidized vitrinite (Senftle, Landis, and McLaughlin 1993; Barker 1996) (figure 6.5). It has been found that at least 20 measurements on indigenous vitrinite are needed to properly evaluate the mean (Barker and Pawlewicz 1993; Senftle, Landis, and McLaughlin 1993). However, it is well recognized that some organic facies are poor in indigenous vitrinite, and even 20 measurements on first-cycle vitrinite may be difficult to obtain. Pseudovitrinite, "recycled" vitrinite, semifusinite, solid bitumen, or other macerals misidentified as first-cycle or indigenous vitrinite were eliminated based on reflectance crossplots (Kilby 1988), which use the maximum and minimum reflectance data measured during rotational polarization to distinguish particles with similar optical characteris-

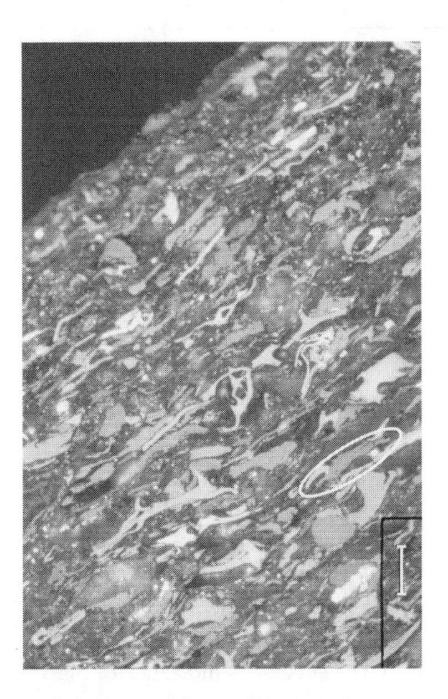

FIGURE 6.5 Very organic-rich black shale with a variety of vitrinite and inertinite macerals. The circled maceral with high length to width ratio, wispy appearance, and low gray reflectance is characteristic of first-cycle vitrinite. Scale bar is 20 µm.

tics. Uniaxial to biaxial negative materials such as vitrinite and other organic macerals have measured apparent maximum reflectances that vary between a true maximum and an intermediate reflectance and have measured apparent minimum reflectances that vary between an intermediate reflectance and a true minimum. These relationships are illustrated with the reflectance indicatrix for a biaxial negative material in figure 6.6. In a reflectance crossplot (figure 6.7A–C), the apparent maximum and minimum of a particle are each plotted against the bireflectance (apparent maximum minus apparent minimum) (Kilby 1988). When data for a number of particles with the same true maximum, true minimum, and intermediate reflectances (i.e., the same reflectance indicatrix) are plotted, the apparent maxima plot in a discrete field from the apparent minima (figure 6.7C). Data from particles with different true maximum, true minimum, and intermediate reflectances should plot outside these fields (figure 6.7D). This provides a higher degree of confidence that one is using optically similar particles in the calculation of the mean random reflectance than is provided by either reflectance histograms (Stach et al. 1982; Senftle, Landis, and McLaughlin 1993) or subjective visual choice of true vitrinite (Barker and Pawlewicz 1993; Senftle, Landis, and McLaughlin 1993).

From the reflectance data in this study, estimates of eroded or missing pre–Coastal Plain strata were made using Dow's (1977) method. For borehole vitrinite reflectance profiles in which the base 10 logarithm ($\log_{10}$) of the percent reflectance ($\%R_0$) increases linearly with depth, the regression line through the reflectance data is extrapolated back to $\%R_0 = 0.2$, the reflectance of recently deposited woody plant matter or low-grade peat (Stach et al. 1982). It is assumed that there is no change in the slope of the reflectance profile in the missing section due to change in conductive geothermal gradient, advective heat flow, or drastic change in thermal conductivity. The difference between the depth intercept at 0.2% reflectance and the erosional unconformity in question is the estimate of missing section (figure 6.8). Even though the reflectance versus depth graphs in this study plot reflectance on the *x*-axis and depth on the *y*-axis, depth in feet was used as the independent variable, and the $\log_{10}$ of

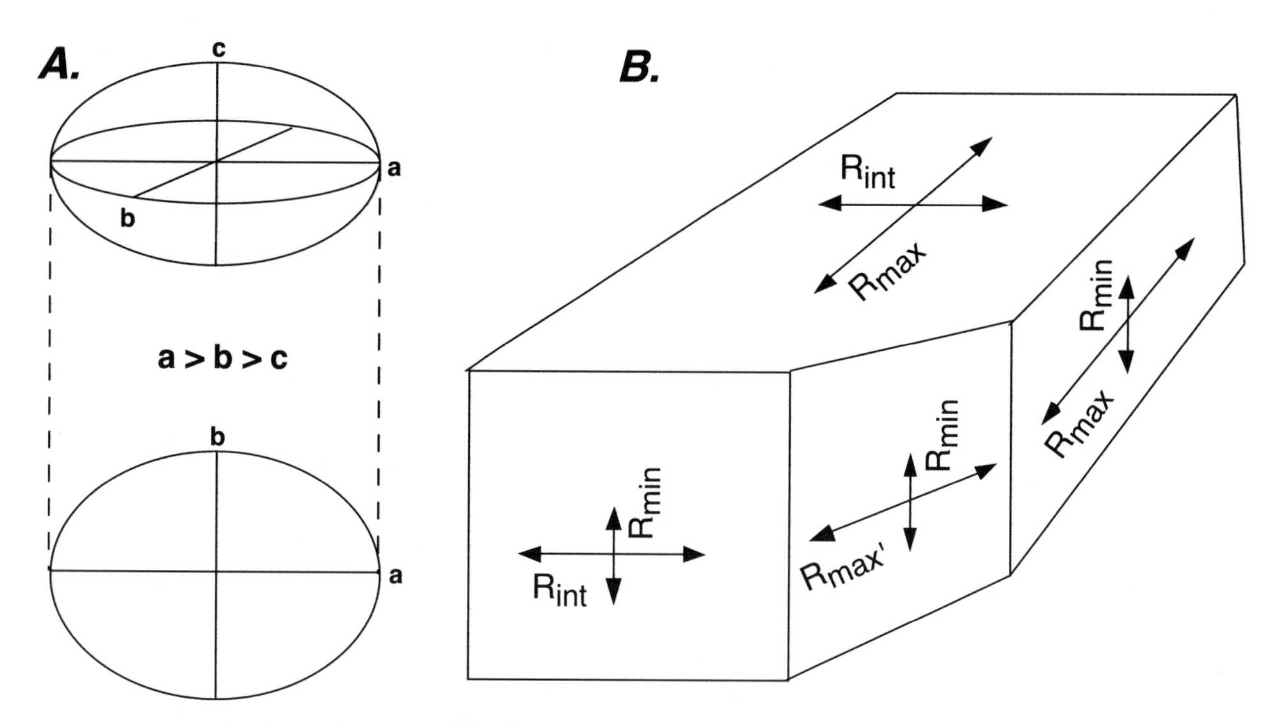

FIGURE 6.6 (*A*) Reflectance indicatrix of biaxial negative material with axes *a*, *b*, and *c*, which correspond to the true maximum reflectance (R_{max}), intermediate reflectance (R_{int}), and true minimum reflectance (R_{min}), respectively. (*B*) Relationship of reflectance axes on different surfaces of a hypothetical piece of biaxial negative vitrinite. $R_{max'}$ is the apparent maximum that will have a value between the true maximum and the intermediate reflectance.

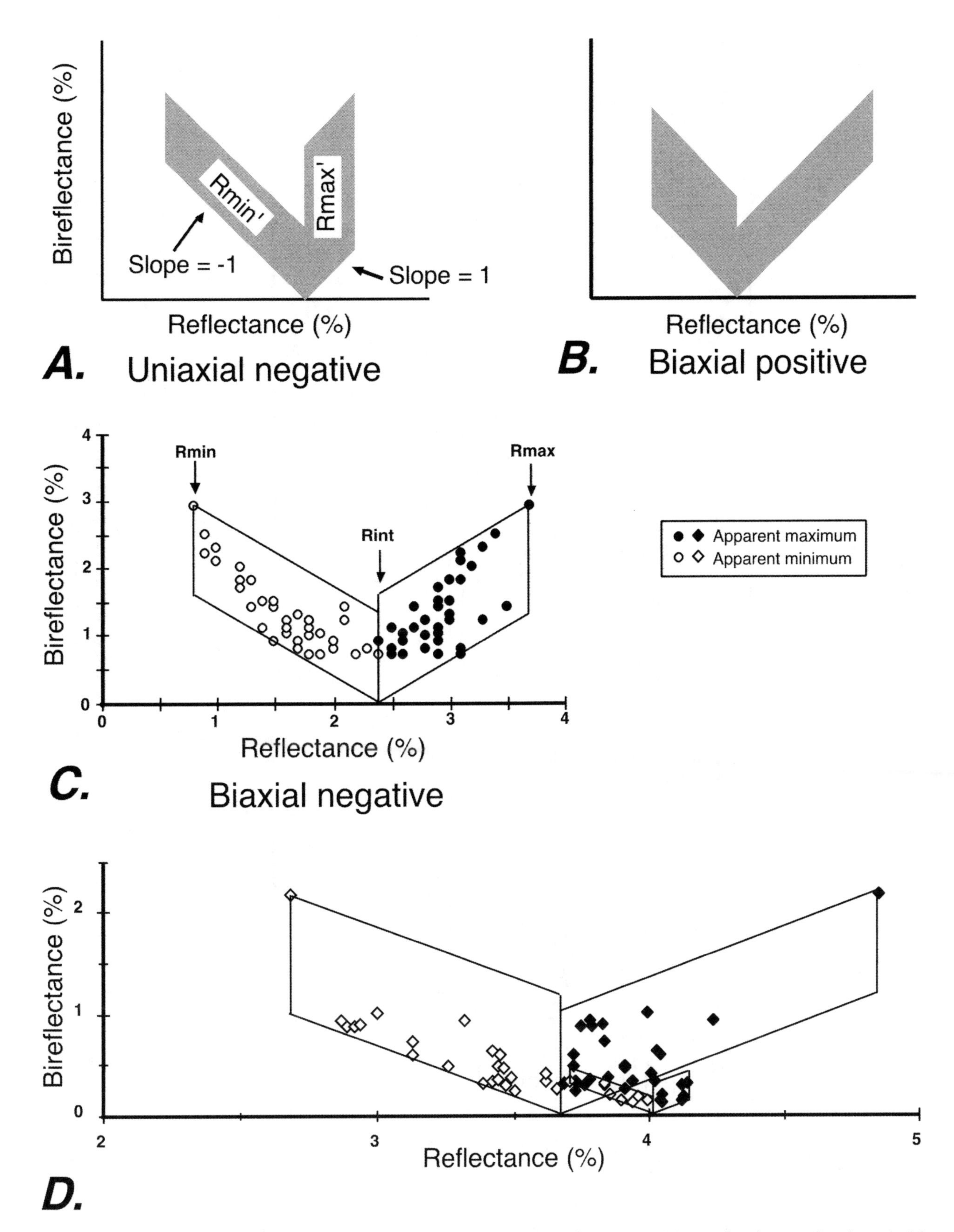

FIGURE 6.7 (*A* and *B*) Examples of field shapes of apparent maximum and apparent minimum data in crossplots for uniaxial negative and biaxial positive reflectance indicatrices. (*C*) Crossplot of apparent maximum and apparent minimum reflectance data (after Malinconico 1993). The lowest minimum and highest maximum are the true minimum and true maximum reflectances for this material; where the apparent minimum and maximum data fields meet is the value of the intermediate reflectance (R_{int}). (*D*) Crossplot of apparent maximum and minimum data from sample BU1218 of the Texaco Butler corehole, Taylorsville basin, Virginia, showing overlapping crossplot fields for vitrinites with different reflectance indicatrices.

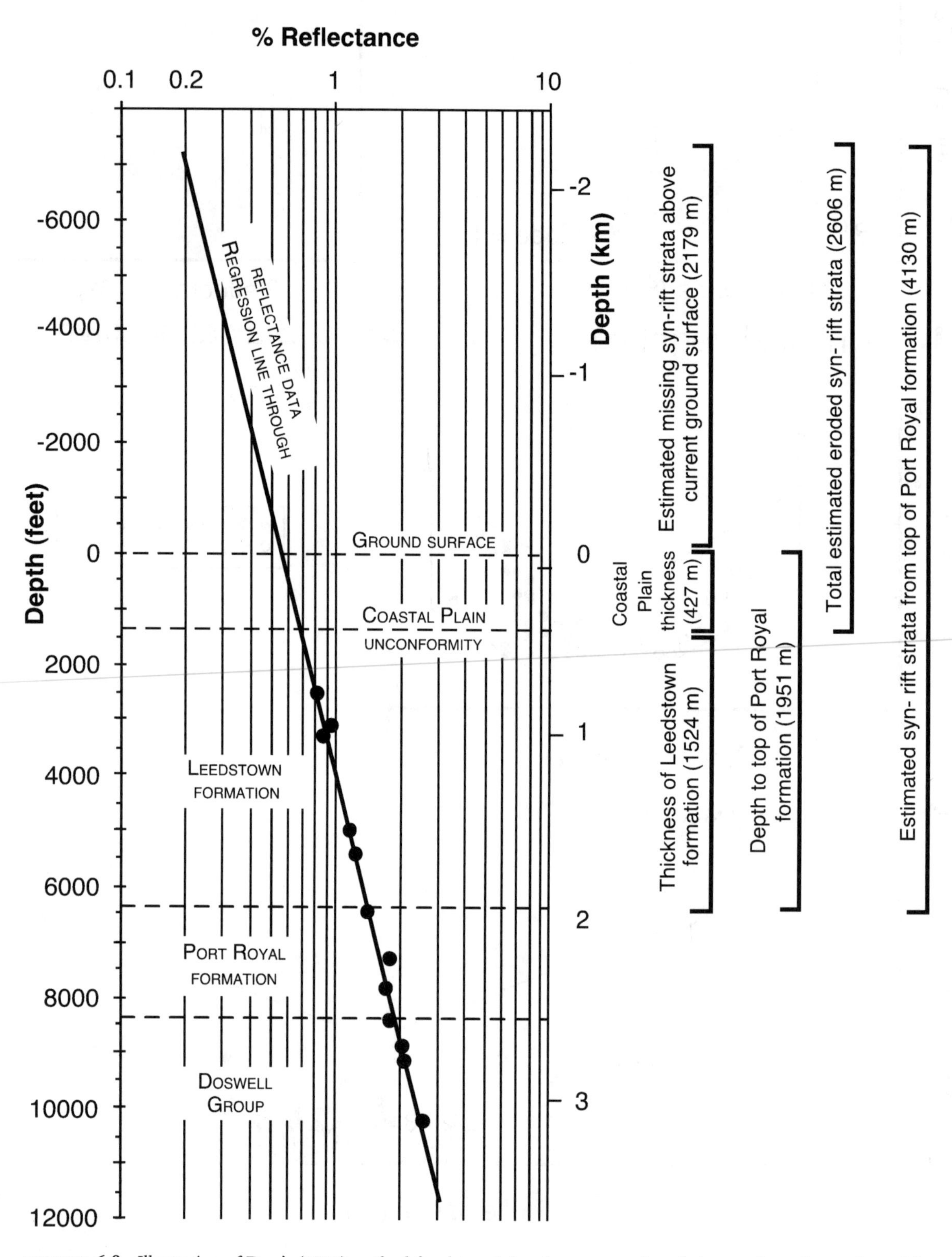

FIGURE 6.8 Illustration of Dow's (1977) method for determining the amount of eroded strata from linear plots of the $\log_{10}$ of vitrinite reflectance versus depth, using data from the Thornhill hole. The regression line through borehole data is extrapolated back to 0.2% R_0, the reflectance of early diagenetic woody plant matter, to find the amount of missing section. The thickness of the eroded section both above the ground surface and above the Coastal Plain (Aptian) unconformity and other relevant stratigraphic intervals are outlined on the left. It is assumed that there was no change in thermal conductivity or geothermal gradient in the eroded section.

the percent reflectance was used as the dependent variable for the regressions.

Results

The mean random reflectances with depth for each of six Taylorsville boreholes (Campbell, Bowie Fogg, Ellis, Wilkins, Gouldman, and Thornhill) are given in table 6.1, and semilog plots of mean random reflectance with depth are shown in figure 6.9. Butler was not included because the data were dominated by high reflectance excursions due to contact metamorphism by igneous dikes, which obscured most of the background conductive burial reflectance profile; the only vitrinite samples in Roberts came from an interval too short (~100 m) to do a meaningful regression; important samples from the middle of the Payne core were not yet available. Gaps in most reflectance profiles are due to the presence of nonvitrinite-bearing strata. The gap in Campbell, however, from 295 to 945 m is due to a high reflectance excursion (up to 1.5 %R_0) attributed to advective heat flow through the basal fluvial stratum of LeTourneau's (1999) Basin-Fill Sequence II and is not used here in the regressions and estimation of eroded section.

The variation in mean random vitrinite reflectance is apparent within a sample from the standard deviation and among downhole samples by the correlation coefficient of linear regression (table 6.2). The standard deviation increases with rank or maturity due to increasing reflectance anisotropy. The correlation coefficient decreases, as expected, with increase in length of the hole and with number of downhole samples. The primary source of the reflectance variation downhole is considered to be variable oxidation of the land-plant woody precursor prior to deposition. In coals, it has long been recognized that vitrinite macerals include vitrinite and pseudovitrinite (Benedict et al. 1968). The pseudovitrinite has a slightly higher reflectance (a couple of tenths of a percent) than normal vitrinite and a lower hydrogen content, and it is not totally reactive in industrial utilization. Nonreactive inertinite macerals, including semifusinite and fusinite, have much higher reflectances than vitrinite in bituminous rank coals and higher oxygen contents. Oxidation of inertinite macerals in coals is due to ancient forest fires, composting, or other nonreducing conditions in the peat swamp (Stach et al. 1982); similar oxidizing conditions during or before early coalification have been suggested for pseudovitrinite (Benedict

TABLE 6.1 Downhole Vitrinite Reflectance

Depth		*Mean Random Reflectance(%)*	*Standard Deviation*
ft	*m*		
Campbell (CH. 1)			
453.0	138.1	0.62	0.04
532.1	162.2	0.52	0.03
598.0	182.3	0.68	0.03
765.7	233.4	0.66	0.03
794.6	242.2	0.67	0.04
799.5	243.7	0.67	0.03
923.0	281.3	0.67	0.03
955.0	291.1	0.74	0.03
956.0	291.4	0.65	0.04
956.1	291.4	0.66	0.02
3,181.5	969.7	1.01	0.04
3,184.5	970.6	0.80	0.03
3,505.0	1,068.3	0.97	0.07
3,512.1	1,070.5	1.08	0.05
3,531.8	1,076.5	0.88	0.03
3,802.6	1,159.0	0.91	0.02
3,867.0	1,178.7	1.08	0.05
4,263.0	1,299.4	1.04	0.09
4,701.5	1,433.0	1.20	0.04
4,891.0	1,490.8	1.36	0.11
5,030.7	1,533.4	1.11	0.04
5,270.0	1,606.3	1.34	0.10
5,278.5	1,608.9	1.38	0.10
5,398.0	1,645.3	1.57	0.12
Ellis (CH. 4)			
2,300.0	701.0	0.48	0.02
2,913.0	887.9	0.54	0.03
3,350.0	1,021.1	0.45	0.02
3,518.6	1,072.5	0.63	0.04
3,523.5	1,074.0	0.71	0.03
4,184.0	1,275.3	0.74	0.04
4,186.0	1,275.9	0.72	0.03
4,228.0	1,288.7	0.62	0.03
4,230.0	1,289.3	0.72	0.02
Bowie-Fogg (CH. 5)			
2,201.2	670.9	0.53	0.03
2,249.5	685.6	0.65	0.05
2,653.5	808.8	0.58	0.02
2,827.4	861.8	0.63	0.03
2,870.4	874.9	0.64	0.01
3,432.3	1,046.2	0.79	0.05
3,433.9	1,046.7	0.87	0.02
3,591.0	1,094.5	0.82	0.05
3,829.9	1,167.4	0.87	0.13
4,296.6	1,309.6	1.04	0.04
4,399.4	1,340.9	0.80	0.03
4,399.5	1,341.0	0.86	0.07
4,410.5	1,344.3	0.89	0.04
4,639.0	1,414.0	0.86	0.04

(*continued*)

TABLE 6.1 Continued.

Depth		*Mean Random Reflectance(%)*	*Standard Deviation*
ft	*m*		
		Wilkins	
2,500.0	762.0	0.78	0.03
2,980.0	908.3	0.75	0.04
4,060.0	1,237.5	1.11	0.08
4,630.0	1,411.2	1.41	0.05
5,700.0	1,737.4	1.55	0.06
6,590.0	2,008.6	1.73	0.10
7,440.0	2,267.7	1.88	0.13
8,140.0	2,481.1	1.92	0.10
8,360.0	2,548.1	1.76	0.07
9,120.0	2,779.8	2.25	0.14
9,520.0	2,901.7	2.70	0.15
9,910.0	3,020.6	2.74	0.16
10,010.0	3,051.0	2.94	0.24
		Thornhill	
2,500.0	762.0	0.81	0.05
3,100.0	944.9	0.92	0.05
3,270.0	996.7	0.86	0.05
4,980.0	1,517.9	1.15	0.07
5,420.0	1,652.0	1.22	0.08
6,460.0	1,969.0	1.37	0.06
7,300.0	2,225.0	1.78	0.20
7,810.0	2,380.5	1.70	0.09
8,400.0	2,560.3	1.77	0.17
8,840.0	2,694.4	2.03	0.19
9,120.0	2,779.8	2.10	0.18
10,200.0	3,109.0	2.49	0.25
		Gouldman	
1,900.0	579.1	0.45	0.03
2,260.0	688.8	0.44	0.03
3,910.0	1,191.8	0.52	0.02
4,120.0	1,255.8	0.52	0.03
4,450.0	1,356.4	0.57	0.03
4,710.0	1,435.6	0.63	0.03
5,230.0	1,594.1	0.64	0.04
7,050.0	2,148.8	0.99	0.03
7,260.0	2,212.8	1.19	0.06
7,400.0	2,255.5	1.14	0.08
7,440.0	2,267.7	1.24	0.07
7,670.0	2,337.8	1.19	0.07
7,720.0	2,353.1	1.23	0.05

et al. 1968). Dispersed vitrinite in lacustrine and marine sediments is allochthonous, and the taphonomy of the woody precursor may include some time under oxidizing conditions that will affect its subsequent coalification path. Variable reflectance, more than could be attributed to normal anisotropy, was seen among and within vitrinite particles from fluvial sandstones in the Campbell, Ellis, and Bowie Fogg coreholes and in black nearshore or shallow-water, organic-rich shales in Campbell. In those shale samples, there was also a large variation in the reflectance and preservation of spores, some being partially to completely fusinized, another indication of variable oxidation of terrestrial organic material. This finding suggests that higher-reflectance vitrinite populations on reflectance histograms, commonly labeled "recycled"—meaning eroded from older sedimentary rocks—more frequently may be first-cycle pseudovitrinite. "Reworked" vitrinite is a term also used for vitrinite oxidized in transport (Creaney 1980).

Because of this variable preservation, although one may be measuring reflectance on the lowest gray vitrinite macerals in a sample, those particles may just be the least-oxidized vitrinites available rather than truly unoxidized normal vitrinite. If the lowest gray macerals are also not as abundant as pseudovitrinite-like macerals and therefore will not form a population peak discrete from more oxidized vitrinite on a reflectance histogram, using the reflectance histogram method of identifying the vitrinite population may result in an overestimation of maturity. In this study, because apparent maximum and minimum reflectance data were available due to rotational polarization, reflectance crossplots (Kilby 1988) were constructed to eliminate pseudovitrinite and to identify low gray particles with a similar reflectance indicatrix and assumably similar chemical composition.

Some noise among samples with vitrinite reflectances greater than 1.3%, where anisotropy is more developed, may be due to a preferred rather than a random orientation among particles measured. Use of the mean rotational reflectance rather than the mean random reflectance will eliminate this noise. For example, in the deepest Campbell sample at 1,645.3 m (5,398 ft.), the mean random reflectance is 1.57%, which is only several hundredths of a percent less than the apparent maximum reflectance for the measurements included in the mean. The mean rotational reflectance is 1.39%, which is closer to the regression line and eliminates any impression that there may be a sudden increase in maturation or metamorphism at the bottom of the core (figure 6.9). This case appears to be, however, an isolated sample in this study, and the mean random reflectance for this depth is reported.

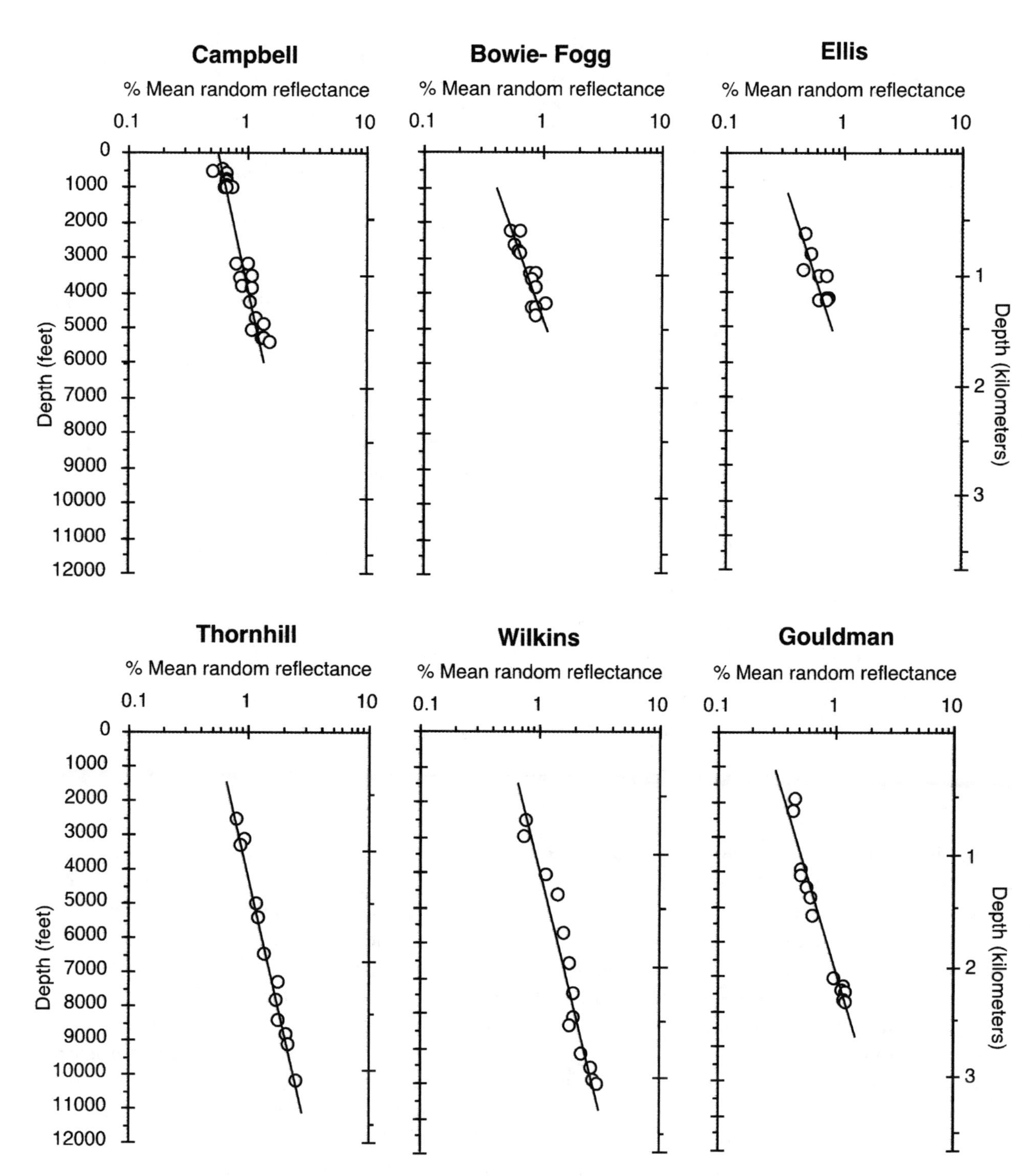

FIGURE 6.9 Semilog plots of reflectance versus depth with regression lines for the six boreholes used in this study. The mean random reflectance values are listed in table 6.1. The correlation coefficients of the regression lines are in table 6.2.

Linear regression of the reflectance profiles from each borehole back to 0.2% reflectance yields an estimated amount of eroded synrift section above the current ground surface and the total estimated eroded synrift strata above the Coastal Plain unconformity (table 6.2). Besides the sources of error discussed in the next paragraph, error in the estimate of eroded synrift strata may also be due both to depth measured,

TABLE 6.2 Estimated Eroded Synrift Overburden and Associated Error

	Linear Regression Results			
Hole	*Estimated Eroded Section Above Ground Level (m)*	*Correlation Coefficient*	*Thickness of Coastal Plain (m)*	*Total Estimated Erosion Above Cretaceous Unconformity (m)*
Campbell	2,024	0.913	131	2,155
Thornhill	2,179	0.987	427	2,606
Wilkins	1,904	0.939	488	2,392
Bowie-Fogg	969	0.753	543	1,512
Gouldman	413	0.950	457	870
Ellis	526	0.525	482	1,008
	Bootstrap Statistics (100 regressions)			
	Mean Eroded Section Above Ground Level (m)	*Standard Deviation (m)*	*Minimum Estimated Erosion (m)*	*Maximum Estimated Erosion (m)*
Campbell	1,994	105	1,738	2,280
Thornhill	2,253	283	1,599	3,318
Wilkins	1,885	167	1,603	2,400
Bowie-Fogg	1,116	265	639	2,235
Gouldman	409	99	214	705
Ellis	543	177	207	970

in some holes, from the derrick floor/kelly bushing and to drill-hole deviation from vertical. Data on these sources of error were not available for all holes. The height of the kelly bushing was found for only two holes, in which it was 7.9 m (26 ft.); for several others, depth, at least in borehole electric logs, was measured from the ground surface. The true depth for three holes in this study, according to continuous verticality logs, was between 3 and 18 m (11 to 57 ft.) shorter than the drilled depth. Any error from not correcting for height of the kelly bushing or for hole deviation from vertical was considered minimal compared with that caused by the standard deviation of the reflectance data for individual samples and scatter among samples for each hole.

The major sources of error in the calculation of the erosion estimate are twofold: one due to uncertainty in the data and linear regression, and the other due to assumptions in the method. Error due to the data and regression was determined using the parametric bootstrap (Efron and Tibshirani 1993), a statistical resampling technique, which takes into account that each mean random reflectance value is a mean with its own standard deviation. Error estimation calculated as the variance of error about the curve was not considered appropriate because the resultant hyperbolic confidence envelope increases away from the mean values of the x, y data set so that boreholes with higher reflectances will have a very large error at the 0.2 %R_0 intercept regardless of correlation coefficient.

For error determination using the parametric bootstrap, because the bootstrap is the $\log_{10}$ that is linear with depth, the log mean and standard deviation for each mean random reflectance were recalculated directly from the logarithms of the measured reflectances. One hundred random numbers subsequently were generated from a range that had a normal (Gaussian) distribution with the same $\log_{10}$ mean random reflectance value and standard deviation. Using the generated random numbers for each reflectance point,

100 linear regressions were performed for each borehole, and the *y* intercept or depth value at $\log_{10}$ of 0.2% vitrinite reflectance was determined. Table 6.2 shows the results of the resampling statistics, including the mean estimated erosion for the 100 regressions per borehole, the standard deviation of the erosion estimate population, and the range of estimates (maximum and minimum). The standard deviation of the resampling estimates ranges from 99 m for Gouldman to 283 m for Thornhill, although the ranges indicate that an error up to 1.1 km is possible. The calculated mean erosion estimates for each borehole from the bootstrap exercise also do not match exactly the estimates from the original linear regressions; however, the means theoretically should be equivalent with larger resampling populations.

The second source of error in the erosion estimates comes from assumptions in the method itself. The Dow (1977) method assumes that thermal conductivity and geothermal gradient in the missing section are the same as in the existing strata. In theory, uncompacted or less compacted sediments buried less than approximately 2 km will have a lower thermal conductivity due to fluids filling that higher porosity and therefore will have a higher geothermal gradient (Sclater and Christie 1980; Holliday 1993). If so, then the Dow method will overestimate the eroded section. However, some published data (Stach et al. 1982:fig. 6.22A) does show linear log reflectance profiles at shallow depths in some young basins, suggesting little or no thermal conductivity contrast due to porosity. In addition, Blackwell and Steele (1989) stated that thermal conductivity in shales (1.1 to 1.3 $Wm^{-1}K^{-1}$) is not affected by compaction: deeply buried shale has a low vertical conductivity due to orientation of platy minerals; at shallow depths, random clay mineral orientation would enhance vertical thermal conductivity, but high porosity and water content negate the mineralogical effects. Studies of diagenesis in the Newark and related Hartford basins indicate that early cementation has resulted in less compaction than expected (Hubert, Feshbach-Meriney, and Smith 1992; Simonson 1996). There is also no increase in the slope of the $\log_{10}$ reflectance data in the least-eroded boreholes (Bowie Fogg, Ellis, Gouldman) that might be attributed to increased porosity. Therefore, it appears that the effects of compaction on thermal conductivity may be variable, and assuming a standard compaction gradient may be just as erroneous as assuming none at all. A comprehensive review of the effects of porosity and compaction on thermal conductivity can be found in Giles, Indrelid, and James (1998).

Based on the low shale thermal conductivities found by Blackwell and Steele (1989), one might expect obvious changes in the slope of reflectance data between shale-dominated and sand/quartz–dominated lithologies. However, the downhole reflectance profiles of this study show no change in slope that can be attributed to material changes in thermal conductivity. For example, an apparent steeper-slope segment in the Wilkins borehole (figure 6.9) from approximately 1,370 m (4,500 ft.) to 2,440 m (8,000 ft.) crosses the lithologic and formational boundary from more sand-rich strata of the Leedstown Formation to the lacustrine black shales of the Port Royal Formation at 1,920 m (6,300 ft.). The thermal conductivity of shales and mudstones of the Taylorsville basin is probably, therefore, closer to that of sandstones, siltstones, and limestones. Lacustrine mudstones of the Newark basin have a notable carbonate content (Olsen 1990), and many Newark basin Jurassic mudstones have a silt component of a few tens percent. The nonclay constituents, including early cements mentioned previously, will increase the bulk thermal conductivity of the shales and diminish any thermal conductivity contrast.

Other methods for estimating eroded overburden include fission track analysis with or without vitrinite reflectance data (Bray, Green, and Duddy 1992; Green, Duddy, and Bray 1995; Lerche 1997), thermal history reconstruction using vitrinite reflectance data (Armagnac et al. 1989), models of various downhole data (sonic and density logs and drilling exponent methods), and Airy isostasy (Lerche 1997). If fission track analysis is used with vitrinite reflectance, these data are first converted to maximum paleotemperatures before regression to a chosen surface temperature in order to estimate missing section. Linearity of geothermal gradient is assumed, as in the Dow method, but the additional step of conversion to temperatures adds a further source of error. Bray, Green, and Duddy (1992) calculated errors of 0.6 to greater than 6.5 km for erosion estimates determined using vitrinite reflectance alone and 0.4 to 1.8 km for apatite fission track data combined with reflectance data. The thermal history reconstruction described in Armagnac et al. (1989)

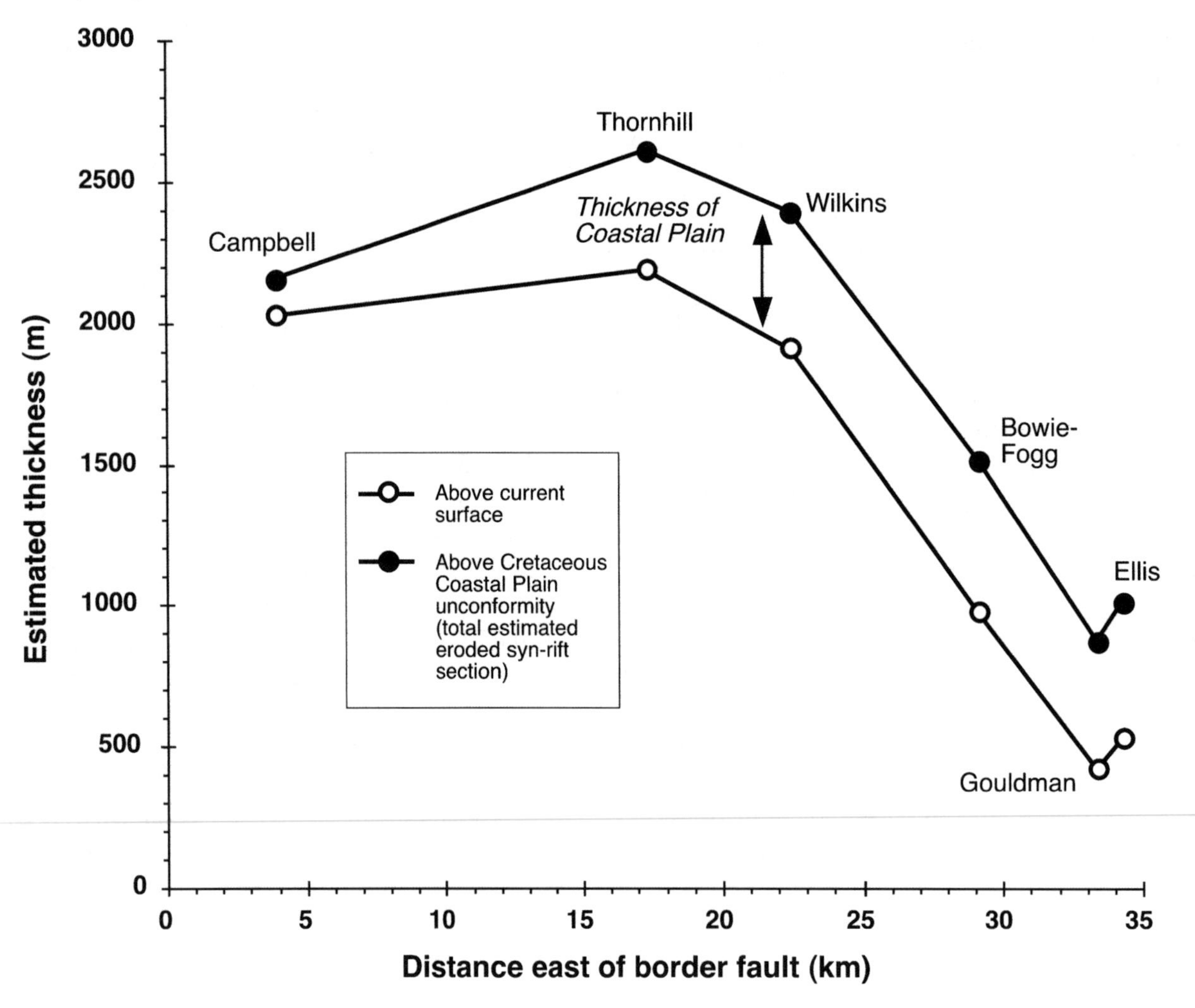

FIGURE 6.10 Change in estimated eroded synrift strata above the current ground surface and above the Coastal Plain unconformity across the basin, based on extrapolation of reflectance versus depth regression lines. The maximum at Thornhill and Wilkins lies over a faulted basement antiform and is attributed to structural inversion.

and based on the physical vitrinite reflectance maturation model of Lerche, Yarzab, and St. C. Kendall (1984) varies heat flux and sediment thickness until the mean-square-residual fit of model reflectance data with measured data is minimized. This method also requires the addition of thermal history parameters. The range of uncertainty in estimates of sediment removal by Armagnac et al. (1989) were up to approximately 800 m. They also used the Dow method and calculated (method not stated) uncertainties of 1.5 to 4 km. Lerche (1997) discussed uncertainty in all these methods and concluded that the minimum uncertainty for estimates of thickness of eroded sediment is ±500m, and most likely ±1 km. These uncertainties (0.5 to 1.0 km) are consistent with errors calculated in this study and with assumptions in the Dow method.

The estimated amount of missing section from the vitrinite data for each borehole is plotted against distance from the western border fault in figure 6.10. The amount of pre–Coastal Plain strata eroded generally increases from east to west, except in Gouldman, with a maximum at the Thornhill and Wilkins localities and then a slight decrease to the west at Campbell. The maximum estimated eroded strata at the Wilkins and Thornhill sites mimics the basement antiform in the faulted hanging wall over which the holes are projected to sit. The minimum in eroded synrift strata at Gouldman corresponds to LeTourneau's (1999, chapter 3 in volume 2) interpretation of that hole's shallowest extant synrift strata as being younger than that in Bowie Fogg, and of its location as being in the center of a synform (figure 6.2).

Discussion

Basement antiforms, like the one over which the Thornhill and Wilkins boreholes are located, are typical of rift basin inversion anticlines (Mitra 1993; Withjack, Olsen, and Schlische 1995). The maximum in postrift basin erosion at Thornhill and Wilkins and the minimum at the Gouldman synform indicate that these structures formed after maximum coalification or heating, which therefore supports an inversion origin for this basement feature.

In the Newark basin, the estimated amount of eroded synrift strata above the youngest existing Jurassic Boonton Formation is 2 to 3 km, based on surface vitrinite reflectance and fission track studies (Pratt, Shaw, and Burruss 1988; Steckler et al. 1993). This missing amount is comparable to that missing at the Thornhill and Wilkins boreholes in the Taylorsville basin, but the approximately 2.5 km missing at the Thornhill and Wilkins sites is the maximum for the basin, whereas the 2 to 3 km missing above the Jurassic Boonton Formation is probably the minimum for the Newark basin. In general in the Newark basin, the oldest exposed stratigraphy and the highest surface vitrinite reflectance values not possibly affected by igneous contact metamorphism are along the eastern edge of the basin. The vitrinite reflectance values there are 2.0 to 2.6% (Katz et al. 1988; Pratt, Shaw, and Burruss 1988), which suggests that approximately 4 to 5 km total eroded there, assuming the same $\log_{10}$ $\%R_0$ versus depth slope in both basins. These differences in basin erosion suggest that the style of inversion is different between basins: the Taylorsville basin is eroded more in the central basin over a basement antiform, whereas the Newark basin generally experienced more erosion and uplift along its eastern (hanging-wall) edge, possibly due to differential reactivation of basement fault blocks.

Duration of Synrift Sedimentation

Methods

To determine the duration of synrift sedimentation, the length of time for deposition of the missing strata plus existing section above an interval of known depositional age can be calculated if the sedimentation rate is known. Estimates of the duration of sedimentation were done for the Thornhill and Wilkins boreholes because of their depth or length of section and low correlation coefficient. The duration of sedimentation was determined from the top of the Port Royal Formation because it is a well-defined correlatible stratigraphic horizon in several wells and because the faulted section in the Port Royal in the Thornhill borehole had to be avoided (LeTourneau 1999). The top of the Port Royal Formation in Thornhill is at 1,951 m (6,400 ft.) and in Wilkins at 1,920 m (6,300 ft.); it is correlated with the middle of the Skunk Hollow Member, Lockatong Formation, in the Newark basin, and is dated at approximately 221 Ma (Kent, Olsen, and Witte 1995; LeTourneau 1999).

Five constant and two decreasing sedimentation rates (compacted) were chosen to estimate the elapsed time required to deposit the extant and estimated missing synrift strata from the deposition of the top of the Port Royal Formation. The five constant sedimentation rates used are 0.35, 0.30, 0.25, 0.20 and 0.14 mm/yr. Paul Olsen (personal communication 1994) determined the rate of 0.35 mm/yr from the average 20,000-year lake-cycle thickness in the Port Royal Formation in both the Campbell core (stored at Lamont) and a short Wilkins core (Texaco). Assuming that 0.35 mm/yr is a maximum compacted sedimentation rate for the basin, 0.30 and 0.25 mm/yr were chosen as average and minimum compacted sedimentation rates for Thornhill and Wilkins, based on the proportional relationships of maximum, average, and minimum cycle thicknesses in the correlative interval of the Newark Basin Coring Project (NBCP) scaled composite stratigraphic section (Olsen and Kent 1996). The rates of 0.14 and 0.20 mm/yr were based on correlative paleomagnetic stratigraphy of the Taylorsville Butler and Bowie Fogg cores with stratigraphy of the Newark basin cores (LeTourneau 1999).

The two decreasing sedimentation rates start with an initial maximum rate of 0.35 or 0.30 mm/yr and then decrease linearly. The rate of sedimentation decrease with time is based on the results of Contreras, Scholz, and King's (1997) rift basin evolution model, which allows for the interplay of faulting, flexure, erosion, sedimentation, and isostasy. In their model, basinal locations adjacent to the border fault have high initial sedimentation rates, with a markedly exponential decrease in sedimentation rate with time; deposition at points far from the border fault have low overall sedimentation rates that are constant or decrease very slowly once a maximum is reached. For their midbasin location 3 (Contreras, Scholz, and King 1997:fig. 6.18b), similar to the setting of the Thornhill and Wilkins boreholes, sedimentation quickly increases to a maximum rate within 5 million years after the initiation of rifting, and then decreases slowly and almost linearly. The rate of change used in this study after a maximum sedimentation rate of 0.35 mm/yr is -1.33×10^{-11} mm/yr^2; for a maximum sedimentation rate of 0.30 mm/yr, the rate of change is -8.9×10^{-12} mm/yr^2. Again, it must be emphasized that all the sedimentation rates are for compacted sediment.

Results and Discussion

The total composite synrift section younger than 221 Ma is 4,130 m (13,549 ft.) for Thornhill and 3,824 m (12,547 ft.) for Wilkins. This total is the sum of the previously estimated missing section above the ground surface (table 6.2) plus the depth of the top of the Port Royal Formation below the ground surface, which in Thornhill is 1,951 m (6,400 ft.) and in Wilkins, 1,920 m (6,300 ft.). The results of the calculations for the five constant and two decreasing rates for the Thornhill and Wilkins drill holes are found in table 6.3 and graphically illustrated in figure 6.11. Estimates of the termination of synrift sedimentation based on these calculations vary from 191.5 to 209.2 Ma in Thornhill and from 193.7 to 210.1 Ma in Wilkins.

The difference in duration of synrift sedimentation between Thornhill and Wilkins is due to the difference in stratigraphic thickness above the Port Royal Formation. This discrepancy may be due to a difference in sedimentation rate. The Leedstown Formation in Thornhill is more fluvial in character (LeTourneau 1999), with a more frequent occurrence of telinite (structured vitrinite), typical of isolated logs or large chunks of wood, throughout the formation, than the equivalent strata in Wilkins, indicating a more proximal position to sources of clastic input. The 300 m difference, however, is within the range of standard

TABLE 6.3 Duration and End of Synrift Sedimentation Calculations

Sedimentation Rate	*Estimated Age of Synrift Sediments at Coastal Plain Unconformity (Ma)*	*Duration of Synrift Sedimentation from 221 Ma (including eroded strata) (my)*	*Estimated End of Synrift Sedimentation (Ma) (Triassic–Jurassic boundary = 201 Ma)*
Thornhill			
0.35 mm/yr constant	216.7	11.8	209.2
0.35 mm/yr and decreasing	216.2	17.9	203.1
0.30 mm/yr constant	215.9	13.8	207.2
0.30 mm/yr and decreasing	215.5	19.3	201.7
0.25 mm/yr constant	214.9	16.5	204.5
0.20 mm/yr constant	213.4	20.7	200.4
0.14 mm/yr constant	210.1	29.5	191.5
Wilkins			
0.35 mm/yr constant	216.9	10.9	210.1
0.35 mm/yr and decreasing	216.5	15.5	205.5
0.30 mm/yr constant	216.2	12.8	208.3
0.30 mm/yr and decreasing	215.8	17.1	203.9
0.25 mm/yr constant	215.3	15.3	205.7
0.20 mm/yr constant	213.9	19.1	201.9
0.14 mm/yr constant	210.8	27.3	193.7

deviation for the eroded section estimates and therefore may not indicate a material difference in accumulation rate between the two boreholes.

Several sources of error can be found in these calculations. There is, of course, error compounded from the erosion estimates, both in the calculations and in the assumptions in the Dow (1977) method, as described earlier. In addition, in the erosion calculations, arguments were made regarding why high porosity of shallowly buried sediments may not result in a higher geothermal gradient and therefore why degree of compaction can be ignored. In calculations on duration of sedimentation, however, ignoring compaction factors skews the duration estimates to longer intervals and younger ages for the end of sedimentation. A unit thickness of compacted sediment represents more time than the same thickness of uncompacted sediment; the apparent accumulation rate of compacted sediment is consequently lower. By how much the duration of synrift sedimentation is overestimated in this study is not clear. Uncompacted sandstones have porosities of approximately 40% (Giles, Indrelid, and James 1998). Hartford basin lacustrine sandstones have intergranular volumes (cement plus current porosity) of up to 42%, indicating early cementation, probably at depths less than 1 km (Hubert, Feshbach-Meriney, and Smith 1992). For that lithology, therefore, the accumulation rate for compacted sediment may not differ much from the uncompacted, assuming that the Taylorsville sandstones experienced a similar diagenetic history. For nonmarine mudstones of the Newark basin, Simonson (1996) also suggested early cementation, but how much compaction or decrease in original volume actually has taken place is not known. Therefore, the degree to which the duration estimates are overestimated is not apparent.

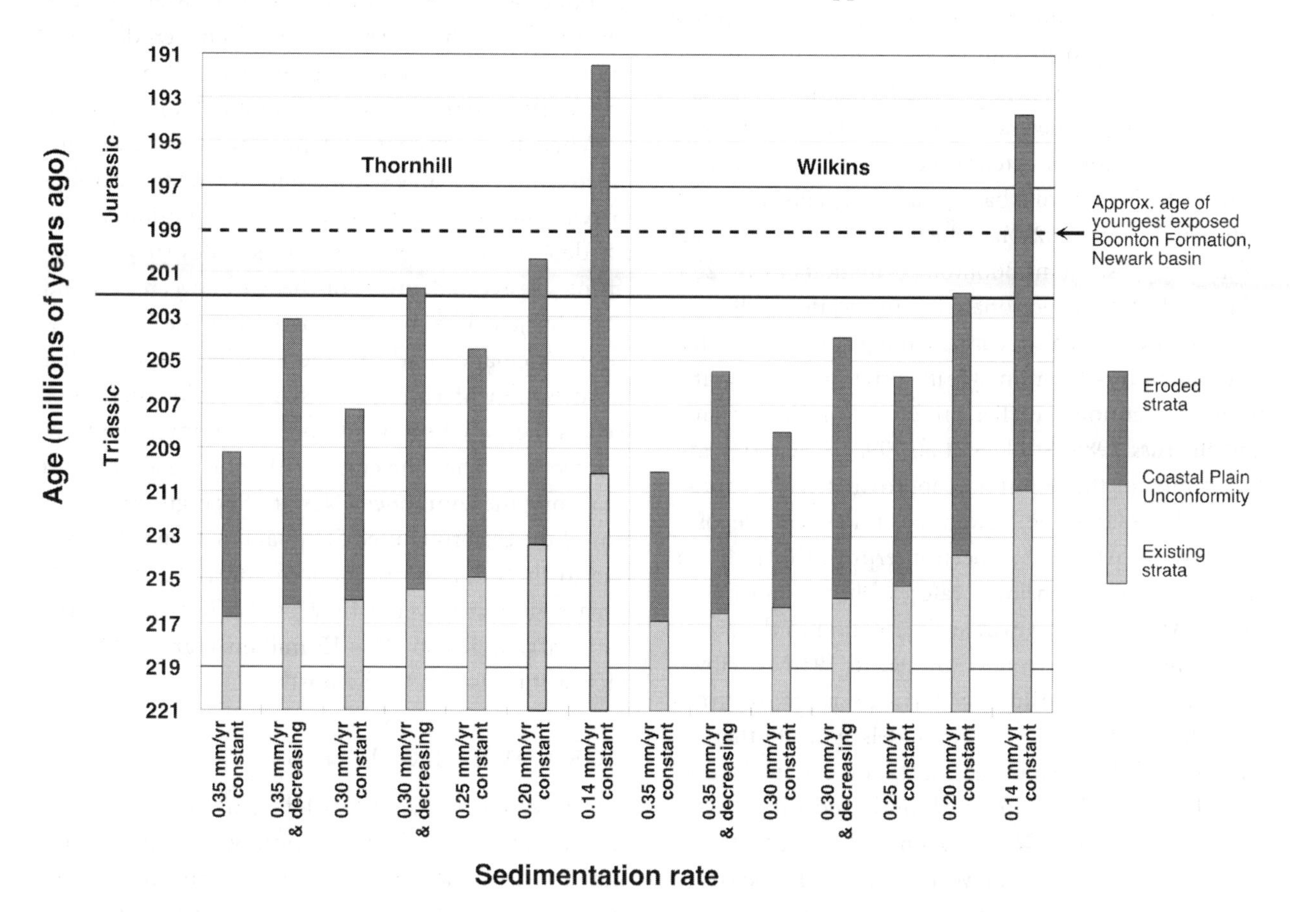

FIGURE 6.11 Estimated end of synrift sedimentation in the Thornhill and Wilkins boreholes, using seven sedimentation rates. The estimated age of the youngest strata in each borehole at the Coastal Plain unconformity is also shown. The youngest extant unit in the Newark basin is the Boonton Formation, whose age is also marked. It is estimated that 2 to 3 km of eroded synrift Jurassic strata are missing above the Boonton in the Newark basin (Pratt et al. 1988; Steckler et al. 1993).

Most error derives, however, from lack of definitive knowledge of appropriate compacted sedimentation rate. The 0.35 mm/yr rate measured in the Wilkins core is in the Port Royal Formation below the datum (the boundary between the Port Royal and Leedstown Formations) from which the sediment duration estimate is calculated and therefore is not appropriate for younger strata. Moreover, the accumulation rate information from magnetic stratigraphy of the Butler and Bowie Fogg cores indicates much lower rates of 0.14 to 0.19 mm/yr. The possible rates of 0.16 to 0.19 mm/yr in the Butler core are higher than those in the same interval in Bowie Fogg, which is reasonable because of Butler's more central basin location. The lower Leedstown Formation in the Thornhill and Wilkins boreholes may have similar or higher rates because they are west of Butler and closer to the border fault, which narrows the possible sedimentation rates for Wilkins and Thornhill in this test to 0.35 mm/yr and decreasing, 0.30 mm/yr and decreasing, or the constant rate of 0.30, 0.25, or 0.20 mm/yr.

Using the parameters and arguments outlined earlier, the cessation of extension and synrift sedimentation in the Taylorsville basin, therefore, falls generally between 208 and 202 Ma, which is older than the age of the youngest extant Boonton Formation (199 to 200 Ma) (P. E. Olsen, personal communication 1997) in the Newark basin, also marked on figure 6.8. Estimates of eroded post-Boonton synrift strata for the Newark basin, as mentioned earlier, are 2 to 3 km (Pratt, Shaw, and Burruss 1988; Steckler et al. 1993). The sedimentation rate for the Boonton is approximately 0.50 mm/yr: the Jurassic of the Newark basin has high depositional rates, which have been interpreted to indicate an increase in extensional rate (Schlische and Olsen 1990). At that rate, synrift sedimentation in the Newark basin could have ended by 195 to 193 Ma. However, all the exposed Jurassic formations are relatively near the border fault, where models indicate that an exponential decrease in sedimentation rate with time would be typical (Contreras, Scholz, and King 1997), so deposition in the Newark basin may have continued for another several million years after that. In the Hartford and Fundy basins, the youngest extant strata are Pliensbachian (Wade et al. 1996; Olsen 1997).

These dates for the end of synrift deposition in the Taylorsville basin significantly add to the growing evidence—as outlined in Schlische and Ackermann (1995) and in Withjack, Schlische, and Olsen (1998)—that the rift–drift transition off the passive margin of the southern United States began by the Triassic–Jurassic boundary, much earlier than that to the north, which is estimated at 185 Ma. However, whether or not Early Jurassic lavas were erupted into the Taylorsville basin cannot be proved. If they were present, the flows would have been at or near the top of the synrift section and would have been among the first strata eroded.

Conclusions

The estimates of eroded synrift strata from Taylorsville basin borehole vitrinite reflectance indicate increasing erosion generally from east to west across the basin, with a maximum of 2.6 km missing in the central portion of the basin over a basement antiform. The pattern of erosion is attributed to postrift inversion. The Taylorsville basin, however, is less deeply eroded than the Newark basin, where the estimated 2 to 3 km of post–earliest Jurassic strata is a minimum for the basin.

When these estimates of missing strata are combined with reasonable sedimentation rates for two Taylorsville boreholes, calculations of the duration of synrift deposition and extensional tectonics suggest that rifting ceased and inversion began between 208 and 202 Ma, just before the Triassic–Jurassic boundary. These calculations add to the growing evidence that the cessation of rifting and synrift sedimentation as well as the initiation of compressional inversion were diachronous along the current Atlantic passive margin and along the component Newark Supergroup rift basins of the eastern United States and maritime Canada due to the initiation of seafloor spreading from Virginia southward by the time of the Triassic–Jurassic boundary, approximately 15 million years before the rift–drift transition to the north.

Acknowledgments

This research was supported in part by student research grants from the Geological Society of America and the American Association of Petroleum Geologists, and by a grant from the student field fund of the Department of Earth and Environmental Sciences, Columbia University, New York. Reflectance data was collected in the Coal Characterization Laboratory, Department of Geology, Southern Illinois University,

Carbondale, and at the Department of Geological Sciences, University of Missouri, Columbia. Photomicrographs were taken at the Coal and Organic Petrology Laboratory of Pennsylvania State University, University Park. I thank Michael Underwood (University of Missouri), John Crelling and William Huggett (Southern Illinois University), and Gareth Mitchell and David Glick (Pennsylvania State University) for their help with this project. I also thank Douglas Martinsen of the Lamont-Doherty Earth Observatory (LDEO) for discussing error determination with me and for suggesting the bootstrap technique, and Paul E. Olsen and Michael Steckler, also of LDEO, for their useful discussions and reviews of the manuscript for this chapter, which is Lamont-Doherty Earth Observatory contribution no. 5939.

Literature Cited

Altamura, R. J. 1996. Tectonics of the Lantern Hill fault, southeastern Connecticut: Evidence for the embryonic opening of the Atlantic Ocean. In P. M. LeTourneau and P. E. Olsen, eds., *Aspects of Triassic–Jurassic Rift Basin Geoscience: Abstracts,* p. 5. State Geological and Natural History Survey of Connecticut Miscellaneous Reports, no. 1. Hartford: Connecticut Department of Environmental Protection.

Armagnac, C., J. Bucci, G. St. C. Kendall, and I. Lerche. 1989. Estimating the thickness of sediment removed at an unconformity using vitrinite reflectance data. In N. Naeser and T. McCulloh, eds., *Thermal History of Sedimentary Basins: Methods and Case Histories,* pp. 217–238. New York: Springer

Barker, C. E. 1996. A comparison of vitrinite reflectance measurements made on whole rock and dispersed organic matter concentrate mounts. *Organic Geochemistry* 24:251–256.

Barker, C. E., and M. J. Pawlewicz. 1993. An empirical determination of the minimum number of measurements needed to estimate the mean random vitrinite reflectance of disseminated organic matter. *Organic Geochemistry* 20:643–651.

Benedict, L. G., R. R. Thompson, J. J. Shigo III, and R. P. Aikman. 1968. Pseudovitrinite in Appalachian coking coals. *Fuel* 47:125–143.

Bensley, D. F., and J. C. Crelling. 1991. The interpretation of vitrinite reflectance measurements using rotational polarization [abstract]. *Proceedings of the Annual Meeting of the Society for Organic Petrology* 8:29.

Benson, R. N. 1992. *Map of Exposed and Buried Early Mesozoic Rift Basins/Synrift Rocks of the U.S. Middle Atlantic Continental Margin.* Delaware Geological Survey Miscellaneous Map Series, no. 5. Newark: Delaware Geological Society.

Blackwell, D. D., and J. L. Steele. 1989. Thermal conductivity of sedimentary rocks: Measurement and significance. In N. Naeser and T. McCulloh, eds., *Thermal History of Sedimentary Basins: Methods and Case Histories,* pp. 13–36. New York: Springer.

Bobyarchick, A. R., and L. Glover III. 1979. Deformation and metamorphism in the Hylas zone and adjacent part of the eastern Piedmont in Virginia. *Geological Society of America Bulletin* 90:737–752.

Bray, R. J., P. F. Green, and I. R. Duddy. 1992. Thermal history reconstruction using apatite fission track analysis and vitrinite reflectance: A case study from the UK East Midlands and the southern North Sea. In R. F. P. Hardman, ed., *Exploration Britain: Geological Insights for the Next Decade,* pp. 3–25. Geological Society Special Publication, no. 67. London: Geological Society.

Contreras, J., C. Scholz, and G. C. P. King. 1997. A model of rift basin evolution constrained by first order stratigraphic observations. *Journal of Geophysical Research,* ser. B4, 102:7673–7690.

Creaney, S. 1980. The organic petrology of the Upper Cretaceous Boundary Creek Formation, Beaufort-Mackenzie basin. *Bulletin of Canadian Petroleum Geologists* 28:112–129.

Dow, W. G. 1977. Kerogen studies and geological interpretations. *Journal of Geochemical Exploration* 7: 79–99.

Efron, B., and R. J. Tibshirani. 1993. *An Introduction to the Bootstrap.* Monograph on Statistics and Applied Probability, no. 57. New York: Chapman and Hall.

Gates, A. E. 1997. Multiple reactivations of accreted terrane boundaries: An example from the Carolina terrane, Brookneal, Virginia. In L. Glover III and A. E. Gates, eds., *Central and Southern Appalachian Sutures: Results of the EDGE Project and Related Studies,* pp. 49–64. Geological Society of America Special Paper, no. 314. Boulder, Colo.: Geological Society of America.

Giles, M. R., S. L. Indrelid, and D. M. D. James. 1998. Compaction: The great unknown in basin modelling. In S. J. Düppenbecker and J. E. Illiffe, eds., *Basin Modelling: Practice and Progress,* pp. 15–43. Geological Society Special Publication, no. 141. London: Geological Society.

Green, P. F., I. R. Duddy, and R. J. Bray. 1995. Applications of thermal history reconstruction in inverted basins. In J. G. Buchanan and P. G. Buchanan, eds., *Basin Inversion,* pp. 149–165. Geological Society Special Publication, no. 88. London: Geological Society.

Hevia, V., and J. M. Virgos. 1977. The rank and anisotropy of anthracites: The indicating surface of reflectivity in uniaxial and biaxial substances. *Journal of Microscopy* 109:23–25.

Hill, R. I. 1991. Starting plumes and continental breakup. *Earth and Planetary Science Letters* 104:398–416.

Holbrook, W. S., and P. B. Kelemen. 1993. Large igneous province on the U.S. Atlantic margin and implications for magmatism during continental breakup. *Nature* 364:433–436.

Holliday, D. W. 1993. Mesozoic cover over northern England: Interpretation of apatite fission track data. *Journal of the Geological Society of London* 150:657–660.

Houseknecht, D. W., D. F. Bensley, L. A. Hathon, and P. H. Kastens. 1993. Rotational reflectance properties of Arkoma basin dispersed vitrinite: Insights for understanding reflectance populations in high thermal maturity regions. *Organic Geochemistry* 20:187–196.

Hubert, J. F., P. E. Feshbach-Meriney, and M. A. Smith. 1992. The Triassic–Jurassic Hartford rift basin, Connecticut and Massachusetts: Evolution, sandstone diagenesis, and hydrocarbon history. *American Association of Petroleum Geologists Bulletin* 76:1710–1734.

Katz, B. J., and X. Liu. 1998. Summary of the AAPG research symposium on lacustrine basin exploration in China and Southeast Asia. *American Association of Petroleum Geologists Bulletin* 82:1300–1307.

Katz, B. J., C. R. Robison, T. Jorjorian, and F. D. Foley. 1988. The level of organic maturity within the Newark basin and its associated implications. In W. Manspeizer, ed., *Triassic–Jurassic Rifting: Continental Breakup and the Origin of the Atlantic Ocean and the Passive Margins,* pp. 683–696. Developments in Geotectonics, no. 22. Amsterdam: Elsevier.

Kelemen, P. B., and W. S. Holbrook. 1995. Origin of thick, high-velocity igneous crust along the U.S. East Coast margin. *Journal of Geophysical Research* 100:10077–10094.

Kent, D. V., P. E. Olsen, and W. K. Witte. 1995. Late Triassic–earliest Jurassic geomagnetic polarity sequence and paleolatitudes from drill cores in the Newark rift basin, eastern North America. *Journal of Geophysical Research* 100:14965–14998.

Kilby, W. E. 1988. Recognition of vitrinite with non-uniaxial negative reflectance characteristics. *International Journal of Coal Geology* 9:267–285.

Lee, J. K. W., and I. S. Williams. 1993. Microstructural controls on U-Pb mobility in zircons [abstract]. *Eos: Transactions of the American Geophysical Union* 74: 650–651.

Lerche, I. 1997. Erosion and uplift uncertainties in the Barents Sea, Norway. *Mathematical Geology* 29:469–501.

Lerche, I., R. F. Yarzab, and C. G. St. C. Kendall. 1984. Determination of paleoheatflux from vitrinite reflectance data. *American Association of Petroleum Geologists Bulletin* 68:1704–1717.

LeTourneau, P. M. 1999. Depositional history and tectonic evolution of Late Triassic age rifts of the U.S. central Atlantic margin: Results of an integrated stratigraphic, structural, and paleomagnetic analysis of the Taylorsville and Richmond basins. Ph.D. diss., Columbia University.

Malinconico, M. L. 1993. Reflectance cross-plot analysis of graptolites from the anchi-metamorphic region of northern Maine, U.S.A. *Organic Geochemistry* 20:197–207.

Milici, R. C., K. C. Bayer, P. A. Pappano, J. K. Costain, C. Coruh, and J. E. Nolde. 1991. *Preliminary Geologic Section Across the Buried Part of the Taylorsville Basin, Essex and Caroline Counties, Virginia.* Virginia Division of Mineral Resources Open-File Report, no. 91-1. Charlottesville: Virginia Division of Mineral Resources.

Mitra, S. 1993. Geometry and kinematic evolution of inversion structures. *American Association of Petroleum Geologists Bulletin* 77:1159–1191.

Olsen, P. E. 1990. Tectonic, climatic, and biotic modulation of lacustrine ecosystems: Examples from the Newark Supergroup of eastern North America. In B. J. Katz, ed., *Lacustrine Basin Exploration: Case Studies and Modern Analogs,* pp. 209–224. American Association of Petroleum Geologists Memoir, vol. 50. Tulsa, Okla.: American Association of Petroleum Geologists.

Olsen, P. E. 1997. Stratigraphic record of the early Mesozoic breakup of Pangea in the Laurasia–Gondwana rift system. *Annual Review of Earth and Planetary Science* 25:337– 401.

Olsen, P. E., and D. V. Kent. 1996. Milankovitch climate forcing in the tropics of Pangea during the Late Triassic. *Palaeogeography, Palaeoclimatology, Palaeoecology* 122:1–26.

Olsen, P. E., D. V. Kent, B. Cornet, W. K. Witte, and R. W. Schlische. 1996. High-resolution stratigraphy of the Newark rift basin (early Mesozoic, eastern North America). *Geological Society of America Bulletin* 108:40–77.

Olsen, P. E., R. W. Schlische, and M. S. Fedosh. 1996. 580 Ky duration of the Early Jurassic flood basalt event in eastern North America estimated using Milankovitch cyclostratigraphy In M. Morales, ed., *The Continental Jurassic*, pp. 11–22. Museum of Northern Arizona Bulletin, no. 60. Flagstaff: Museum of Northern Arizona.

Olsen, P. E., R. W. Schlische, and P. J. W. Gore, eds. 1989. *Tectonic, Depositional, and Paleoecological History of Early Mesozoic Rift Basins, Eastern North America.* International Geological Congress Field Trip no. T-351. Washington, D.C.: American Geophysical Union.

Pratt, L. M., C. A. Shaw, and R. C. Burruss. 1988. Thermal histories of the Hartford and Newark basins inferred from maturation indices of organic matter. In A. J. Froelich and G. R. Robinson Jr., eds., *Studies of the Early Mesozoic Basins of the Eastern United States*, pp. 58–62. U.S. Geological Survey Bulletin, no. 1776. Washington, D.C.: Government Printing Office.

Ragland, P. C., L. E. Cummins, and J. D. Arthur. 1992. Compositional patterns for early Mesozoic diabases from South Carolina to central Virginia. In J. H. Puffer and P. C. Ragland, eds., *Eastern North American Mesozoic Magmatism*, pp. 301–331. Geological Society of America Special Paper, no. 268. Boulder, Colo.: Geological Society of America.

Ratcliffe, N. M., W. C. Burton, R. M. D'Angelo, and J. K. Costain. 1986. Low-angle extensional faulting, reactivated mylonites, and seismic reflection geometry of the Newark basin margin in eastern Pennsylvania. *Geology* 14:766–770.

Robinson, G. R., Jr. 1979. Pegmatite cutting mylonite: Evidence supporting pre-Triassic faulting along the western border of the Danville Triassic basin, southern Virginia. *Geological Society of America, Abstracts with Programs* 11:210.

Schamel, S., R. Ressetar, S. Gawarecki, G. K. Taylor, A. Traverse, H. F. Houghton, and P. M. LeTourneau. 1986. Early Mesozoic rift basins of the eastern United States [abstract]. *American Association of Petroleum Geologists Bulletin* 70:644.

Schlische, R. W. 1993. Anatomy and evolution of the Triassic–Jurassic continental rift system, eastern North America. *Tectonics* 12:1026–1042.

Schlische, R. W., and R. V. Ackermann. 1995. Rift basin inversion around the margins of the North Atlantic Ocean: Chronology, causes, and consequences. *Geological Society of America, Abstracts with Programs* 27:80.

Schlische, R. W., and P. E. Olsen. 1990. Quantitative filling models for continental extensional basins with application to the early Mesozoic rifts of eastern North America. *Journal of Geology* 98:135–155.

Sclater, J. G., and P. A. F. Christie. 1980. Continental stretching: An explanation of the post–mid-Cretaceous subsidence of the Central North Sea basin. *Journal of Geophysical Research* 85:3711–3739.

Senftle, J. T., C. R. Landis, and R. L. McLaughlin. 1993. Organic petrographic approach to kerogen characterization. In M. H. Engel and S. A. Macko, eds., *Organic Geochemistry*, pp. 355–374. New York: Plenum.

Simonson, B. M. 1996. Vein formation in non-marine mudstones of the Newark basin. In P. M. LeTourneau and P. E. Olsen, eds., *Aspects of Triassic–Jurassic Rift Basin Geoscience: Abstracts*, p. 44. State Geological and Natural History Survey of Connecticut Miscellaneous Reports, no. 1. Hartford: Connecticut Department of Environmental Protection.

Stach, E., M.-Th. Mackowsky, M. Teichmüller, G. H. Taylor, D. Chandra, R. Teichmüller, D. G. Murchison, and F. Zierke. 1982. *Stach's Textbook of Coal Petrology.* 3d ed. Berlin: Gebruder Borntraeger.

Steckler, M. S., G. I. Omar, G. D. Karner, and B. P. Kohn. 1993. Pattern of hydrothermal circulation with the Newark basin from fission-track analysis. *Geology* 21:735–738.

Wade, J. A., D. E. Brown, A. Traverse, and R. A. Fensome. 1996. The Triassic–Jurassic Fundy basin, eastern Canada: Regional setting, stratigraphy, and hydrocarbon potential. *Atlantic Geology* 32:189–231.

Weems, R. E. 1980. Geology of the Taylorsville basin, Hanover County, Virginia. In *Contributions to Virginia Geology*, vol. 4, pp. 23–28. Charlottesville: Virginia Division of Mineral Resources.

Withjack, M. O., P. E. Olsen, and R. W. Schlische. 1995. Tectonic evolution of the Fundy rift basin, Canada: Evidence of extension and shortening during passive-margin development. *Tectonics* 14:390–405.

Withjack, M. O., R. W. Schlische, and P. E. Olsen. 1998. Diachronous rifting, drifting, and inversion on the passive margin of central eastern North America: An analog for other passive margins. *American Association of Petroleum Geologists Bulletin* 82:817–835.

7

Stress Regimes in the Newark Basin Rift: Evidence from Core and Downhole Data

David Goldberg, Tony Lupo, Michael Caputi, Colleen Barton, and Leonardo Seeber

Analysis of composite geophysical and borehole televiewer logs in the Newark basin drill holes indicate basinwide changes in its physiochemical properties and stress regimes from the Late Triassic to the present day. During its evolution as an enclosed lake basin, approximately 2% organic content was deposited in gray-black shales, which we have estimated from the resistivity log. The basin was extensively deformed throughout its history, and we consequently observed significant fracturing in acoustic images recorded by the borehole televiewer. Both open and filled fractures having greater than 0.5 cm aperture are identified in two well-defined populations, one dipping steeply and striking NE–SW and the other subhorizontal. The subhorizontal fractures increase in number toward the western basin boundary, whereas the steeply dipping fractures are more numerous toward the interior. The formation of these two fractures sets is attributed to initial extension in the NW–SE direction (intrabasin fractures) and to subsequent compression and exhumation of the basin along the same principal direction (subhorizontal fractures). A N47°E orientation of the present-day principal compressive stress is observed by borehole breakouts only at the westernmost site of the basin. This direction is subparallel to the strike of regional faults and to the N68°E average orientation of seismic focal solutions in the area. The mismatch between the regional seismic average and the breakout directions is attributed to the influence of local variations in stress near the western border-fault system. The lack of shallow earthquakes within the Newark basin may be due to reverse/strike-slip displacements along the preexisting fractures within the basin and possibly along deep faults below that were formed during extension, accommodating NE–SW shortening through time.

General Geology of the Newark Basin

Continental rifting along the eastern seaboard of North America during the Late Triassic and Early Jurassic has been associated with the breakup of the supercontinent of Pangea. Such large-scale extensional forces resulted in the formation of a series of rift basins that presently are located along the eastern seaboard (Olsen et al. 1996). As in typical half-graben rifts, the Newark basin is fault bounded on one side only and was filled with sedimentary rocks and volcanics as it subsided. Geologic evidence suggests that preexisting Paleozoic thrust faults were reactivated extensionally in the Newark basin during the Late Triassic and earliest Jurassic (Schlische and Olsen 1990). Thereafter, protracted extension, basin subsidence, and continuing sedimentation of fluvial, lacustrine, and playa deposits thickened most toward the western basin-bounding fault system. The extension of the half-graben in a NW–SE direction produced many deformational features such as normal faults, joints, and fractures perpendicular to the principal horizontal stress direction (S_{Hmax}). All the sediments and volcanics filling the Newark basin are crosscut by faults and fractures.

After a period of quiescence when the rift made the transition to active seafloor spreading in the Early Jurassic, compression along S_{Hmax} has been associated with tectonic shortening and a number of deformational features, such as reverse faulting, folding, and uplift in the Newark basin. Withjack, Olsen, and Schlische (1995) determined that the smaller Fundy basin, a northern analog, also underwent a distinct episode of compressional deformation after rifting. This postrift compressional regime has been attributed to the breakup of the continent and the initiation of seafloor spreading. For the Newark basin, Schlische (1999) reviewed the evidence for inversion of the stress regime and suggested that it initiated when sedimentation and subsidence ceased, reversing motion along the border-fault system and uplifting and unloading the sedimentary sequence. The subsequent rotation of the principal compressive stress to its present direction may be related to tectonic opening of the Labrador Sea and to changes in spreading direction in the northern Atlantic.

The rocks in the Newark basin are well exposed and have been studied in great detail for more than 140 years (e.g., Redfield 1856; Schlische and Olsen 1990). The stratigraphy is dominated by nine nonmarine formations that dip approximately 10° toward the western boundary fault, suggesting that the basin has subsided. Paleomagnetic evidence shows that it has drifted northward during sediment deposition (Kent, Olsen, and Witte 1995; Olsen et al. 1996). The five oldest basin-filling sequences are discussed in this chapter. The Stockton Formation consists of yellow-brown and red conglomerate, arkose, and sandstone units. These coarse-grained units are indicative of a fluvial environment, which requires high energy to erode and transport large clasts, and they are believed to have been deposited during the early phases of rifting. The overlying Lockatong Formation consists of mostly fine-grained gray-black mudstone that was deposited in a moderate-deep lacustrine environment. The Passaic Formation is dominated by gray and red mudstone and sandstone, interpreted to be a progradation from lacustrine- to playa-dominated deposits. The Lockatong Formation and the lower part of the Passaic Formation are rich in carbon, suggesting that they were emplaced when the basin environment was reducing. To produce this environment, it is likely that the basin was hydrologically closed at that time, meaning that the primary input and output components of the water budget were precipitation and evaporation, not fluvial inflow and runoff. Repeated depositional cycles observed in these formations have been attributed to Milankovitch period oscillations in the Earth's orbit that affect variations in global climate (Van Houten 1962; Olsen and Kent 1996). At the top of the sequence, the Orange Mount Basalt is an Early Jurassic sill, and the Feltville Formation is believed to have been emplaced syndepositionally as the rift widened.

The Newark Basin Coring Project

During an eight-month period in 1990/1991 and two months in early 1993, a series of seven deep coreholes were drilled through the Triassic–Jurassic deposits in the Newark basin. The prime objectives of this effort were to refine the region's climate and tectonic history and to correlate orbital and geomagnetic polarity timescales from around 200 to 230 Ma (Olsen et al. 1996). This chapter focuses on the downhole logs acquired during the Newark Basin Coring Project (NBCP), particularly the unpublished borehole televiewer (BHTV) data, and examines the variation in deformational structures that reflect the evolution of stress regimes across the basin.

The stratigraphic sequence in the basin was drilled continuously in seven offset holes, exploiting the eroded half-graben geometry of the rift (figure 7.1). Offset drilling of these dipping beds resulted in an efficient recovery of the entire stratigraphic sequence, where the youngest units are exposed at the northwest end and the oldest units at the southwest end of the basin. The drilling sites were placed close to mappable and easily identifiable lithologic units, which helped to position the next offset site, to establish correlations

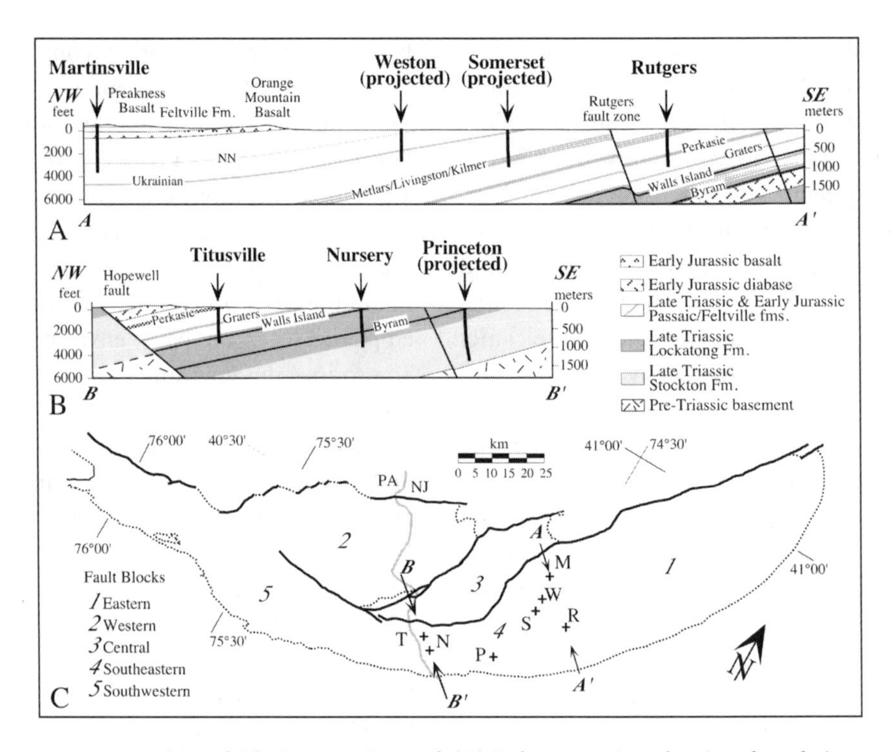

FIGURE 7.1 (*A* and *B*) Cross section and (*C*) index-map view showing the relative position of the seven offset coreholes, fault blocks, and principal rock units in the Newark basin. Coring sites: *M,* Martinsville; *N,* Nursery; *P,* Princeton; *R,* Rutgers; *S,* Somerset; *T,* Titusville; *W,* Weston Canal. (From Olsen et al. 1996)

between holes, and to ensure that the entire Newark basin section was cored. Drilling at the seven sites recovered approximately 6,770 m (22,200 ft.) of core, penetrating more than 4,725 m (15,500 ft.) of the contiguous basin sequence when overlapping sections between sites are excluded.

After each site was cored, a series of downhole logging tools was run to collect continuous geophysical data. The technology and functionality of various logging tools for scientific applications have been described in works by Ellis (1987) and Goldberg (1997). Goldberg, Reynolds, and colleagues (1994) discussed the operations and results of geophysical logging at the Newark basin sites. A variety of formation properties—including the electrical resistivity, natural gamma radiation, bulk density, and neutron porosity—indicated that relatively constant physical properties were measured through significant portions of the Newark basin (Goldberg, Reynolds, et al. 1994). A formation dip log confirmed that the conformal, clay-rich sequences dip approximately 10° NW. Results of the composite geophysical and BHTV logs in these holes are presented in the rest of the chapter.

Data Analysis

Geophysical Logs

A composite of the offset stratigraphic section can be made by appending the geophysical log data from the seven Newark basin holes at the geological tie points used by Olsen et al. (1996). Reynolds (1994) generated a composite of the sonic log in this manner to investigate the cyclicity of seismic signals deep in the basin. In figure 7.2, the gamma ray, resistivity, bulk density, and neutron porosity logs are displayed as a composite of relative depth. Data from each hole were smoothed using a 0.60 m (2 ft.) moving depth window. The logs were appended at the published tie points such that the overlapping section of the shallower hole was truncated. In general, the caliper logs indicate that shallower intervals were more affected by rough hole conditions and have lower data quality. The deeper intervals nominally have a 0.10 m (4 in.) diameter and are smooth walled (Goldberg, Reynolds, et al. 1994). As a result, the truncated sections eliminated the poorer data, and corrections for hole-size variation in the composite logs are not significant. The vertical

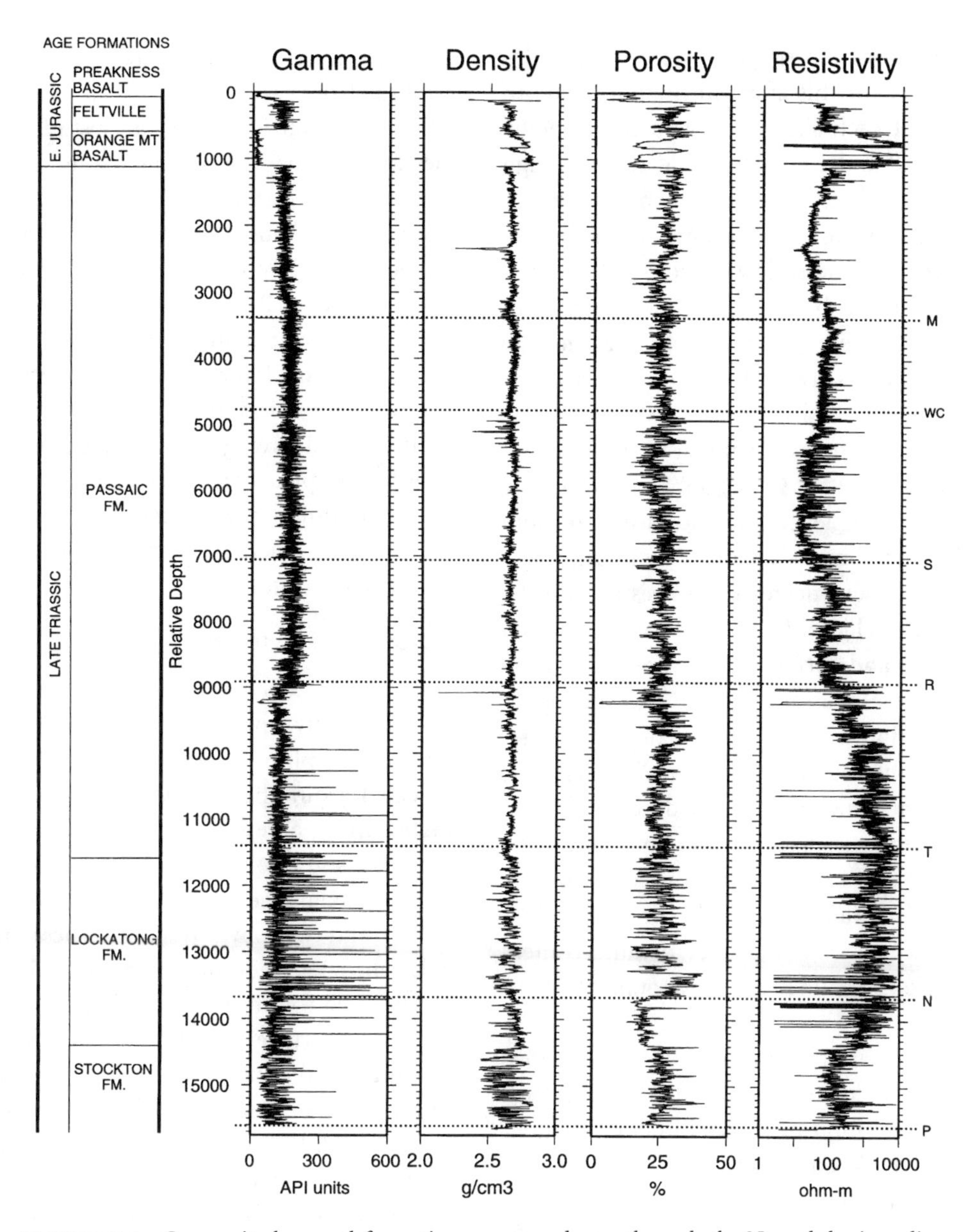

FIGURE 7.2 Composite logs and formation names and ages through the Newark basin sedimentary sequence. Gamma Ray (API units), bulk density (g/cm^3), neutron porosity (%), and electrical resistivity (Ω-m) logs from the seven holes are appended at depths based on the geologic tie points (*dashed lines*) identified by Olsen et al. (1996). Neutron porosity and bulk density logs were linearly detrended over the upper 300 ft. to remove the effects of exhumation. Relative depth is indicated in feet, referenced to the top of the Martinsville hole.

continuity of the composite logs allows for an interpretation of the basinwide variation in sediment and volcanic physical properties.

The gamma ray log has a mean and standard deviation through most of the sequence of approximately 150 and 50 API units, respectively, but shows thin layers up to 1,000 API units in the gray-black shales of the Lockatong Formation. Average gamma ray values decrease by approximately a factor of two in the Orange Mount Basalt, having a much lower concentration of radioisotopes than the underlying clay-rich sediments. The concentration of open fractures and secondary mineralization in the upper 300 m (1,000 ft.) of each hole (El-Tabakh, Riccioni, and Schreiber 1997) causes some systematic variation in microcrack and secondary porosity due to uplift and erosion. The density and neutron porosity logs were linearly detrended to correct for this effect in the upper 200 to 300 m interval at each site. Density values generally vary between 2.60 and 2.80 g/cm³, and reach as high as 2.90 g/cm³ in the lower Passaic and Lockatong, but on average remain relatively constant throughout the entire sequence. The fine-scale variations in bulk density probably indicate borehole washouts related to fracturing; the greater variability observed in the Stockton Formation is associated with thin bedding (Goldberg, Reynolds, et al. 1994).

The resistivity log shows the greatest variations, averaging approximately 20 Ω-m (ohm-meter) in the upper part of the sequence and increasing to more than 5,000 Ω-m in some of the older rocks. Although the sharp and frequent decreases in resistivity in the deeper sequence are likely due to fracturing, its overall readings are high and correlate with large gamma ray values in the lower Passaic, Lockatong, and upper Stockton Formations. Thin layers of organic matter identified in this interval replace electrically conductive pore water with nonconductive hydrocarbons that were deposited in gray-black shales. Accordingly, the increases in electrical resistivity and the concentration of natural U and Th radioisotopes can be measured by log responses to indicate the total organic content (TOC) in various sedimentary environments (e.g., Messnier 1978).

The basinwide variation in TOC can be observed from the composite log. Passey et al. (1990) generated a model based on empirical and laboratory relationships (Archie 1942; Waxman and Smits 1968) to estimate TOC from resistivity and porosity. This approach is based on the logarithmic increase in electrical resistivity through organic source rocks above a baseline in a particular reservoir, proportionally scaled to its response to the porosity. In both the Lockatong (source) and Passaic (baseline) Formations, the average porosity observed in the logs is essentially constant, eliminating a significant porosity scaling effect. Using average resistivity values of 2,000 Ω-m in the Lockatong and 20 Ω-m in the Passaic, Passey's approach yields an estimate of 1.3 to 2.0% (weight) TOC. Olsen et al. (1996) also estimated an average TOC value of approximately 2% from outcrop samples in the deep lacustrine sediments, with individual measurements as high as 7% TOC. Laboratory analyses of samples from the moderately reducing environments in Nursery Road cores yielded average TOC values of 1.9% (unpublished data [depth rank = 3], Geochem Group, USA, 1993). Vitrinite reflectance studies on core samples indicate that the organic matter in the Lockatong Formation is overmature (LOM = 15 to 17) (Malinconico 2002). This overmaturity unfortunately results in more poorly constrained parameters for the log-based estimate (Passey et al. 1990). However, considering the greater than 20-fold difference in their resolution of thin layers and the uncertainties in their parametric assumptions, the agreement of log- and sample-based averages is notable. Depth profiles of densely sampled core tests compared with the log-based computation will serve to calibrate the estimates of TOC in thinner layers and over a broader range of reducing environments.

Borehole Televiewer

The BHTV is a downhole logging device that provides a representative image of formation properties derived from the amplitude and travel time of acoustic signals reflected from the borehole wall. As the BHTV is pulled up the hole and the rotating acoustic transceiver records a spiral of data returned through the borehole fluid, digital images can be constructed from their amplitude and two-way travel time. These images provide a flattened 360° view of a length of the borehole, which can be used to estimate fracture location, orientation, dip, and aperture as well as to indicate stress-related spalling of the borehole wall, hole ellipticity, and drilling-induced damage (e.g., Zemanek et al. 1970; Zoback et al. 1985). An example of an amplitude image from the Martinsville hole over a 3 m (10 ft.) interval is shown in figure 7.3. The sinusoidal feature identified near the top of the image is a steeply dipping planar fracture; below this, several patchy zones of borehole wall spalling are observed. These patchy zones are stress-induced features, commonly called *breakouts*, that form in vertical holes at azimuths perpendicular to the principal horizontal stress direction when the strength of the rock surrounding the borehole breaks under tangential shear forces (Bell and Gough 1979; Zoback et al. 1985). BHTV image analysis in six Newark basin holes indicates a small population ($n = 98$) of stress-induced breakouts toward the bottom of the Martinsville hole. No breakouts are observed in the other holes. Figure 7.4 shows the location and orientation of these breakouts, which have a consistent width of approximately 33° and an azimuth of 137.2°

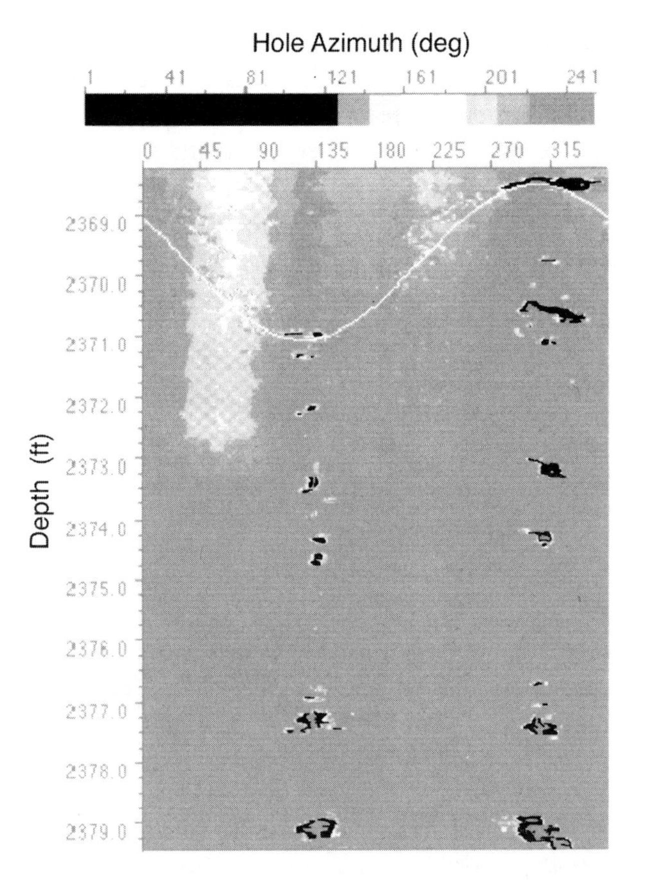

FIGURE 7.3 Example image from the borehole televiewer (BHTV) between 2,269 and 2,379 ft. depth (vertical axis) in the Martinsville hole. Borehole azimuth is the horizontal axis (N = 0°), and the gray scale indicates signal amplitude. The sinusoidal feature highlighted at 2,370 ft. depth is a fracture that approximately dips 80°SE and strikes N35°E. The dark patchy zones are caused by wall spalling (breakouts) that form along the direction of least-compressive stress (NW–SE). Similar images were generated from BHTV logs in six Newark basin holes (Weston Canal was not logged with the BHTV).

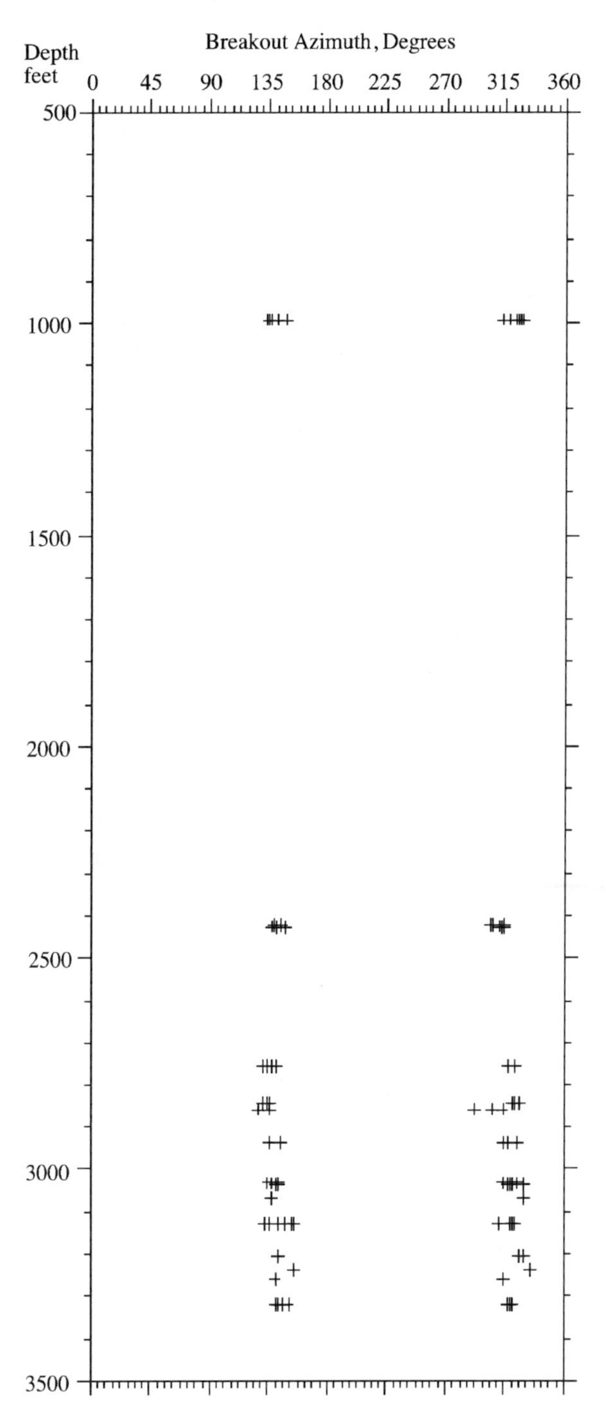

FIGURE 7.4 Location and azimuth of the breakouts identified from BHTV images in the Martinsville hole. Breakouts were observed below 2,400 ft. depth at this site only and indicate that the present-day principal compressive stress direction is NE–SW, normal to the breakout direction.

(±3.5° standard deviation). These features indicate a N47.2°E direction of the present-day principal horizontal stress S_{Hmax}, perpendicular to the breakout orientation, which is subparallel to the strike of the sediment bedding and regional faulting, and to geologic structures in and around the Newark basin (Goldberg, Caputi, et al. 1994).

Although a limited depth interval of borehole breakouts can be observed only in the Martinsville hole, the BHTV data acquired in all six holes identified extensive fracturing throughout the basin, with dips ranging from subhorizontal to vertical. BHTV amplitude images were used to identify fractures by their low acoustic impedance relative to the high-density competent rock that makes up most of the borehole wall. Many fractures are observed in BHTV images as conjugate populations of steeply dipping (60 to 85°) fractures that strike NW–SE. Subhorizontal fractures are identified predominantly at the Martinsville site. The total fracture count is also three times greater at Martinsville than in any other individual hole. Figure 7.5 shows a compilation of all the fractures identified from the BHTV data at six sites, presented as histograms of the fracture count over 15 m depth intervals. A modest decrease in the number of fractures toward the bottom of each hole and across the basin (Martinsville to Princeton) can be qualitatively observed.

In order to interpret and "ground-truth" the BHTV fracture logs, direct observations of fractures in approximately 0.3 m (1 ft.) sections of the recovered core were examined in sequential photographs (figure 7.6). Numerical estimates of fracture location, dip, and aperture were quantitatively made from these photos, which also provided an indication of the mineralized infilling of fractures. Core data sets were compiled for

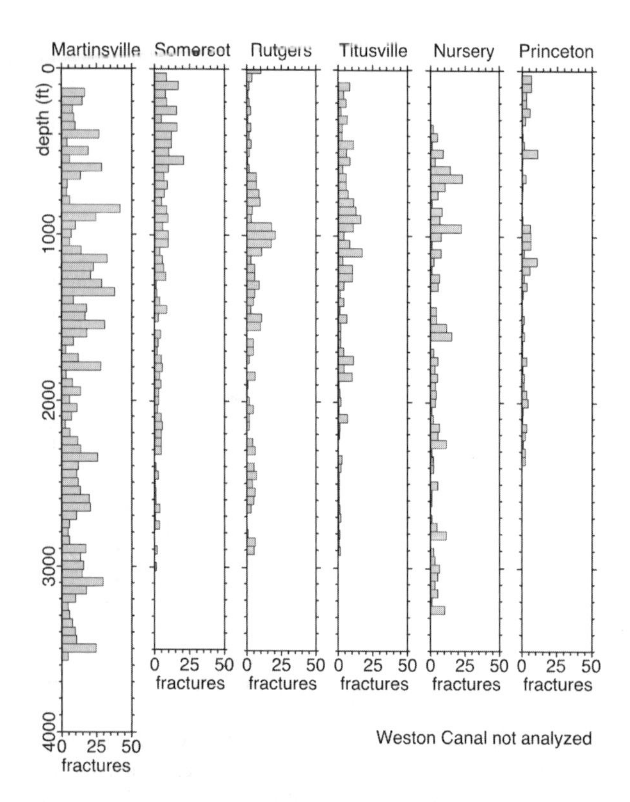

FIGURE 7.5 Histograms of fracture count identified over 50 ft. intervals in the BHTV logs at six Newark basin sites, west to east, from Martinsville to Princeton. The fracture count decreases slightly with depth in each hole and eastward toward the interior of the Newark basin.

the Martinsville and Nursery Road sites, which represent the stratigraphy of the basin from northeast to southwest. The fracture counts from core observations are 552 and 3,415 at Nursery Road and Martinsville, respectively. These totals are 1.9 to 3.4 times greater than those from the BHTV data. Parkinson, Dixon, and Jolley (1999) similarly reported a ratio of core to BHTV fracture count on the order of five to one in a North Sea sandstone reservoir with a high correlation in more intensively fractured zones.

Histograms comparing the core with the BHTV fracture count clearly indicate the disparities between the two data sets as a function of depth (figures 7.7 and 7.8). In figure 7.7, the three columns of core data distinguish open and filled fractures for fracture apertures of greater than 0.0 cm, greater than 0.305 cm, and greater than 0.610 cm. The fracture count from core is reduced considerably with increasing fracture aperture. The total fracture count measured by the BHTV shows a similar profile with depth but falls between the 0.305 and 0.610 cm curves in terms of total fracture count. A precise match between the BHTV and the core fracture count can be quantified by the simple proportional equation,

$$D_a = \left(\frac{|F_3 - F_{BHTV}|}{|F_3 - F_6|}\right) \times |a_3 - a_6|$$

where

D_a = aperture difference

F_{BHTV} = total fracture count from BHTV

F_3 = total fracture count from core with apertures greater than 0.305 cm

F_6 = total fracture count from core with apertures greater than 0.610 cm

a_3 = fracture aperture equal to 0.305 cm

a_6 = fracture aperture equal to 0.610 cm

Adding the difference D_a to $|a_3 - a_6|$ provides an estimate of apparent aperture resolution of the BHTV. In the Martinsville hole, the resolution is 0.52 cm. The greatest proportion of filled fractures are observed in the interval from approximately 335 to 520 m (1,100 to 1,700 ft.). The fracture count observed from the BHTV over this interval is high, suggesting that it successfully differentiates filled fractures from the surrounding clay formations, even though their acoustic

FIGURE 7.6 Example photograph of Martinsville cores at various depths in red and gray mudstone. Each core shown is 1 ft. long and oriented top upward. Open (O) and filled (F) fractures are annotated. Similar photos were used to count fractures and estimate aperture, dip magnitude, and mineral infilling for the Martinsville and Nursery Road cores.

contrast is less than in open fractures. Overall, the BHTV resolves approximately 30% of the fractures observed in the core at Martinsville; these fractures predominantly have apparent apertures greater than approximately 0.5 cm.

In figure 7.8, both core and BHTV fracture histograms at Nursery Road also indicate similar profiles with depth but fewer filled fractures than at Martinsville. The proportional fracture counts yield essentially an equivalent apparent aperture resolution of 0.55 cm in this hole for the BHTV. Because of the greater variation in fracture dip observed at Nursery Road, the total fracture count also may be affected by dip magnitude. In figure 7.9, histograms for fractures having dips greater than 45° are compared with all fractures observed in the core, with fractures having apertures greater than 0.610 cm, and with the BHTV fracture count. Both the dip and aperture histograms correlate well and are similar to the BHTV fracture profile. The comparison indicates that fractures large enough to be resolved by the BHTV ($n = 217$) generally have dips greater than 45° ($n = 296$). As a consequence, many of the subhorizontal fractures in the Nursery Road hole are too thin to be resolved. Compared with those at Martinsville, however, subhorizontal fractures at Nursery Road account for only approximately 50% of the fractures observed in the core and are significantly thinner, having apparent apertures less than approximately 0.5 cm.

Fracture Dips

The BHTV image analysis illustrates different distributions of fracture dip and orientations across the Newark basin. In figures 7.10 and 7.11, stereographic Wulff projections of the poles normal to fracture planes are presented for Nursery Road and Martinsville, respectively. Fractures are plotted for Nursery Road ($n = 283$) and contoured to show the conjugate set of fractures that dip 60 to 85° and strike NE–SW (figure 7.10). In contrast, the distribution of poles to 986 fractures observed at Martinsville are nearly vertical with a slight orientation NW–SE (figure 7.11).

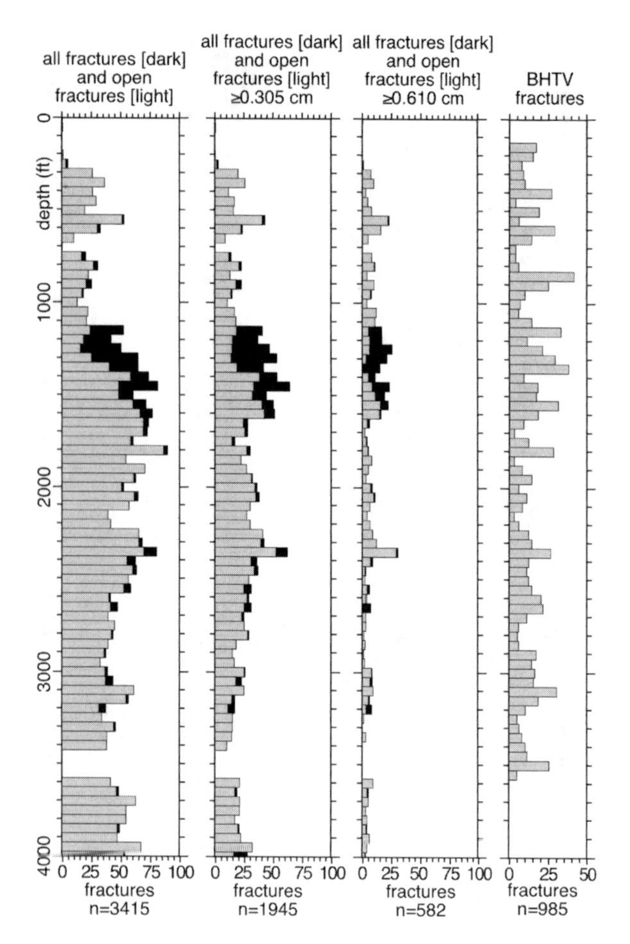

FIGURE 7.7 Histograms of all fractures identified in the Martinsville core at 50 ft. intervals (Track 1), fracture with aperture ≥ 0.305 cm (Track 2), fractures with aperture ≥ 0.610 cm (Track 3), and BHTV fractures (Track 4). Dark shading indicates filled fractures. The core and BHTV fracture data have similar profiles versus depth. The aperture resolution of the BHTV is estimated to be 0.52 cm for both open and filled fractures. For further discussion, see the text.

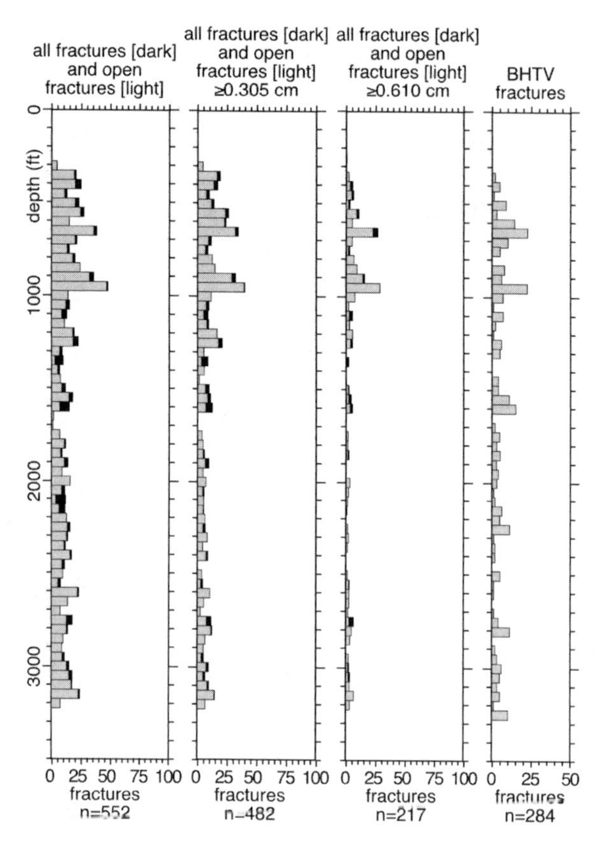

FIGURE 7.8 Histograms of all fractures identified in the Nursery Road core at 50 ft. intervals (Track 1), fracture with aperture ≥ 0.305 cm (Track 2), fractures with aperture ≥ 0.610 cm (Track 3), and BHTV fractures (Track 4). Dark shading indicates filled fractures. The core and BHTV fracture data have similar profiles versus depth. The aperture resolution of the BHTV is estimated to be 0.55 cm for both open and filled fractures. For further discussion, see the text.

The N47°E orientation of the principal horizontal stress S_{Hmax} inferred from the breakouts is also shown. Some steeply dipping fractures may exist in the Martinsville hole, and it is likely as well that subhorizontal fractures exist in the Nursery Road (and other) holes, but they are thinner than approximately 0.5 cm and are thus not resolved by the BHTV. Figure 7.12 illustrates a summary of dip of the large-aperture fractures at sequential sites from west to east (Martinsville to Princeton). In general, the number of steeply dipping fractures increases toward the interior of the basin (Princeton), and the number of subhorizontal fractures increases toward the western basin boundary (Martinsville). In the following discussion, we interpret the significance of the large-aperture fractures observed by the BHTV in terms of the evolution of the stress regimes in the Newark basin.

Discussion

The two populations of fracture orientations in the Newark basin are consistent with the principal stress directions formed during extension and rifting in the Late Triassic, followed by their subsequent inversion to a compressional regime in the Early Jurassic. Initial rifting and subsidence caused the development of the population of steeply dipping extensional fractures that are oriented NE–SW, normal to the direction of principal extension in the Late Triassic; these fractures proportionally increase in number and widen in ap-

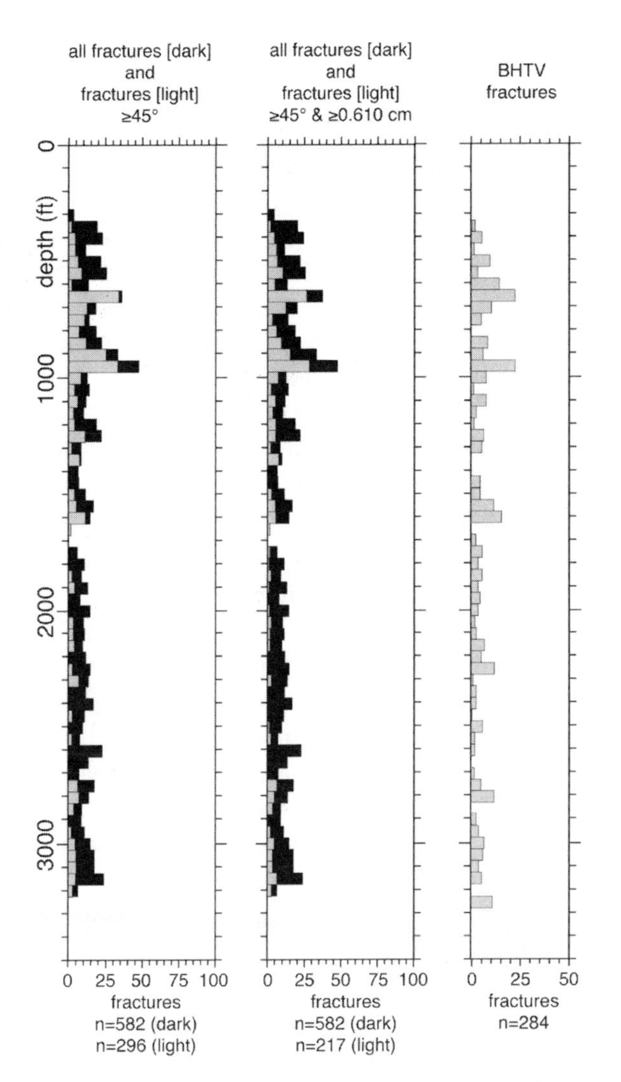

FIGURE 7.9 Histograms of fractures with dips ≥ 45° (*light,* Track 1) and aperture ≥ 0.610 cm (*light,* Track 2) versus all fractures identified in the Nursery Road core (*dark*) and the BHTV fractures (Track 3). Fractures with dips ≥ 45° and aperture ≥ 0.610 cm have similar profiles versus depth and approximately match the BHTV fracture count. For further discussion, see the text.

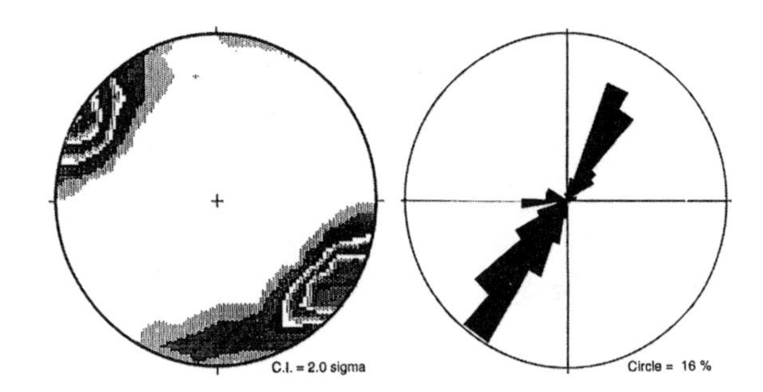

FIGURE 7.10 (*Left*) Kamb contour stereogram of poles to fracture planes and (*right*) rose diagram of fracture orientations identified by the BHTV (n = 283) in the Nursery Road hole. These displays indicate that the population of fractures at Nursery dip steeply and are oriented NE–SW. This orientation is consistent with an extensional stress regime generated during initial rifting of the Newark basin.

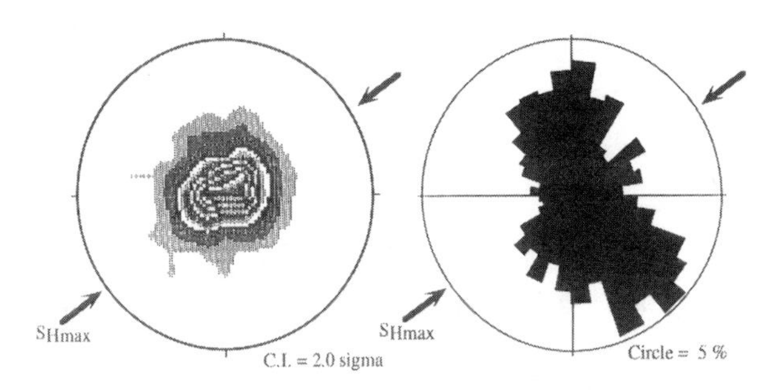

FIGURE 7.11 (*Left*) Kamb contour stereogram of poles to fracture planes and (*right*) rose diagram of fracture orientations identified by the BHTV (n = 986) in the Martinsville hole. These displays indicate that the population of fractures at Martinsville are primarily subhorizontal. The direction of the present-day maximum horizontal compressive stress (S_{Hmax}) inferred from breakout orientation is also shown. This orientation is consistent with a reverse/strike-slip stress regime, where tensile fracturing likely generated by unloading of the overburden occurred during basin uplift.

erture at Nursery Road and Princeton sites toward the interior of the basin. Such features may have formed to accommodate protracted extension away from the border-fault system. Near the western boundary of the basin, however, synrift extension may have been accommodated largely by activation of the border fault itself; hence large, steeply dipping fractures are not observed near Martinsville.

The population of subhorizontal fractures and cracks most likely developed during uplift and exhumation of the basin-filling sediments after inversion of the stress regime in the Early Jurassic. Subhorizontal fractures can form by tensional failure when the least-compressive stress is vertical, as is likely to occur under a reverse/strike-slip stress regime. Such fractures are most concentrated where the greatest overburden is released, and, consequently, the largest population of subhorizontal fractures in the Newark basin is observed where the synrift sedimentation is thickest, near Martinsville. Progressively fewer and thinner subhorizontal fractures are observed in both the BHTV and core data at the more eastern sites because erosion of

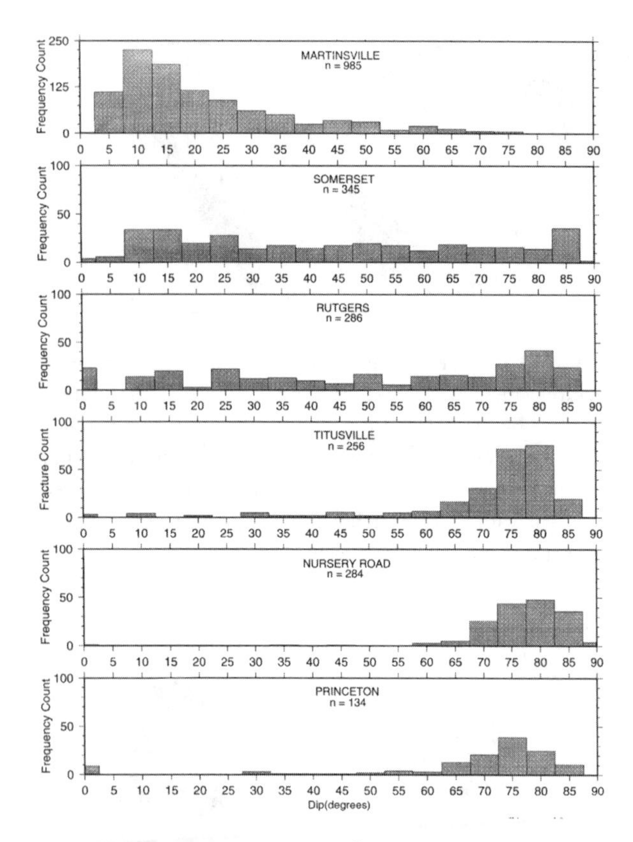

FIGURE 7.12 Histograms over 5° intervals of the dip magnitude of all fractures identified by the BHTV at six Newark basin sites, west to east, from Martinsville to Princeton. The histograms show two superimposed populations of fractures: (1) a subhorizontal fracture set that decreases in number eastward toward the interior of the basin (Princeton), and (2) a steeply dipping fracture set that decreases in number westward toward the boundary of the basin (Martinsville).

a thinner sediment cover released less overburden. In addition to the effects of exhumation, some concentration of subhorizontal fractures near Martinsville may be related to damaged zones in hanging-wall rocks close to the border-fault system.

El-Tabakh, Schreiber, and Warren (1998) studied core samples and observed that a zone of shallow subhorizontal fractures (100 to 300 m depth) filled with fibrous gypsum transect stratigraphic boundaries across the Newark basin. The gypsum formed by groundwater dissolution and recrystallization of anhydrite in tensional cracks that opened by the release of overburden after uplift, tilting, and erosion of the half-graben. The gypsum is undeformed, thereby postdating other tectonic features, and is less frequent toward the interior of the basin. This result is similar to evidence from the BHTV that the population of subhorizontal fractures formed after the sediment overburden was released.

Our evidence from borehole breakouts indicates that the current direction of S_{Hmax} is NE–SW. This direction has changed since the Jurassic so that it is now subparallel to the strike of the basin and to its preexisting faults and fractures. Figure 7.13 illustrates a compilation of the orientation of the present-day stress direction from breakouts and the alignment of earthquake focal solutions around the Newark basin over the past 200 years. Few earthquakes are located within the basin itself, suggesting that present-day strain within it is largely aseismic. One explanation is that the population of preexisting fractures reflects deep basement faults toward the interior of the basin that developed during formation of the initial rift graben. Reverse/strike-slip displacement along these faults and fractures, which are subparallel to the current NE–SW direction of S_{Hmax}, may accommodate regional strain within the basin. Such displacements release stresses below the critical failure criterion for unfractured rock and do not cause intrabasin earthquakes. If fault reactivation in this direction is physically prevented, as may be the case along the western border fault, tectonic shortening related to present-day strains cannot be accommodated in a similar manner. The absence of fault reactivation below the sediments at Martinsville may have resulted in the formation of borehole breakouts after the hole was drilled into regionally stressed rock.

Conclusions

We used acoustic images from the BHTV to observe borehole breakouts, which indicate the present-day stress regime as well as two distinct populations of fractures within the Newark rift basin sediments. These data indicate the evolution of three distinct stress regimes around the Newark basin, as summarized in figure 7.14. In the Late Triassic, protracted extension in the NW–SE direction and subsidence reactivated the basin-bounding fault at its western edge and caused the formation of a dense population of steeply dipping fractures in the interior of the basin that strike NE–SW, normal to the extension. Subsequent shortening in the Early Jurassic associated with

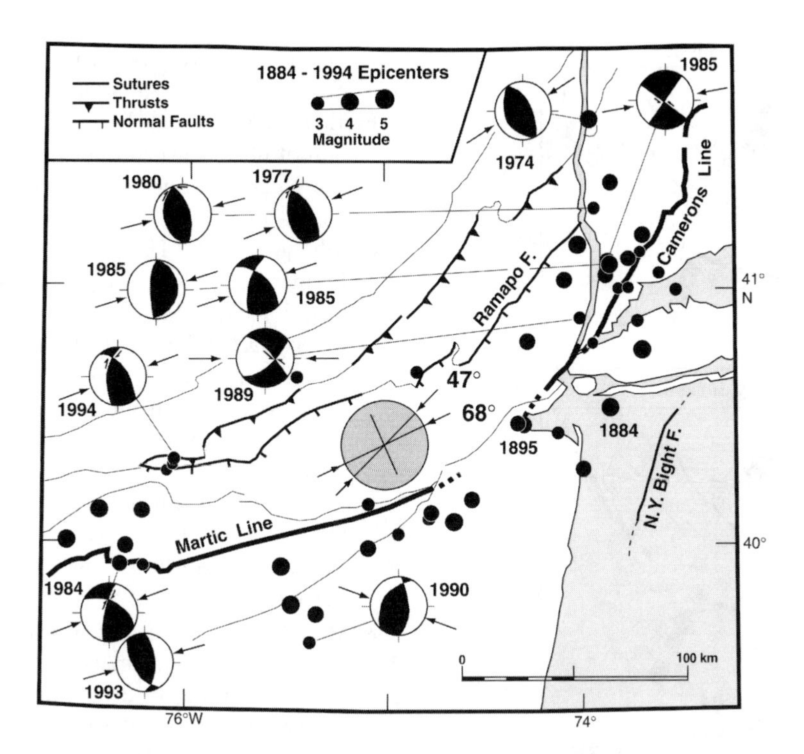

FIGURE 7.13 Earthquake epicenters from 1884 to 1994 and focal plane solutions around the Newark basin. The direction of the present-day principal horizontal compressive stress inferred from BHTV breakouts (N47.2°E) is subparallel to the average direction of earthquake compression (N68°E). Few earthquakes have occurred within the Newark basin itself between the Martic line and the Ramapo thrust fault.

the initiation of seafloor spreading in the Atlantic reversed the axis of extension to compression along the same direction and uplifted the basin. Sediments throughout the basin were exhumed and eroded, forming a population of subhorizontal fractures concentrated toward the border fault where the greatest overburden was released. The change in direction of S_{Hmax} to its present-day NE–SW orientation, as observed by breakouts, is subparallel to the preexisting extensional fractures and to the NE–SW strike of major geologic features within and around the Newark basin. These fractures may reflect deeper faults and contribute to the present-day reverse/strike-slip stress regime in the basin, accommodating shear strain and causing a quiescence of seismic activity within the basin itself. Further work investigating slickensides observed on exposed faults, petrographic studies of microfracturing on samples, and microseismic monitoring in the region may provide important new evidence about the continuing evolution of the stress regime in and around the Newark rift basin.

Acknowledgments

We gratefully acknowledge the financial support from the Continental Dynamics Program of the National Science Foundation for acquisition and preparation of these data. The National Science Foundation summer intern program supported T. Lupo and M. Caputi at the Lamont-Doherty Earth Observatory while they analyzed the core and BHTV data. T. Lupo and A. Martino prepared, reviewed, and loaded the geophysical log data from the Newark Rift Basin Coring Project online, which is currently available at the URL address http://www.ldeo.columbia.edu/BRG/Newark/. Technical assistance was provided by D. Moos and E. Scholz during field operations and for calibration of the BHTV logs. S. Brower and K. Nagao assisted in preparation of several figures presented in this chapter. Critical reviews by D. Kent and J. Contreraz also contributed substantially to improve the paper. This study is Lamont-Doherty Earth Observatory contribution no. 6306.

Late Triassic

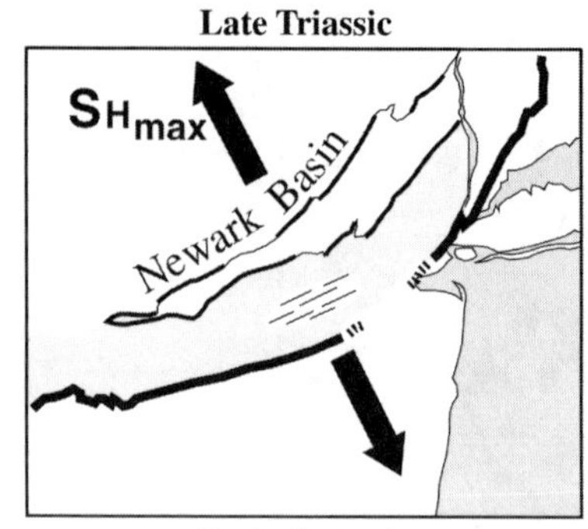

Early Jurassic

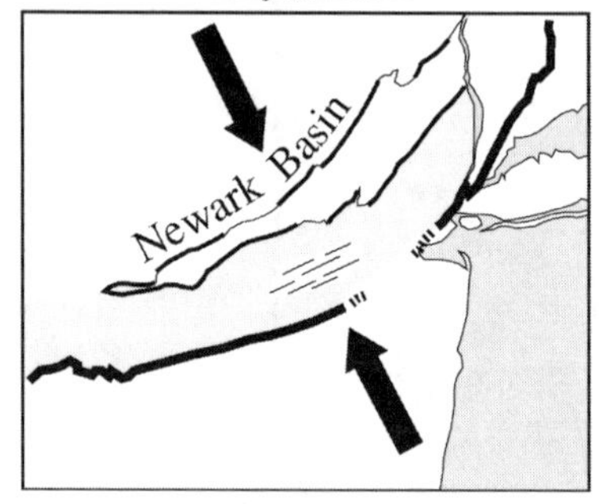

Present

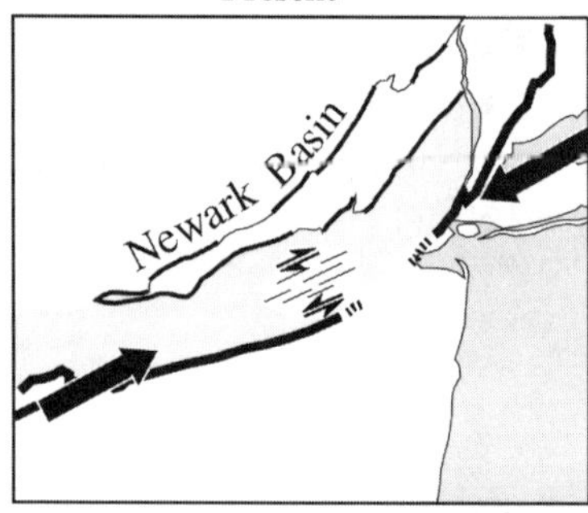

FIGURE 7.14 Schematic diagram of the evolution of principal horizontal stress S_{Hmax} direction in the Newark basin: Late Triassic extension, Early Jurassic compression and uplift, and present-day reverse/strike-slip regimes. Possible shearing along preexisting fractures and faults may accommodate strain and moderate seismic activity within and below the basin itself.

Literature Cited

Archie, G. E. 1942. The electrical resistivity log as an aid in determining some reservoir characteristics. *Transcripts of the Society of Petroleum Engineers* 146:54–62.

Bell, J. S., and D. I. Gough. 1979. Northeast–southwest compressive stress in Alberta: Evidence from oil wells. *Earth and Planetary Science Letters* 45:475–482.

Ellis, D. 1987. *Well Logging for Earth Scientists.* New York: Elsevier.

El-Tabakh, M., R. Riccioni, and B. C. Schreiber. 1997. Evolution of Late Triassic rift basin evaporites (Passaic Formation): Newark basin, eastern North America. *Sedimentology* 44:767–790.

El-Tabakh, M., B. C. Schreiber, and J. Warren. 1998. Origin of fibrous gypsum in the Newark rift basin, eastern North America. *Journal of Sedimentary Research* 68:88–99.

Goldberg, D. 1997. The role of downhole measurements in marine geology and geophysics. *Review of Geophysics* 35:315–342.

Goldberg, D., M. Caputi, C. Barton, and L. Seeber. 1994. Preliminary analysis and interpretation of BHTV and wireline log data from the Newark Rift Basin Coring Project. *Transactions of the Geological Society of America* 26:336.

Goldberg, D., D. Reynolds, C. Williams, W. Witte, P. Olsen, and D. Kent. 1994. Well logging results from the Newark Rift Basin Coring Project. *Scientific Drilling* 4:267–279.

Kent, D. V., P. E. Olsen, and W. Witte. 1995. Late Triassic–earliest Jurassic geomagnetic polarity sequence and paleolatitudes from drill cores in the Newark rift basin, eastern North America. *Journal of Geophysical Research* 100:14965–14998.

Malinconico, M. L. 2002. Lacustrine organic sedimentation, organic metamorphism and thermal history of selected early Mesozoic Newark Supergroup basins, eastern U.S.A. Ph.D. diss., Columbia University.

Meissner, F. F. 1978. Petroleum geology of the Bakken Formation, Williston basin, North Dakota and Montana. In *The Economic Geology of the Williston Basin,* pp. 207–227. Williston Basin Symposium. Billings: Montana Geological Society.

Olsen, P. E., and D. V. Kent. 1996. Milankovitch climate forcing in the tropics of Pangea during the Late Triassic. *Palaeogeography, Palaeoclimatology, Palaeoecology* 122:1–26.

Olsen, P. E., D. V. Kent, B. Cornet, W. K. Witte, and R. W. Schlische. 1996. High-resolution stratigraphy of the Newark rift basin (early Mesozoic, eastern North America). *Geological Society of America Bulletin* 108: 40–77.

Parkinson, D. N., R. J. Dixon, and E. J. Jolley. 1999. Contributions of acoustic imaging to the development of the Bruce Field, northern North Sea. In M. Lovell, G. Williamson, and P. Harvey, eds., *Borehole Imaging: Applications and Case Histories*, pp. 259–270. Geological Society Special Publication, no. 159. London: Geological Society.

Passey, Q., S. Creaney, J. Kulla, F. Moretti, and J. Stroud. 1990. A practical model for organic richness from porosity and resistivity logs. *American Association of Petroleum Geologists Bulletin* 74:1777–1794.

Redfield, W. C. 1856. On the relations of the fossils fishes of the sandstone of Connecticut and the Atlantic states to the Liassic and Oolitic periods. *American Journal of Science* 22:357–363.

Reynolds, D. J. 1994. Sedimentary basin evolution: Tectonic and climatic interaction. Ph.D. diss., Columbia University.

Schlische, R. W., and P. E. Olsen. 1990. Quantitative filling model for continental extensional basins with applications to early Mesozoic rifts of eastern North America. *Journal of Geology* 98:135–155.

Van Houten, F. B. 1962. Cyclic sedimentation and the origin of analcime-rich Upper Triassic Lockatong Formation, west-central New Jersey and adjacent Pennsylvania. *American Journal of Science* 260:561–576.

Waxman, M. H., and L. Smits. 1968. Electrical conductivities in oil-bearing shaly sands. *Society of Petroleum Engineers Journal* 8:107–122.

Withjack, M. O., P. E. Olsen, and R. W. Schlische. 1995. Tectonic evolution of the Fundy rift basin, Canada: Evidence of extension and shortening during passive-margin development. *Tectonics* 14:390–405.

Zemanek, J. E., E. Glenn, L. Norton, and R. Caldwell. 1970. Formation evaluation by inspection with the borehole televiewer. *Geophysics* 35:254–269.

Zoback, M. D., D. Moos, L. Mastin, and R. N. Anderson. 1985. Well bore breakouts and in situ stress. *Journal of Geophysical Research* 90:5523–5530.

8

A Lagerstätte of Rift-Related Tectonic Structures from the Solite Quarry, Dan River–Danville Rift Basin

Rolf V. Ackermann, Roy W. Schlische, Lina C. Patiño, and Lois A. Johnson

The Solite Quarry within the Dan River–Danville basin contains an extensive suite of rift-related structures. The cyclical upper member of the Cow Branch Formation has been deformed both in continuous fashion and via three brittle failure modes, exhibiting fracture partitioning such that failure mode is lithologically dependent. All structures are tectonic; extension estimates are roughly comparable for all failure modes; and there is an absence of bedding-parallel detachment horizons with normal separation. All extensional structures formed in response to Triassic rifting. These observations imply that different beds failed coevally or semicoevally in extension. All contractional structures are consistent with earliest Jurassic inversion.

The small normal faults in the Solite Quarry are in most ways like larger faults and occur both as isolated features and as segments of relay systems. The faults exhibit slickensided, mineralized fault surfaces, footwall uplift, hanging-wall subsidence, relay ramps, and elliptical fault surfaces, with maximum displacement occurring at fault centers and tapering to zero at the tips. Detailed analysis of these structures and integration with other data sets suggest that faults exhibit linear length-displacement scaling over nine orders of magnitude of fault length. These small faults can be divided into two subsets based on length and on their spatial distribution within the rock volume, with the set of smaller structures exhibiting anticlustering with respect to the larger structures (called *master faults*), forming fault shields due to the presence of stress-reduction shadows around the master faults.

A field locality in the Dan River–Danville basin known as a major Lagerstätte—or a location that contains an exceptional suite of fossils, minerals, and the like—for Triassic terrestrial arthropods and flora (e.g., Fraser et al. 1996) also contains an exceptionally well exposed suite of rift-related discontinuous (brittle) and continuous (ductile) deformation features. Detailed study of the exquisite array of structures found in the Solite Quarry permits verification or nullification of a range of empirical relationships related to fault-population studies, a topic of considerable interest in recent years (e.g., Cowie, Knipe, and Main 1996). The goal of these studies is to unravel the scaling laws of fractures—including their size and spatial distribution—the relationship between displacement or aperture and fracture length, and the strain accommodated by fractures. In addition to advancing our understanding of the mechanics of fracturing and the evolution of fracture systems (e.g., Cowie and Scholz 1992b; Cowie 1998), these scaling relations also have more practical applications. For example, studies of small-scale systems potentially can provide information relevant to an understanding of large-scale systems such as rift basins (studies of which commonly are hampered by poor exposure and complex basin architecture)—if the scaling laws and the scale range over which these systems operate are known. In addition, models of fractured bedrock systems, such as the eastern North American rift system, require detailed information on the size and spatial distribution of fractures in order to be effective for groundwater prospecting, petroleum exploration and exploitation, and the remediation of contaminated aquifer systems. To address these issues, we require high-quality data sets from well-exposed regions with relatively simple strain histories in which the fractures span at least two orders of magnitude of size. The fractures in the Solite Quarry meet or exceed these requirements.

In this chapter, we systematically discuss the rift-related structures present within the Solite Quarry. We describe the features first by category (brittle versus ductile; extension versus shear fractures) and then by type of strain accommodated (extensional versus contractional). We then examine the size and spatial distribution of some of the fractures within the quarry and discuss the implications of these observations for fault scaling relationships, lithologically dependent failure modes (mechanical stratigraphy), and the existence of stress-reduction shadows (regions of lower shear stress compared with the remote stress) around normal faults.

Geologic Setting

Outcrops discussed in this chapter are located in the main and new quarries of the Virginia Solite Corporation located in the Dan River–Danville basin, directly on the border between Virginia and North Carolina (figure 8.1). The Dan River–Danville basin is part of the Mesozoic rift system on the eastern coast of North America, which formed in response to the Triassic initiation of breakup of the supercontinent Pangea. At the time of rifting, σ_1 and σ_3 are inferred to have been vertical and NW–SE directed, respectively (Schlische, chapter 4 in this volume). The basin itself is a highly elongate half-graben with a SE-dipping border-fault system (Chatham fault zone) (figure 8.1). The basin experienced inversion during earliest Jurassic time, most likely due to reorientation of the principal stresses at the onset of seafloor spreading in the central Atlantic Ocean (Withjack, Schlische, and Olsen 1998).

The 4,000 m of basin fill (Dan River Group) rest in depositional unconformity over basement and dip antithetically to the border-fault system at angles from 20 to 50° (Kent and Olsen 1997). The Solite Quarry lies within the upper member of the Cow Branch Formation, as defined by Olsen (1997). Kent and Olsen (1997) have described this member as an approximately 1,950 m thick sequence of cyclic gray to black mudstones of deep-water lacustrine facies. Within the quarry, bedding dip angles range from 30 to 40°. In subsequent sections of this chapter, we have corrected all structural orientations for bedding dips.

Structures Within the Quarry

Brittle Deformation-Extension

Extension Fractures. Extension fractures are common within the Solite Quarry and exhibit in both cross section and plan view the classic "penny-shaped crack" described by Pollard and Aydin (1988) (figure 8.2a). Layers containing extension fractures are typically medium- to fine-grained feldspathic sandstones. The vast majority of extension fractures in the quarry are calcite veins with cockscomb texture (e.g., Davis 1984). They are restricted to discrete lithologies, are oriented subnormal to bedding, terminate at sharp lithologic boundaries, and have a median fracture spacing (at this scale range) that is roughly proportional to mechanical layer thickness (e.g., Gross et al. 1995) (figure 8.2a). The fractures can be divided into two sets based on dominant orientations. The dominant set is subvertical, striking approximately 045° (figure 8.3), suggesting that σ_1 and σ_3 were vertical and NW–SE directed, respectively, at the time of formation, which is consistent with stress directions inferred from basin geometry and intrabasinal faults. The subordinate set is also subvertical, striking approximately 005°, subparallel to a regional (subordinate) dike set (~358°) (Schlische, chapter 4 in this volume) (figure 8.4) that postdates the initiation of basin inversion in earliest

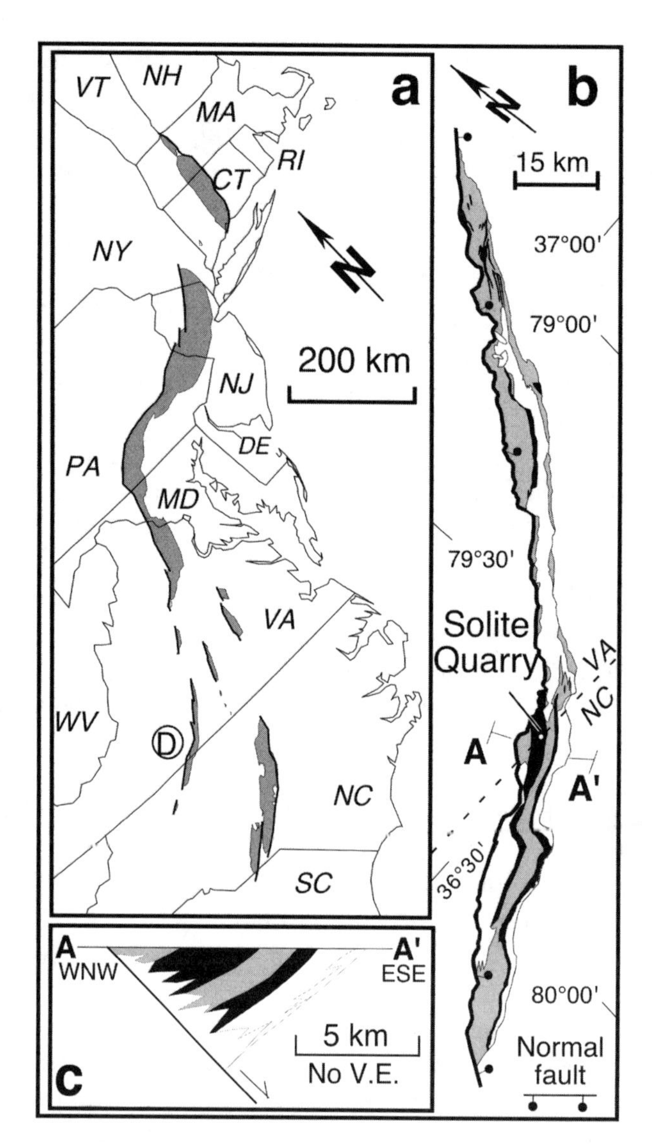

FIGURE 8.1 (*a*) Exposed Mesozoic rift basins of the eastern United States; *D*, Danville basin. (*b*) Simplified geologic map of the Danville basin, showing the location of the Solite Quarry. Black represents the lower and upper members of the Cow Branch Formation. (*c*) Simplified geologic cross section of the Danville basin (for location, see *b*), illustrating half-graben geometry modified by inversion. (Modified from Schlische et al. 1996; Kent and Olsen 1997; and Schlische, chapter 4 in this volume)

Jurassic time (~200 Ma). The dominant regional dike set (~320°) (Schlische, chapter 4 in this volume) (figure 8.4) is believed to be coeval with the initiation of basin inversion (Withjack, Schlische, and Olsen 1998).

Strain accommodated by extension fractures was calculated using the sum of bed segments between veins along a scan line and the length of the scan line. The average strain estimate is 5.4%. As yet, no quantitative thin section work has been done on the Solite rocks. Hence our average strain estimate for extension fractures should be regarded as a minimum at this time. A histogram of vein spacings (figure 8.4) for a bed in the main Solite Quarry resembles a skewed, log-normal distribution of spacings that is commonly observed where extension fractures are restricted to lithologically controlled mechanical layers (e.g., Ladeira

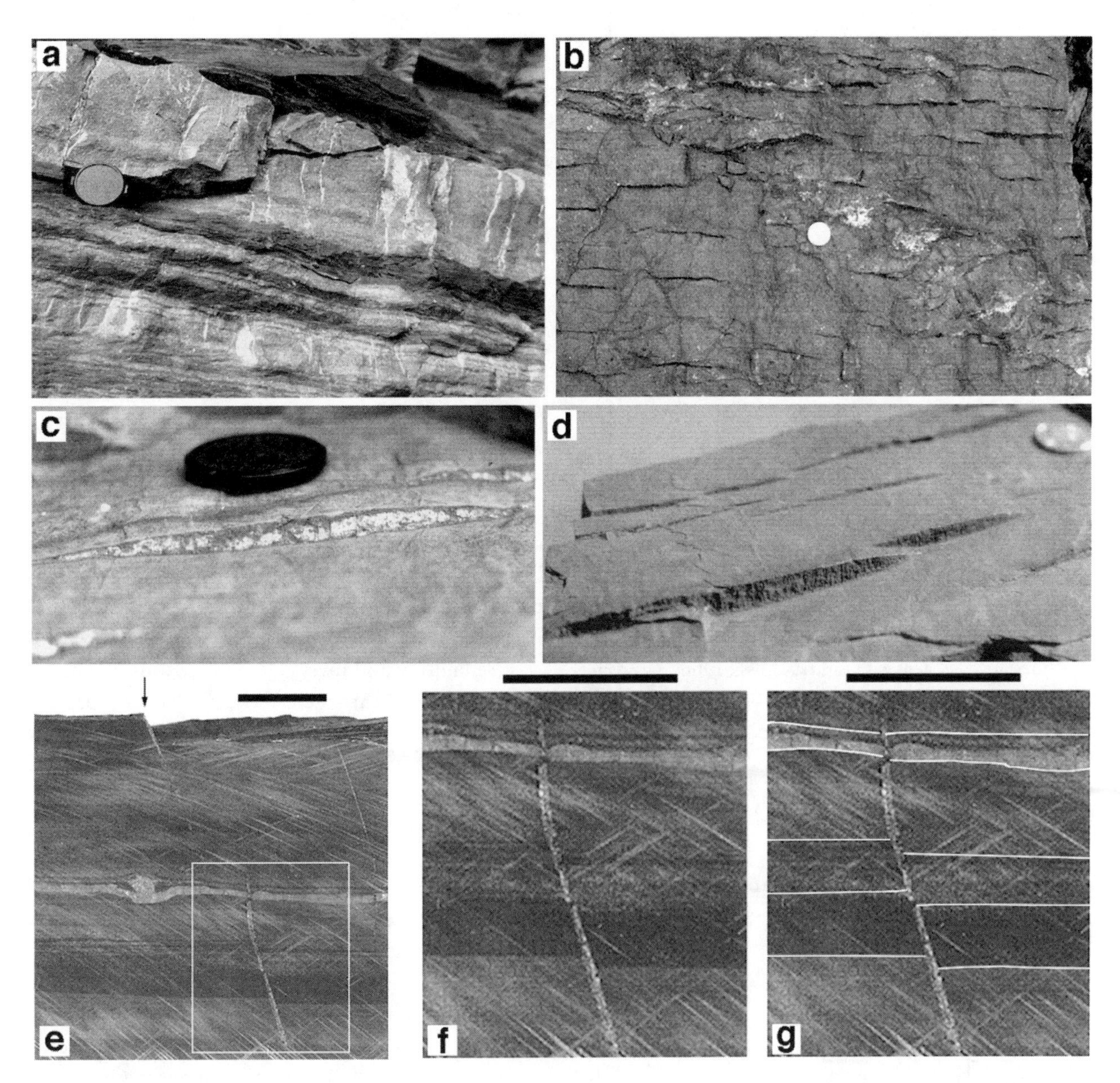

FIGURE 8.2 Brittle extensional features of the Solite Quarry. (*a*) Medium-fine-grained sandstones that contain calcite veins, interbedded with black shales that failed continuously. Lens cap for scale. (*b*) Hybrid fractures that formed in medium-coarse-grained feldspathic sandstones. The feature trending from top-left to bottom-right of the photo is a shear zone. Coin is 2.3 cm in diameter. (*c*) Single small normal fault that formed in fine-grained massive siltstones. The fault surface contains fibrous slickenlines that indicate predominantly dip-slip motion. Lens cap for scale. (*d*) Small normal faults separated by a relay structure. The fault surfaces contain tool-and-groove slickenlines. Coin is approximately 1 cm in diameter. (*e*) Sawed section of a slab that contains small faults. The arrow points to a prime example of footwall uplift and hanging-wall subsidence (reverse drag) along a bedding-plane parting cut by a normal fault. (*f* and *g*) Close-ups of a boxed fault; image *g* is interpreted. Notice that displacement along the fault is at a maximum toward the center of the fault, tapering to zero at the tips, where the structure terminates in zero-displacement cracks.

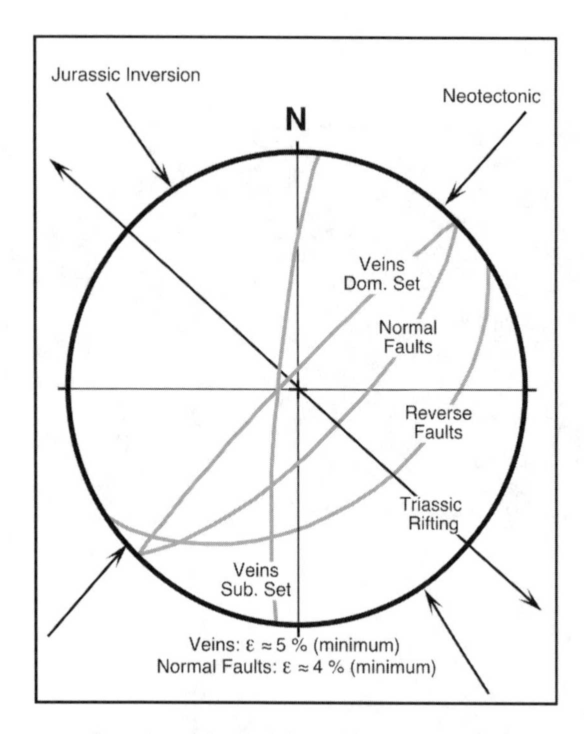

FIGURE 8.3 Synoptic diagram and stress-orientation analysis for brittle features present at the Solite Quarry. Extensional structures are consistent with Triassic rifting; contractional structures are consistent with earliest Jurassic rifting. Great circles are means based on multiple structures: 100+ normal faults, 50+ veins, one reverse fault, one imbricate thrust stack, and five quarry buckles.

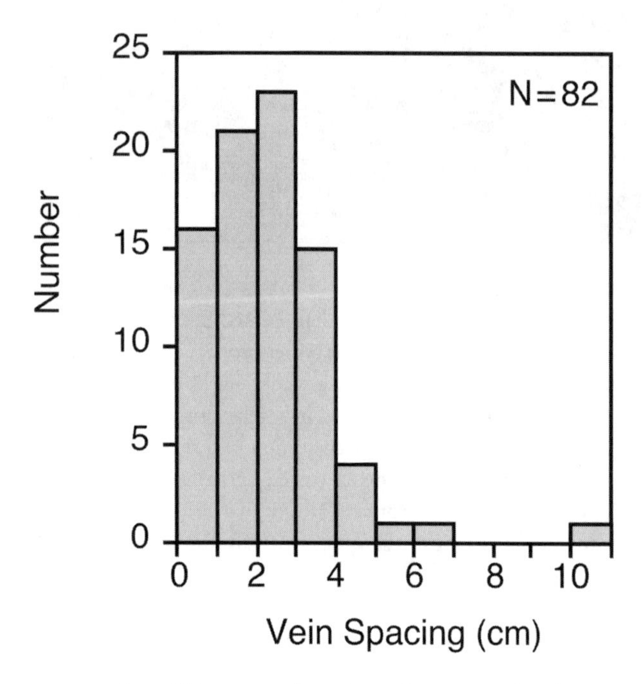

FIGURE 8.4 Histogram of spacings of macroscopic veins.

and Price 1981; Narr and Suppe 1991; Gross 1993); this distribution occurs in part because stress-reduction shadows around existing fractures inhibit the formation of other confined fractures and in part because of undersampling of the smallest structures.

Hybrid Fractures. Hybrid fractures are shear fractures with a substantial extensional component (e.g., Dunne and Hancock 1994). Within the Solite Quarry, hybrid fractures formed within medium- to coarse-grained feldspathic sandstones (figure 8.2b) and are restricted to discrete lithologic intervals in the same manner as extension fractures. Hybrid fractures are the least-common fracture type in the Solite Quarry.

Shear Fractures (Small Normal Faults). Shear fractures—the third brittle failure mode of fractures—are represented by very small, newly formed normal faults (figure 8.2c). They are restricted to fine-grained massive siltstones that occasionally separate cleanly along bedding-plane partings. These structures formed prior to basin-scale reverse drag associated with hanging-wall subsidence, as well as normal drag associated with reverse faulting and inversion, and are thus best called early tectonic shear fractures. Such fractures are the smallest normal faults studied in detail to date and are of particular interest because they are exceptionally well exposed and better constrain the scale range over which faulting takes place. They exhibit normal stratigraphic separation in both plan (figure 8.2c) and cross-section views (figure 8.2e–g), as well as footwall uplift and hanging-wall subsidence (figure 8.2e). Isostasy is not a consideration at this scale range; hence the footwall uplift must have been caused by coseismic elastic deformation of the volume surrounding the fault (e.g., Gupta and Scholz 1998). The bedding surfaces shown in figure 8.2c can be traced continuously around the tips of the faults and thus can be used as an offset marker. These very small faults ($\sim$0.5 cm $< L$ [length] $< \sim$300 cm) are planar and (originally) blind, dip at 70° to bedding (figure 8.2d), and are synthetic to the border-fault system of the basin. They are in many ways like their larger cousins, although they are unaffected by factors such as isostasy, erosion, and sedimentation. Some ($\sim$50%) of the structures exhibit a slight extensional component. The orientations of the small faults are consistent with Triassic rifting (figure 8.3). Bedding thickness remains constant across the faults, indicating that they are not syndepositional.

Their elliptical fault surfaces (Gupta and Scholz 1998) contain either tool-and-groove (figure 8.2d) or fibrous (calcite) slickensides (figure 8.2c), ensuring their brittle origin. Slickenlines rake at high angles, indicating predominantly dip-slip motion. The faults occur both as isolated features and as segments of relay systems, with relay ramps between overlapping fault segments in map view (Schlische et al. 1996; Gupta and Scholz 2000) (figure 8.2c and d). Relay structures are common to faults at a variety of scales (e.g., Larsen 1988; Peacock and Sanderson 1991; Dawers, Anders, and Scholz 1993). The faults also overlap in vertical section.

Extensional strain in beds containing small faults was calculated using the sum of heaves collected along scanlines oriented normal to the fault traces. The average extensional strain is approximately 4%. Along the length of the faults, displacement is at a maximum at the center of the fault and tapers to zero at the tips (figure 8.2c), consistent with displacement profiles observed on larger structures (e.g., Dawers, Anders, and Scholz 1993). This is the geometry predicted by models of fault growth that incorporate a process zone, according to which inelastic and nonbrittle processes such as plastic deformation, frictional wear, and mechanical breakdown occur at the fault tip (Cowie and Scholz 1992b). Displacement also varies along the height of the faults, such that larger displacements occur toward the center of the structures, tapering to zero toward the upper and lower tips (figure 8.2e–g), where the faults terminate into zero-displacement cracks. Far-field deformation affecting the volume surrounding the fault (required to maintain geometric coherence [e.g., Barnett et al. 1987]) is expressed as bedding deflection (reverse drag) that decreases away from the faults, generally consistent with elastic models of fault growth (Gupta and Scholz 1998) and with observations of faults at a variety of scales (e.g., Barientos, Stein, and Ward 1987; Barnett et al. 1987; King, Stein, and Rundle 1988; Schlische, chapter 4 in this volume).

The relation between maximum observed displacement, *D*, and trace length, *L*, for 201 isolated or completely linked faults is linear over approximately 2.5 orders of magnitude of fault length (Schlische et al. 1996) (figure 8.5). Thus in the *D-L* scaling relation

$$D = cL^n$$

the value of the scaling exponent $n = 1$ (*c* is a constant related to rock properties and tectonic environment). The Solite faults extend the global *D-L* data set by two orders of magnitude to a total of eight orders of magnitude of fault length and indicate that there is no significant change in the linear *D-L* scaling relation between small and large faults. Although the value of *n* has been controversial in the past (cf. Cowie and Scholz 1992a, 1992b; Gillespie, Walsh, and Watterson 1992), the Solite data set and others that have appeared in recent years (e.g., Dawers, Anders, and Scholz 1993; Villemin, Angelier, and Sunwoo 1995) indicate that $n = 1$ and that fault growth is largely self-similar. The determination of $n = 1$ supports Cowie and Scholz's (1992b) elastic-plastic model of fault growth.

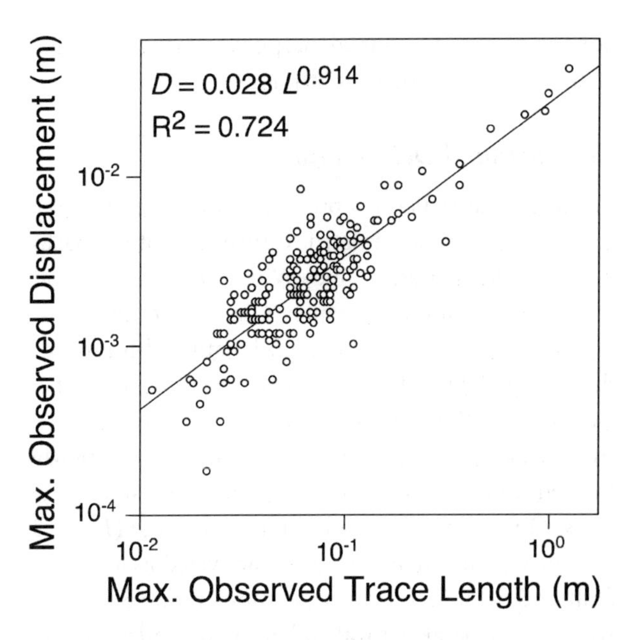

FIGURE 8.5 Log-log plot of displacement versus length for 201 small faults in the Solite Quarry. The relationship between displacement and length is approximately linear ($n = 1$).

Ductile Deformation-Extension

The fourth style of extensional deformation present within the section is ductile. Black shales are interbedded with the three aforementioned lithologies but do not exhibit brittle failure (figure 8.2a). There are no bedding-parallel extensional detachments within the section: all beds are welded together. Thus in order to maintain strain compatibility, the black shale layers deformed ductilely. Deformation of this type is manifested as deformed mudcracks and stretching lineations or microfolds that are perpendicular to the strike of small faults in adjacent lithologies.

Contractional Deformation

Contractional structures present in the Solite Quarry (figure 8.3) are likely related to inversion of the basin during earliest Jurassic time (Withjack, Schlische, and Olsen 1998; Schlische, chapter 4 in this volume). Inversion is indicated by the high length:width ratio of the basin (11:1) and the steep bedding dips (Schlische, chapter 4 in this volume). Some of the contractional structures in the quarry are thought to be newly formed, whereas others are reactivated extensional features. One such structure is a relatively large (L = 15 m) normal fault that has been reactivated as a reverse fault and exhibits reverse separation (figure 8.6a); the structure is parallel to all other small, intermediate, and large normal faults within the quarry. Small faults ($\sim$2 cm $< L <$ $\sim$10 cm) located in the hanging wall of the inverted fault continued to slip as normal faults during inversion but have anomalously high displacements for their lengths (figure 8.6b). This anomaly suggests that these faults did not lengthen according to the same scaling relation they followed during their initial formation. We attribute this difference to changes in the rock properties and, thus, c, between the time of formation of the extensional features and the time of initiation of inversion, which corresponds to the burial and thermal maximums for these rocks (Withjack, Schlische, and Olsen 1998).

Several individual shale layers were shortened during inversion of the Dan River–Danville basin. Black shales occasionally serve as compressional *décollement* surfaces with multiple imbricate thrust sheets (figure 8.6c). Microlaminated organic-rich shale layers known for their terrestrial fossils have also been shortened, forming very small folds in advance of propagating reverse faults (figure 8.6d). The fossils themselves have been deformed (Olsen, Schlische, and Gore 1989; Fraser et al. 1996).

Neotectonic Structures

Quarry buckles present within the Solite Quarry (figure 8.6e) suggest a NE–SW-directed maximum principal horizontal stress (figure 8.3). The buckles formed due to unloading as material was removed during the quarrying process. The resultant folds (buckles) formed such that their fold axes are perpendicular to the maximum horizontal principal stress direction (Stewart and Hancock 1994). Extension fractures associated with the buckles parallel the axes of the buckles.

Vertical Distribution of Structures

Within the Solite Quarry, deformation style and failure mode correlate closely with lithology, such that extension fractures, hybrid fractures, and small normal faults are restricted to discrete lithologic layers; there is no vertical connection among most of the fractures. In order to quantify this relation, we analyzed three samples of rock exhibiting each extensional deformational style (extension, hybrid, shear, ductile) to constrain their composition using a combination of geochemical x-ray diffraction techniques (Ackermann 1997). Because quantitative x-ray diffraction was not possible within the financial and time constraints of the study, we combined the oxide and phase data to generate the bulk mineral compositions of the samples. We did this using a modified least-squares–based magma-mixing model; such models combine the same types of data to determine the bulk compositions of igneous rocks.

X-ray diffraction indicates that all samples have a strong peak at 28° 2θ (albite); all samples are at least 50% albite. Albite is likely the cement for these rocks, consistent with their saline lacustrine origin (P. E. Olsen, personal communication 1996). Other minerals present in the samples are phyllosilicates (biotite), tectosilicates (Na- and K-feldspars, quartz, analcime), carbonates (dolomite, calcite), and inosilicates (riebekite).

Figure 8.7a presents the results of the oxide analysis in terms of general carbonate versus sandstone compositions, following Brownlow (1979). The data are sorted by failure mode and show a correlation between increasing shear-failure component and decreasing carbonate component (carbonates are generally considered to be "strong" [Suppe 1985]). There does not appear to be a correlation between failure mode and sandstone component within the samples. In figure 8.7b, the oxide data are shown in terms of $CaO:Al_2O_3$ ratios, with decreasing CaO and increasing Al_2O_3 trends following phyllosilicate ("weak" platy minerals) composition trends. The Solite data show a correlation between increasing shear component to the deforma-

FIGURE 8.6 Contractional structures present at the Solite Quarry. Parts *a* through *d* are related to inversion in earliest Jurassic time. (*a*) Large normal fault reactivated as a reverse fault. The person's hand is on the hanging wall, and the feet are on the footwall. (*b*) Small faults present in the hanging wall of the fault in *a*, where they were reactivated as normal faults. (*c*) Black shale layer that serves as a *décollement* surface, forming a series of imbricate thrust slices. Lens cap for scale. (*d*) Thin section of one of the arthropod beds, showing small reverse faults with less than 1 mm of separation (photograph courtesy of Nick Fraser). *e*) Neotectonic quarry buckles. Arrows indicate the buckle axes, which are perpendicular to neotectonic σ_1.

tion and increasing phyllosilicate composition trends. Notice that the outlying point in figure 8.7a and b is sample S2395BT. Some vein material (calcite) may have been incorporated erroneously into the sample during crushing of this sample, leading to an anomalously high CaO content.

Figure 8.7c summarizes the results of the least-squares regression analysis, where the major oxide and relative mineral abundance data were combined. The data are presented in terms of percentages of "weak" and "strong" minerals. "Weak" minerals are defined as phyllosilicates; all other minerals present in the sam-

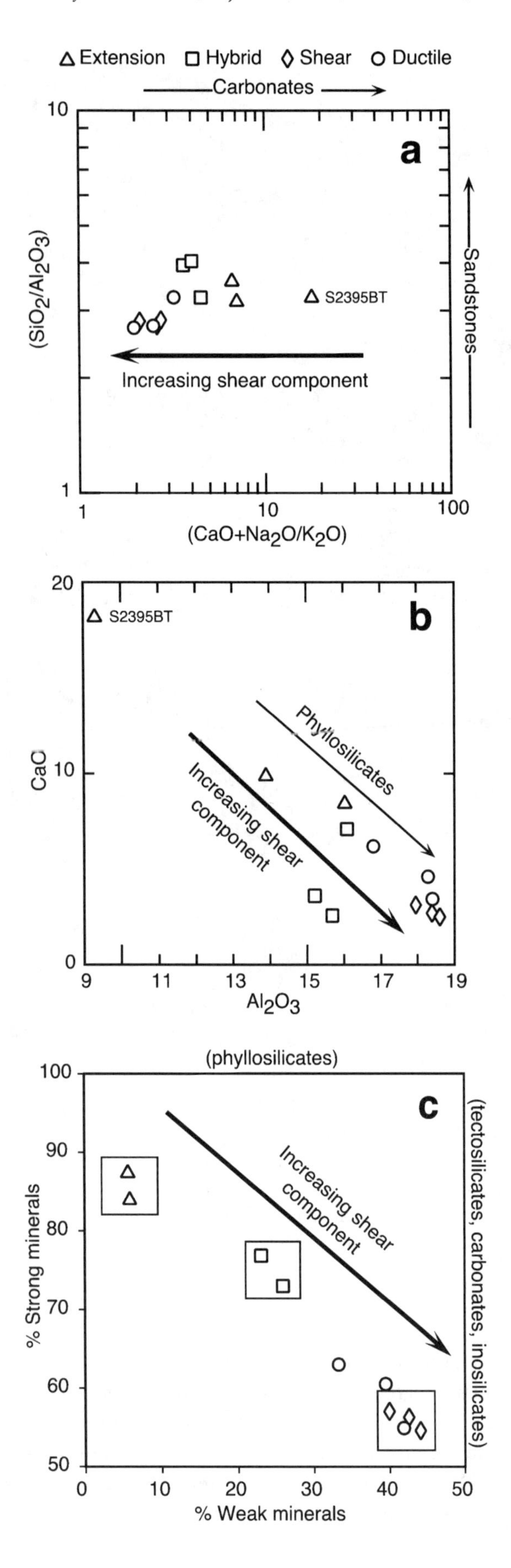

FIGURE 8.7 Sedimentary geochemistry of the rocks of the Solite Quarry. (*a*) Increasing carbonate components correlate with decreasing shear-failure components. Note that the sandstone component does not appear to correlate with the failure mode (characterization adopted from Brownlow 1979). *b*) $CaO:Al_2O_3$ ratios are another way of presenting sedimentary geochemical data, with decreasing CaO and increasing Al_2O_3 trends following phyllosilicate (platy minerals) composition trends. For the Solite data, increasing shear-failure components correlate with increasing phyllosilicates. (*c*) Summary of bulk mineralogy versus failure mode, in terms of weak minerals (phyllosilicates) versus strong minerals (tectosilicates, carbonates, inosilicates). Increasing shear-failure components correlate with increasing proportions of weak minerals.

ples are "strong." There is a very good correlation between increasing proportions of "weak" minerals and increasing shear-failure components (figure 8.7c). An interesting fact is that the hybrid samples contain hydrous inosilicates (riebekite), which may have made them more prone to shear failure.

The correlation between failure mode and bulk mineralogy, the comparable strain estimates for extension fractures and faults, and the distinct absence of extensional bedding-parallel detachments within the section suggest that all the brittle structures failed approximately coevally in response to the same remote applied stress. This phenomenon is known as *fracture partitioning* and has been described by Gross (1995) for the Monterey Formation in California. Based on the approach outlined by Gross (1995), Ackermann (1997) constructed a two-component (extension and shear) macroscopic failure model based on the boundary conditions outlined earlier and on average mechanical properties derived from the rock mechanics literature. This failure model indicates that brittle failure occurred at a burial depth of approximately 4 km in the presence of a pore fluid pressure of approximately 15 MPa.

The macroscopic failure models of Gross (1995) and Ackermann (1997) invoke a specific set of boundary conditions and potentially may be used to construct a predictive mechanical stratigraphy based on the mechanical properties of the rocks. Such a predictive tool will be useful for studies of groundwater and hydrocarbon accumulation and migration. However, these models have limitations. A fundamental weakness is that the mechanical properties used in the models are based on values published in the literature and not the actual samples themselves. Published data listing mechanical properties, bulk mineralogy, sedimentary geochemistry, and good lithologic descriptions are lacking, thus making these failure models more conceptual than case specific. This lack, in turn, points to another weakness: it is not clear if the samples we have now are mineralogically, geochemically, and mechanically similar to how they were when they failed at depth, making load tests questionable even if they are feasible. In the Solite example, the rocks have been metamorphosed to zeolite facies and are thermally mature. The thermal maximum for the basin occurred after the initial formation of the fractures and just before inversion of the basin in earliest Jurassic time (Malinconico 1996, chapter 6 in this volume; Withjack, Schlische, and Olsen 1998). In addition, the stress orientation analysis for the Solite rocks suggests that failure occurred prior to significant tilting of the basin fill, some of which is due to hanging-wall subsidence and reverse drag associated with rifting, but most of which is due to inversion (Schlische, chapter 4 in this volume). Whether or not the details of the failure model are correct, it is clear that the different types of fractures are controlled by lithology and that they formed semicoevally.

Lateral Distribution of Structures

Extension Fractures

A bedding-plane exposure of extension fractures (gashes) from the Solite Quarry is shown in figure 8.8. There is a uniform distribution of features across the surface. The average extension accommodated across the slab (measured using apertures along a series of scanlines) is approximately 3%, with a standard deviation of 0.5%; this estimate is lower than those determined using bed lengths along a cross-section scanline. The reason for this lower estimate is that the bed shown in figure 8.8 has broken at a mechanical layer boundary—the view is thus of the "top" of the mechanical layer, where extension fractures terminate. Thus the apertures measured were not maximum apertures, which would have required sampling at the level in the bed where the fractures nucleated, if there is a single level. Nonetheless, the strain appears to be evenly distributed across the sampled surface.

Figure 8.9 is a cumulative frequency plot (number of features greater than a given size plotted versus size) in log-log space for the lengths of the extension fractures shown in figure 8.8. The entire distribution is best fit using an exponential function rather than a power-law function (which applies to only a very limited scale range [figure 8.9, dark circles]). The definition of a mean value or characteristic size is a diagnostic feature of exponential distributions (Cowie et al. 1994). The exponential distribution is probably a result of a lack of spatial correlation in the system due to the effect of mechanical layer thickness on lateral positioning of fractures, which in turn suppresses the short- and long-range interactions of structures that ordinarily result in power-law (fractal) distributions (e.g., Cowie et al. 1994).

The population of extension fracture lengths shown in figure 8.8 exhibits a characteristic length of approximately 2.60 cm, based on the slope of the best-fit exponential curve in figure 8.9. The slope (0.384) is the reciprocal of $\langle L \rangle$, the mean value of L (e.g., Cowie et al. 1994). This estimate of 2.60 cm contrasts with an average length of approximately 4.25 cm, based on field observations. The difference likely stems from the structural level where fractures were sampled: lengths were sampled at the mechanical layer boundary (a limiting level for fracture height), not at the nucleation level or levels. As a result, small fractures were undersampled, either because they do not intersect the sampled level or because they fall below the resolution of the sampling level (naked eye). The estimate of 4.25 cm is derived from a sample in which small fractures were underrepresented, thus overestimating the average or characteristic length. The estimate of characteristic length of 2.60 cm is likely to be more accurate because it relies on the distribution of fracture sizes. The distribution of sizes (the slope of the line in figure 8.9) will not vary appreciably with structural level, provided most fractures have achieved a height equal to mechanical layer thickness. Some lengths sampled at the layer boundary will be less than the maximum lengths. In other words, depending on the structural level sampled, the intercepts of the cumulative frequency plot will change, but the slope will not, and it is the slope that provides the characteristic length.

FIGURE 8.8 Spatial distribution of extension fractures across a bedding surface. View is of the top of the bed, which is a mechanical layer boundary. Lines are scan lines along which one-dimensional strain data were collected.

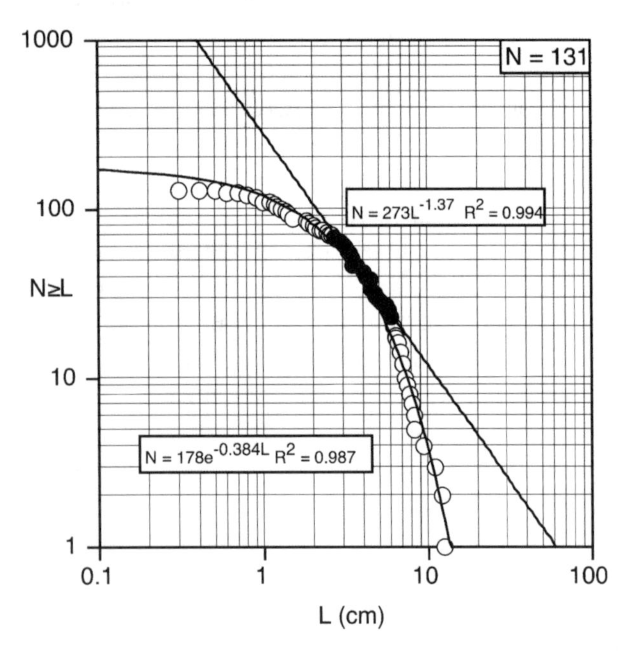

FIGURE 8.9 Cumulative frequency diagram of extension fracture lengths for the bedding surface shown in figure 8.8. It is not clear if there is a "flat" central segment suggestive of a power-law distribution of sizes. The data as a whole are better fit using an exponential function. If the data do indeed follow a power-law distribution, then truncation and censoring effects are severe for this sample.

Small Normal Faults

Over large areas of the Solite Quarry, the small normal faults appear to be distributed fairly uniformly within the units that fail in shear. These structures range in size (0.1 cm $< L <$ 200 cm), although the vast majority of them are less than 10 cm long. A division based on fault size (length) can be made (Ackermann and Schlische 1997) (figure 8.10): there are small faults ($L \leq$ 20 cm) and larger faults ($L \geq$ 20 cm, usually $L \geq$ 100 cm). The larger faults are fairly uncommon and appear to be uniformly distributed within the rock volume. Around these larger faults (master faults), there appear to be ellipsoidal zones devoid of smaller brittle structures (figure 8.10), which are otherwise ubiquitous in the surrounding volume. The smaller faults are thus anticlustered around the master faults. These regions lacking smaller faults are called *fault shields.* Master faults are defined by the presence of a fault shield around them, regardless of their size. The width of the fault shield in plan view scales linearly with the displacement on the master fault (Ackermann and Schlische 1997) and is geometrically similar to the deformation field surrounding a normal fault (e.g., Gibson, Walsh, and Watterson 1989) (figure 8.11a).

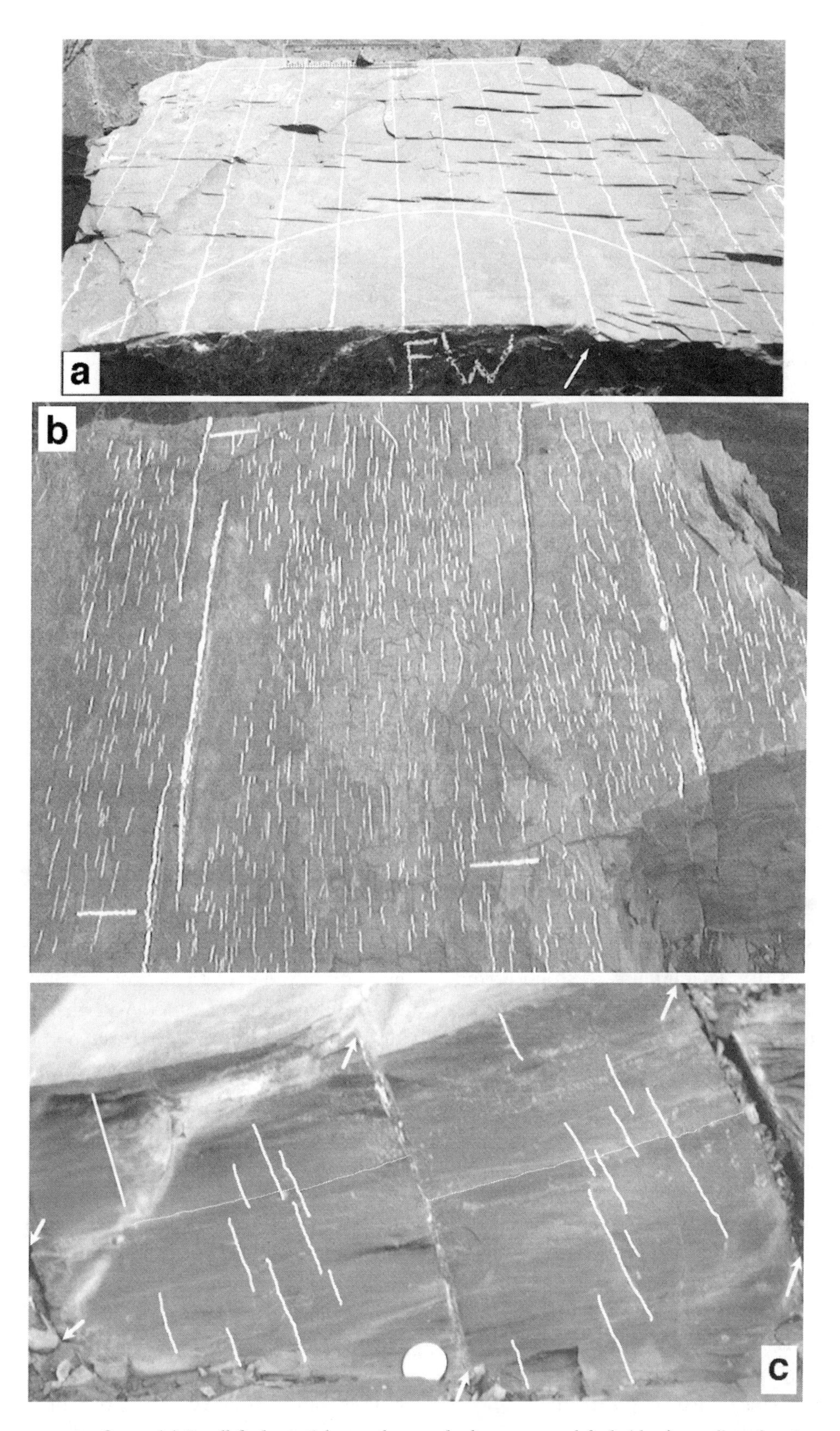

FIGURE 8.10 (*a*) Small faults anticlustered around a larger normal fault (the footwall surface is shown in foreground). Notice the absence of faults within an elliptical region around the larger structure, except for where there is a breached ramp structure in the lower right (*arrow*). Ruler at the top of image is 15 cm wide. (*b*) Distribution of faults exposed on the bedding surface of a quarried boulder. Faults are highlighted with chalk. Scale bars normal to fault traces are 10 cm long. (*c*) Fault shields in cross section, with faults highlighted. Arrows point to tips of master faults. Master faults and small faults offset bedding. Coin for scale.

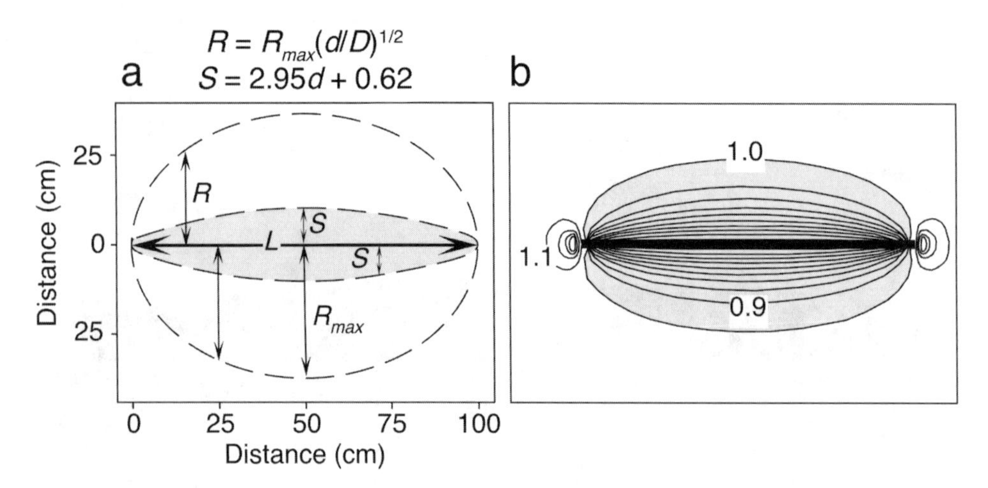

FIGURE 8.11 Comparison of fault shields, deformation fields, and stress-reduction shadows around a normal fault. (*a*) Relationship between fault shield width (*S*) and deformation field width (i.e., reverse drag width, *R* [Gibson, Walsh, and Watterson 1989]) for an idealized fault of length 100 cm and displacement 3 cm. *S* is based on the empirical relationship in Ackermann and Schlische (1997). R_{max} is equal to the mean of the radii of the fault-surface ellipse (in this case, the radii are 50 cm and 25 cm, and thus the aspect ratio is 2:1). *D*, maximum displacement; *d*, displacement. (*b*) Perturbation of the stress field surrounding a vertical fault (aspect ratio 2:1) following a uniform stress drop. The shear-stress component acting parallel to the slip vector is reduced in the shaded areas and enhanced in the unshaded areas. (Modified from Willemse 1997)

The shields are apparent in plan view (figure 8.10a and b) and in cross section (figure 8.10c). The smaller faults appear to follow a semiregular arrangement around the master faults, such that they change step near the center of the fault along its length and around the tips (figure 8.10). The amount of strain accommodated within a rectilinear two-dimensional area around a master normal fault, including smaller faults outside the fault shield, varies along the length of the larger fault, mirroring a length-displacement profile (figure 8.12a). When the master faults present are not included in the strain calculation, strain on the smaller structures does not vary appreciably across the study area, and the amount of strain accommodated is substantially less. The master faults are dominating the strain. If one considers an area without a master fault present (figure 8.12b), the small faults do not exhibit stepping patterns, and the amount of strain accommodated is fairly uniform and substantially less than in areas with master faults. As documented fully in Ackermann and Schlische (1997), anticlustering also affects the distribution of fault sizes, such that the population as a whole is best described by two power-law curves: one covering the smaller faults, which tend to be unbounded structures, and the other covering the master faults, which have completely spanned the mechanical layer containing the faults.

The fault shields are interpreted to correspond to a critical stress-reduction shadow, which prevented the nucleation and growth of smaller faults around the earlier formed master faults (Ackermann and Schlische 1997). This interpretation is based on the nearly complete absence of brittle structures in the fault shields; the geometric similarity between fault shields and the deformation fields around normal faults (which represent regions where the rocks have been strained and the stresses relaxed [Gibson et al. 1989]); and the similarity between the shield geometry and regions of stress reduction derived from numerical models (e.g., Willemse 1997) (figure 8.11).

The Solite Quarry represents the first definitive field documentation of an anticlustered spatial distribution of faults and the existence of stress-reduction shadows around faults (Ackermann and Schlische 1997). In contrast, several workers (e.g., Gillespie et al. 1993; Little 1996) have noted that smaller faults most often are clustered positively around larger faults. The two observations need not be mutually exclusive: numer-

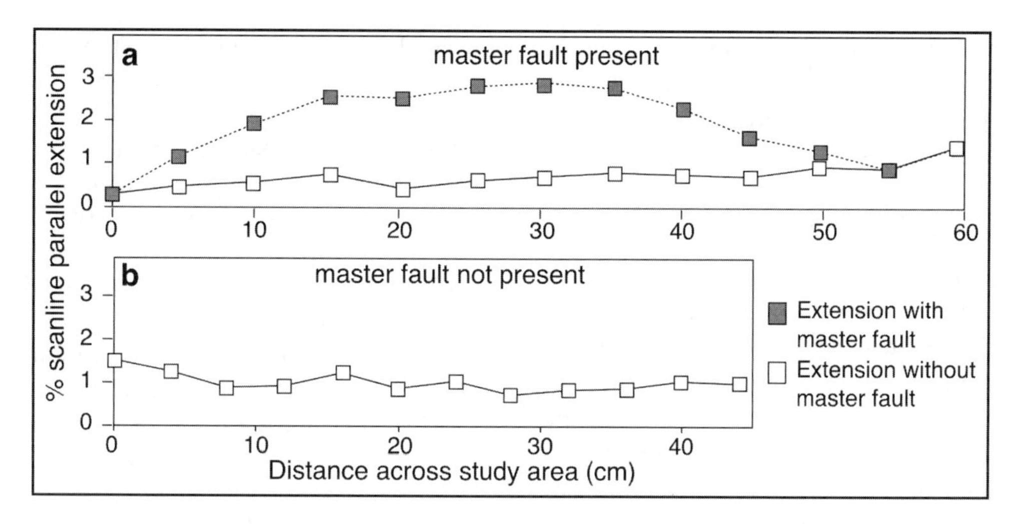

FIGURE 8.12 Distance-strain profiles for two study areas. Diagrams show the distance across the study area (parallel to fault traces) on the ordinate and percent scan-line parallel (fault-orthogonal) extension on the abscissa. (*a*) Profile of an area with a master fault. Curves show total brittle strain. Estimates including the master fault are shown with shaded boxes; estimates not including the master fault contribution are shown with open boxes. (*b*) Profile of an area without a master fault.

ical modeling conducted by Willemse (1997) shows higher shear stresses in the regions surrounding the fault tips (figure 8.11b). It is thus possible that both positive clustering (figure 8.10a, arrow) and anticlustering (in the form of fault shields) can occur along the same fault, with anticlustering dominating near the center of the fault and positive clustering dominating near the tips. A better understanding of the spatial distribution of fractures, made possible by exceptional field localities such as the Solite Quarry, ultimately will lead to the refinement of fractured-rock models, which are so important for groundwater and hydrocarbon prospecting.

Conclusions

The Solite Quarry contains an exquisite suite of rift-related deformation structures that provide a unique opportunity to study fracture geometries, kinematics, size and spatial distributions, and population systematics in great detail.

Extension fractures, hybrid fractures, and shear fractures (small normal faults) are the dominant brittle structures related to Triassic rifting. The faults and extension fractures are geometrically and kinematically distinct, and it is therefore unlikely that the faults originated as macroscopic extension fractures. Also present are reactivated normal faults, imbricate thrust faults, and microfolds—all of which are late-stage features related to inversion of the Dan River–Danville basin in earliest Jurassic time.

All these features are restricted to certain interbedded lithologies, such that beds with greater percentages of weak minerals failed with increasing shear components. There is a distinct absence of extensional detachment horizons between mechanical layers. The extension fractures and the small normal faults have undergone comparable one-dimensional strains. All observations suggest coeval brittle failure of interbedded lithologies under the same remote applied stress.

The Solite faults extend the global *D-L* data set by two orders of magnitude to a total of nine orders of magnitude of fault length and indicate that there is no significant change in the linear *D-L* scaling relation. The faults can be subdivided into two subsets based on their spatial distribution. Larger (master) normal faults accommodate the majority of the strain but, compared with the smaller faults, are relatively uncommon. The other subset of smaller faults is ubiquitous within the rock volume but exhibits anticlustering with respect to the larger structures, forming ellipsoidal fault shields around the larger faults. The shields are geometrically similar to the deformation fields of the master faults, corresponding to a stress-reduction shadow that prevented the nucleation of smaller faults around the master faults.

Acknowledgments

We thank C. H. Gover and the Virginia Solite Corporation for their support of this and other research conducted at the Solite Quarry. Our research was supported generously by Mobil Technology Company, the Virginia Museum of Natural History, Sigma Xi, and the Department of Geological Sciences at Rutgers University. Alison Lighthart kindly and expertly made numerous thin sections. Anu Gupta and MaryAnn Malinconico generously shared unpublished data. Mike Carr, Jeff Niemitz, and Gene Yogodzinski provided advice regarding the analytical geochemistry and x-ray diffraction. We thank Karen Bemis, Jim Carpenter, Amy Clifton, Patience Cowie, Nancye Dawers, Gloria Eisenstadt, Nick Fraser, Mike Gross, Anu Gupta, Peter Hennings, Paul Olsen, Chris Scholz, ShayMaria Silvestri, Martha Withjack, and Scott Young for many engrossing and enlightening discussions regarding the features at Solite, fault-population systematics, and the Mesozoic rift system. Finally, we appreciate Mike Gross's and Chris Scholz's helpful reviews of the manuscript for this chapter.

Literature Cited

Ackermann, R. V. 1997. Spatial distribution of rift-related fractures: Field observations, experimental modeling, and influence on drainage networks. Ph.D. diss., Rutgers University.

Ackermann, R. V., and R. W. Schlische. 1997. Anticlustering of small faults around larger normal faults. *Geology* 25:1127–1130.

Barnett, J. A. M., J. Mortimer, J. H. Rippon, J. J. Walsh, and J. Watterson. 1987. Displacement geometry in the volume containing a single normal fault. *American Association of Petroleum Geologists Bulletin* 71:925–937.

Barrientos, S. E., R. S. Stein, and S. N. Ward. 1987. Comparison of the 1959 Hebgen Lake, Montana, and the 1983 Borah Peak, Idaho, earthquakes from geodetic observations. *Seismological Society of America Bulletin* 77:784–808.

Brownlow, A. H. 1979. *Geochemistry.* Englewood Cliffs, N.J.: Prentice Hall.

Cowie, P. A. 1998. Normal fault growth in three-dimensions in continental and oceanic crust. In *Faulting and Magmatism at Mid-Ocean Ridges,* pp. 325–348. Geophysical Monograph, no. 106. Washington, D.C.: American Geophysical Union.

Cowie, P. A., R. Knipe, and I. G. Main. 1996. Introduction to *Scaling Laws for Fault and Fracture Populations: Analyses and Applications.* Special issue of *Journal of Structural Geology* 18:v–xi.

Cowie, P. A., A. Malinverno, W. B. F. Ryan, and M. H. Edwards. 1994. Quantitative fault studies on the East Pacific Rise: A comparison of sonar imaging techniques. *Journal of Geophysical Research* 99:15205–15218.

Cowie, P. A., and C. H. Scholz. 1992a. Displacement-length scaling relationship for faults: Data synthesis and discussion. *Journal of Structural Geology* 14:1149–1156.

Cowie, P. A., and C. H. Scholz. 1992b. Physical explanation for displacement-length relationship of faults using a post-yield fracture mechanics model. *Journal of Structural Geology* 14:1133–1148.

Davis, G. H. 1984. *Structural Geology of Rocks and Regions.* New York: Wiley.

Dawers, N. H., M. H. Anders, and C. H. Scholz. 1993. Fault length and displacement: Scaling laws. *Geology* 21:1107–1110.

Dunne, W. M., and P. L. Hancock. 1994. Paleostress analysis of small-scale brittle structures. In P. L. Hancock, ed., *Continental Deformation,* pp. 101–120. New York: Pergamon.

Fraser, N. C., D. A. Grimaldi, P. E. Olsen, and B. Axsmith. 1996. A Triassic Lagerstätte from eastern North America. *Nature* 380:615–619.

Gibson, J. R., J. J. Walsh, and J. Watterson. 1989. Modelling of bed contours and cross-sections adjacent to planar normal faults. *Journal of Structural Geology* 11:317–328.

Gillespie, P. A., C. B. Howard, J. J. Walsh, and J. Watterson. 1993. Measurement and characterisation of spatial distributions of fractures. *Tectonophysics* 226:113–141.

Gillespie, P. A., J. J. Walsh, and J. Watterson. 1992. Limitations of dimension and displacement data from single faults and the consequences for data analysis and interpretation. *Journal of Structural Geology* 14:1157–1172.

Gross, M. R. 1993. The origin and spacing of cross joints: Examples from the Monterey Formation, Santa Bar-

bara coastline, California. *Journal of Structural Geology* 15:737–751.

Gross, M. R. 1995. Fracture partitioning: Failure mode as a function of lithology in the Monterey Formation of coastal California. *Geological Society of America Bulletin* 107:779–792.

Gross, M. R., M. P. Fischer, T. Engelder, and R. J. Greenfield. 1995. Factors controlling joint spacing in interbedded sedimentary rocks: Integrating numerical models with field observations from the Monterey Formation, USA. In M. S. Ameen, ed., *Fractography: Fracture Topography as a Tool in Fracture Mechanics and Stress Analysis,* pp. 215–233. Geological Society Special Publication, no. 92. London: Geological Society.

Gupta, A., and C. H. Scholz. 1998. Utility of elastic models in predicting fault displacement fields. *Journal of Geophysical Research* 103:823–834.

Gupta, A., and C. H. Scholz. 2000. A model of normal fault interaction based on observations and theory. *Journal of Structural Geology* 22:865–879.

Kent, D. V., and P. E. Olsen. 1997. Paleomagnetism of Upper Triassic continental sedimentary rocks from the Dan River–Danville rift basin (eastern North America). *Geological Society of America Bulletin* 109:366–377.

King, G. C. P., R. S. Stein, and J. B. Rundle. 1988. The growth of geological structures by repeated earthquakes. 1. Conceptual framework. *Journal of Geophysical Research* 93:13307–13318.

Ladeira, F. L., and N. J. Price. 1981. Relationships between fracture spacing and bed thickness. *Journal of Structural Geology* 3:179–183.

Larsen, P. H. 1988. Relay structures in a Lower Permian basement-involved extension system, East Greenland. *Journal of Structural Geology* 10:3–8.

Little, T. A. 1996. Faulting-related displacement gradients and strain adjacent to the Awatere strike-slip fault in New Zealand. *Journal of Structural Geology* 18:321–340.

Malinconico, M. L. 1996. Paleo-maximum thermal structure of the Triassic Taylorsville (Virginia) basin: Evidence for border fault convection and implications for duration of syn-rift sedimentation and long-term elevated heat flow. In P. M. LeTourneau and P. E. Olsen, eds., *Aspects of Triassic–Jurassic Rift Basin Geoscience: Abstracts,* pp. 25–26. State Geological and Natural History Survey of Connecticut Miscellaneous Reports, no. 1. Hartford: Connecticut Department of Environmental Protection.

Narr, W., and J. Suppe. 1991. Joint spacing in sedimentary rocks. *Journal of Structural Geology* 13:1037–1048.

Olsen, P. E. 1997. Stratigraphic record of the early Mesozoic breakup of Pangea in the Laurasia–Gondwana rift system. *Annual Review of Earth and Planetary Science* 25:337–401.

Olsen, P. E., R. W. Schlische, and P. J. W. Gore, eds. 1989. *Tectonic, Depositional, and Paleoecological History of Early Mesozoic Rift Basins, Eastern North America.* International Geological Congress Field Trip no. T-351. Washington, D.C.: American Geophysical Union.

Peacock, D. J. P., and D. J. Sanderson. 1991. Displacements, segment linkage, and relay ramps in normal fault zones. *Journal of Structural Geology* 13:721–733.

Pollard, D. D., and A. Aydin. 1988. Progress in understanding jointing over the past century. *Geological Society of America Bulletin* 100:1181–1204.

Schlische, R. W., S. S. Young, R. V. Ackermann, and A. Gupta. 1996. Geometry and scaling relations of a population of very small rift-related normal faults. *Geology* 24:683–686.

Stewart, I. S., and P. L. Hancock. 1994. Neotectonics. In P. L. Hancock, ed., *Continental Deformation,* pp. 370–410. New York: Pergamon.

Suppe, J. 1985. *Principles of Structural Geology.* Englewood Cliffs, N.J.: Prentice Hall.

Villemin, T., J. Angelier, and C. Sunwoo. 1995. Fractal distribution of fault length and offsets: Implications of brittle deformation evaluation—Lorraine coal basin. In C. C. Barton and P. R. La Pointe, eds., *Fractals in the Earth Sciences,* pp. 205–226. New York: Plenum.

Willemse, E. J. M. 1997. Segmented normal faults: Correspondence between three-dimensional mechanical models and field data. *Journal of Geophysical Research* 102:675–692.

Withjack, M. O., R. W. Schlische, and P. E. Olsen. 1998. Diachronous rifting, drifting, and inversion on the passive margin of eastern North America: An analog for other passive margins. *American Association of Petroleum Geologists Bulletin* 82:817–835.

PART II

THE CENTRAL ATLANTIC LARGE IGNEOUS PROVINCE

9
Introduction

Paul E. Olsen and J. Gregory McHone

Large igneous provinces (LIPs) comprise enormous edifices of basaltic lava and associated igneous rocks emplaced over a relatively brief time interval (Coffin and Eldholm 1994). Two of the largest terrestrial LIPs, the Siberian Traps ($\sim 2.5 \times 10^6$ km^3) and Deccan Traps ($\sim 2.6 \times 10^6$ km^3), are continental flood basalts associated in time with a mass extinction—the Siberian Traps with the end-Permian extinction at 250 Ma and the Deccan Traps with the end-Cretaceous extinction at 65 Ma (McLean 1985; O'Keefe and Ahrens 1989; Caldeira and Rampino 1990; Courtillot et al. 1994) (figure 9.1). The early Mesozoic basaltic rocks of eastern North America recently have been recognized as part of a third giant continental LIP closely associated in time with a mass extinction, this time the Triassic–Jurassic ($\sim$201 Ma) boundary (Marzoli et al. 1999; Olsen 1999) (figure 9.2). These lavas and associated igneous intrusions are now called the Central Atlantic Magmatic Province (CAMP) (Marzoli et al. 1999). The CAMP, the topic of part II of this volume, covers major parts of at least four tectonic (continental) plates and is integral to the rifting of the supercontinent of Pangea and formation of the Atlantic Ocean.

Although CAMP has very familiar geologic features with a venerable literature (e.g., Percival 1842; Davis 1883; Darton 1890; Walker 1940; May 1971), the fact that it may well be the largest continental LIP on Earth has been recognized within only the past few years (Marzoli et al. 1999). According to J. Gregory McHone and John H. Puffer (chapter 10), CAMP originally may have produced lavas extending over more than 7×10^6 km^2 prior to the formation of the Atlantic. In their chapter, McHone and Puffer review the distribution, chemistry, age, and stratigraphic setting of the CAMP, and they suggest that the eruption of the lavas could have triggered ecologically catastrophic climate change through massive input of volatiles into the atmosphere, as has been suggested for several other LIPs (McLean 1985; Rampino and Stothers 1988; Courtillot et al. 1994; Renne et al. 1995; McHone 1996, 2000). Recent direct fossil evidence does suggest a link to the CAMP (McElwain, Beerling, and Woodward 1999; Palfy et al. 2000).

To one extent or another, the other chapters in part II consider the somewhat controversial origin of the CAMP and the geodynamic conditions necessary for the production of LIPs. The controversy centers on the "deep-mantle plume" model versus the "shallow-mantle convection" model for generating large amounts of basalt (essentially Campbell and Griffiths's [1990] model versus King and Anderson's [1995]). Wilson (1997) and McHone (2000), among others, have suggested applications of these contrasting models to the CAMP. Although both models are essentially variations on the geometry of mantle upwelling (figure 9.3), there are fundamental differences in how each can cause continents to rift and massive basalts to erupt, so both deserve serious attention. Geophysical data that can conclusively demonstrate the requisite mantle geodynamics unfortunately are exceedingly difficult to collect and afterward to interpret. Needed are testable

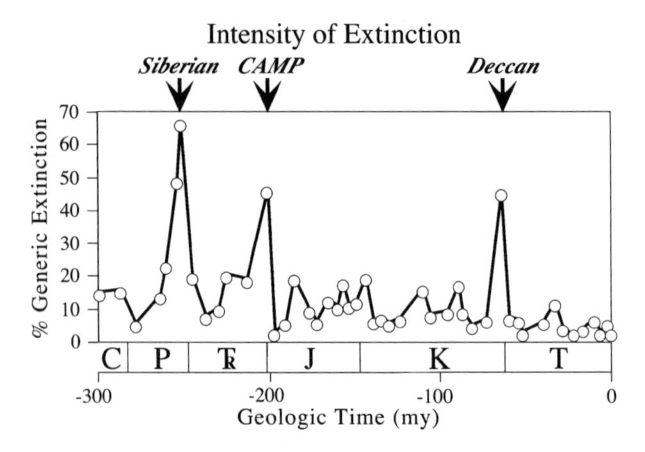

FIGURE 9.1 Extinction rate of "shelly" marine invertebrates through the Phanerozoic, showing the major continental LIPs. (Based on Sepkoski 1997, with timescale modified according to Kent and Olsen 1999)

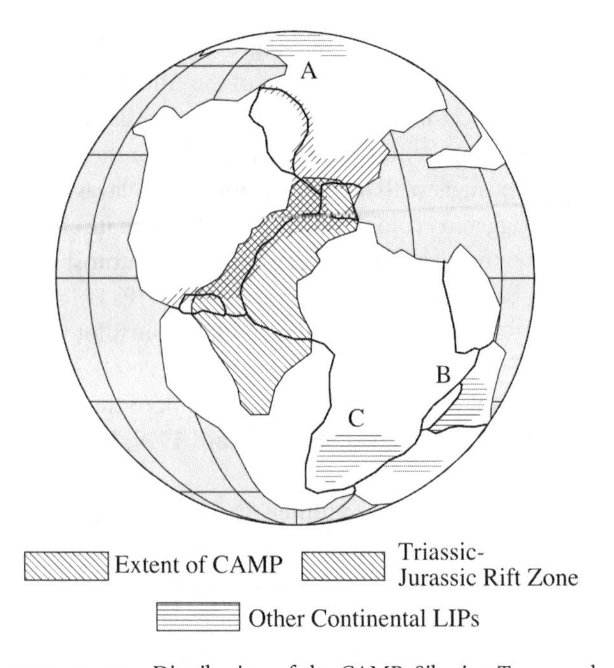

FIGURE 9.2 Distribution of the CAMP, Siberian Traps, and Deccan Traps with a Late Triassic plate configuration. (Based on Olsen 1999)

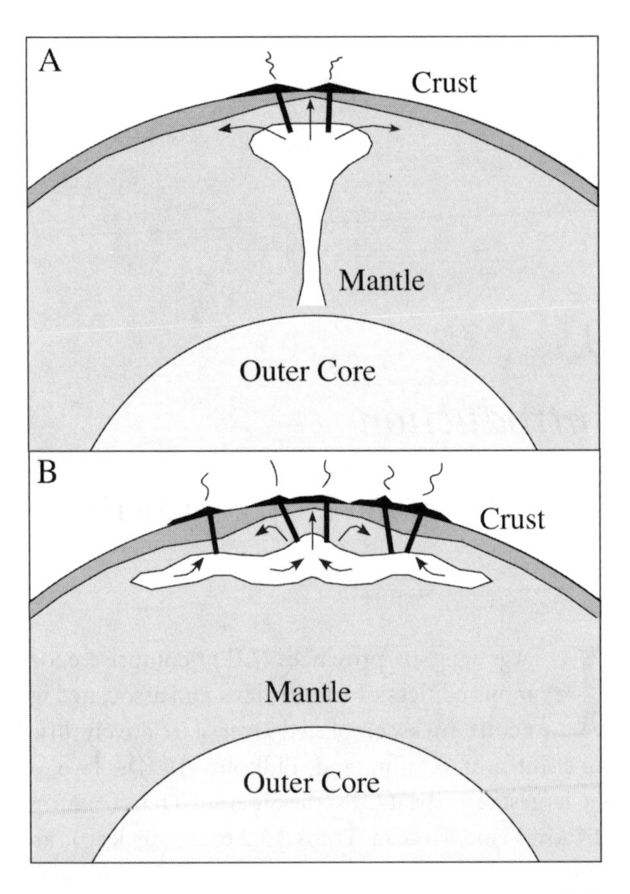

FIGURE 9.3 General geodynamic patterns envisioned for (*A*) the deep-mantle plume model and (*B*) the upper-mantle convection model. Note the more narrowly focused volcanism attributed to the plume model versus the wider-ranging volcanism from upper-mantle convection.

hypotheses that predict associated chemical, physical, dimensional, and temporal characteristics unique to each model (or, quite possibly, unique to some variant model yet to be realized).

In chapter 11, John H. Puffer places the CAMP flood basalts in a larger petrological context with various other continental flood basalt LIPs, including those of the Rodinian and pre-Rodinian supercontinent, the Siberian Traps, the Karoo Province, the Deccan Traps, and the Columbia River Basalts. Surprisingly, CAMP basalts appear similar to arc volcanics, and Puffer concludes that global plate reorganization was the motive force behind the emplacement of the gigantic igneous province, unlike some other continental LIPs that may have been produced by hot spots. Paul C. Ragland, Vincent J. M. Salters, and William C. Parker (chapter 12) examine a massive database of major-oxide analyses from the southeastern U.S. portion of the CAMP and hypothesize that the observed chemical trends indicate melting at deeper levels toward the southwestern portion of the southeastern United States in an area of thicker crustal derivation of the magma from more "fertile" mantle influenced by a hot spot. The different models for origin of LIPs, especially the CAMP, make different prediction of the mechanism of emplacement though dikes. In chapter 13, Jelle Zeilinga de Boer, Richard E. Ernst, and Andrew G. Lindsey describe anisotropy of magnetic susceptibility (AMS) data from the northeastern United States that they conclude indicates northeasterly directed lateral flow in dikes. This flow would be more compatible with a southeastern United States mantle plume source than with any other mechanism.

McHone and Puffer (chapter 10) and Ragland and colleagues (chapter 12) also briefly discuss one of the largest and most critical unanswered questions about the CAMP: What is the relationship between the CAMP and the seaward-dipping reflectors off the eastern United States (Holbrook and Kelemen 1993; Talwani et al. 1995; Withjack, Schlische, and Olsen 1998)? This question, unlikely to be dealt with seriously without extensive scientific drilling (Olsen, Kent, and Raeside 1999), bears directly on the mechanism of supercontinent breakup, the origin of initial oceanic crust, and the magnitude of the CAMP and its environmental effects.

Literature Cited

Caldeira, K. G., and M. R. Rampino. 1990. Deccan volcanism, greenhouse warming, and the Cretaceous/Tertiary boundary. *Geological Society of America Bulletin* 247:117–123.

Campbell, I. H., and R. W. Griffiths. 1990. Implications of mantle plume structure for the evolution of flood basalts. *Earth and Planetary Science Letters* 99:79–93.

Coffin, M. F., and O. Eldholm. 1994. Large igneous provinces: Crustal structure, dimensions, and external consequences. *Review of Geophysics* 32:1–32.

Courtillot, V., J. J. Jaeger, Z. Yang, G. Firaud, and G. Hoffman. 1994. The influence of continental flood basalts on mass extinctions: Where do we stand? In G. Ryder, D. Fastovsky, and S. Gartner, eds., *The Cretaceous–Tertiary Event and Other Catastrophes in Earth History,* pp. 513–525. Geological Society of America Special Paper, no. 107. Boulder, Colo.: Geological Society of America.

Darton, N. H. 1890. The relations of the traps of the Newark system in New Jersey. *U.S. Geological Survey Bulletin* 67:1–82.

Davis, W. M. 1883. On the relations of the Triassic traps and sandstones of the eastern United States. *Bulletin of the Museum of Comparative Zoology* 7:249–309.

Holbrook, W. S., and P. B. Kelemen. 1993. Large igneous province on the U.S. Atlantic margin and implications for magmatism during continental breakup. *Nature* 364:433–436.

Kent, D. V., and P. E. Olsen. 1999. Astronomically tuned geomagnetic polarity timescale for the Late Triassic. *Journal of Geophysical Research* 104:12831–12841.

King, S. D., and D. L. Anderson. 1995. An alternative mechanism of flood basalt formation. *Earth and Planetary Science Letters* 136:269–279.

Marzoli, A., P. R. Renne, E. M. Piccirillo, M. Ernesto, G. Bellieni, and A. De Min. 1999. Extensive 200-million-year-old continental flood basalts of the Central Atlantic Magmatic Province. *Science* 284:616–618.

May, P. R. 1971. Pattern of Triassic–Jurassic diabase dikes around the North Atlantic in the context of predrift position of the continents. *Geological Society of America Bulletin* 82:1285–1291.

McElwain, J. C., D. J. Beerling, and F. I. Woodward. 1999. Fossil plants and global warming at the Triassic–Jurassic boundary. *Science* 285:1386–1390.

McHone, J. G. 1996. Broad-terrane Jurassic flood basalts across northeastern North America. *Geology* 24:319–322.

McHone, J. G. 2000. Non-plume magmatism and rifting during the opening of the central Atlantic Ocean. *Tectonophysics* 316:287–296.

McLean, D. M. 1985. Deccan Traps mantle degassing in the terminal Cretaceous marine extinctions. *Cretaceous Research* 6:235–259.

O'Keefe, J. D., and T. J. Ahrens. 1989. Impact production of CO_2 by the Cretaceous/Tertiary extinction bolide and the resultant heating of the Earth. *Nature* 338:247–249.

Olsen, P. E. 1999. Giant lava flows, mass extinctions, and mantle plumes. *Science* 284:604–605.

Olsen, P. E., D. V. Kent, and R. Raeside. 1999. International workshop for a climatic, biotic, and tectonic pole-to-pole coring transect of Triassic–Jurassic Pangea. *Newsletter, International Continental Drilling Program* (Potsdam) 1:16–20.

Palfy, J., J. K. Mortensen, E. S. Carter, P. L. Smith, R. M. Friedman, and H. W. Tipper. 2000. Timing the end-Triassic mass extinction: First on land, then in the sea? *Geology* 28:39–42.

Percival, J. G. 1842. *Report on the Geology of the State of Connecticut.* New Haven, Conn.: Osborn and Baldwin.

Rampino, M. R., and R. B. Stothers. 1988. Flood basalt volcanism during the past 250 million years. *Science* 241:663–668.

Renne, P. R., Z. Zhang, M. A. Richards, M. T. Black, and A. R. Basu. 1995. Synchrony and causal relations between Permian–Triassic boundary crises and Siberian flood volcanism. *Science* 269:1413–1416.

Sepkoski, J. J., Jr. 1997. Biodiversity: Past, present, and future. *Journal of Paleontology* 71:533–539.

Talwani, M., J. Ewing, R. E. Sheridan, W. S. Holbrook, and L. Glover III. 1995. The EDGE experiment and the U.S. East Coast Magnetic Anomaly. In E. Banda, M. Torne, and M. Talwani, eds., *Rifted Ocean–Continent Boundaries,* pp. 155–181. Dordrecht: Kluwer.

Walker, F. 1940. The Palisade sill of New Jersey. *Geological Society of America Bulletin* 51:1059–1105.

Wilson, M. 1997. Thermal evolution of the central Atlantic passive margins: Continental break-up above a Mesozoic super-plume. *Journal of the Geological Society of London* 154:491–495.

Withjack, M. O., R. W. Schlische, and P. E. Olsen. 1998. Diachronous rifting, drifting, and inversion on the passive margin of central eastern North America: An analog for other passive margins. *American Association of Petroleum Geologists Bulletin* 82:817–835.

10

Flood Basalt Provinces of the Pangean Atlantic Rift: Regional Extent and Environmental Significance

J. Gregory McHone and John H. Puffer

The original extent of Hettangian Pangean rift basalts is estimated from maps of feeder dikes and Mesozoic basins that contain remnants of the basalts. Dikes and basalts across the initial Pangean rift zone are correlated by radiometric dates near 200 Ma, stratigraphy of associated basin sediments, and chemical characteristics. Intermediate-Ti quartz-normative tholeiites in northeastern North America and Morocco were derived from large NE-trending dikes that define a northern subprovince over much of modern northeastern North America, northwestern Africa, and the Iberian Peninsula, with an area approximately 2.8 × 10^6 km^2. Other quartz and olivine tholeiites comprise 2 × 10^5 km^2 of flood basalts that remain beneath the southern U.S. Coastal Plain and continental shelf, as derived from large N–S-trending and NW–SE-trending dike swarms. The southern subprovince originally extended more than 3.2 × 10^6 km^2 across the present southeastern United States, western Africa, and northern South America. Gaps in the lava sheets were likely, and the relationship of these continental basalts with subsequent ocean-crust magmatism remains unclear. The environmental impact from such enormous volumes of basalt includes cooling or greenhouse effects or both from the potential liberation in the order of 10^{12} metric tons each of CO_2 and SO_2 aerosols and of proportionally large amounts of water vapor, halides, and ash—all produced in a brief volcanic event.

Recent work has demonstrated the presence of large volcanic provinces and wedge-shaped basalt bodies that may exceed 10^6 km^3 along portions of the eastern continental margin of North America (Austin et al. 1990; Oh et al. 1995). Such basalts appear to be associated with the initial production of ocean crust during the Jurassic opening of the central Atlantic Ocean, and they explain geophysical features such as the East Coast Magnetic Anomaly (Holbrook and Keleman 1993). Although the Atlantic margin basaltic wedge is a major igneous feature, it has been difficult to discern beneath thick covers of sediment and ocean water. Possible landward counterparts to the Pangean final-rift magmas are exposed mainly as diabase dikes within the circum-Atlantic continental regions (figure 10.1) and as tholeiitic lavas preserved within sections of some early Mesozoic basins (Manspeizer 1988).

Because diabase dikes and sills are prominent locally in Triassic strata that underlie the basin basalts, stratigraphic and tectonic models commonly have assumed

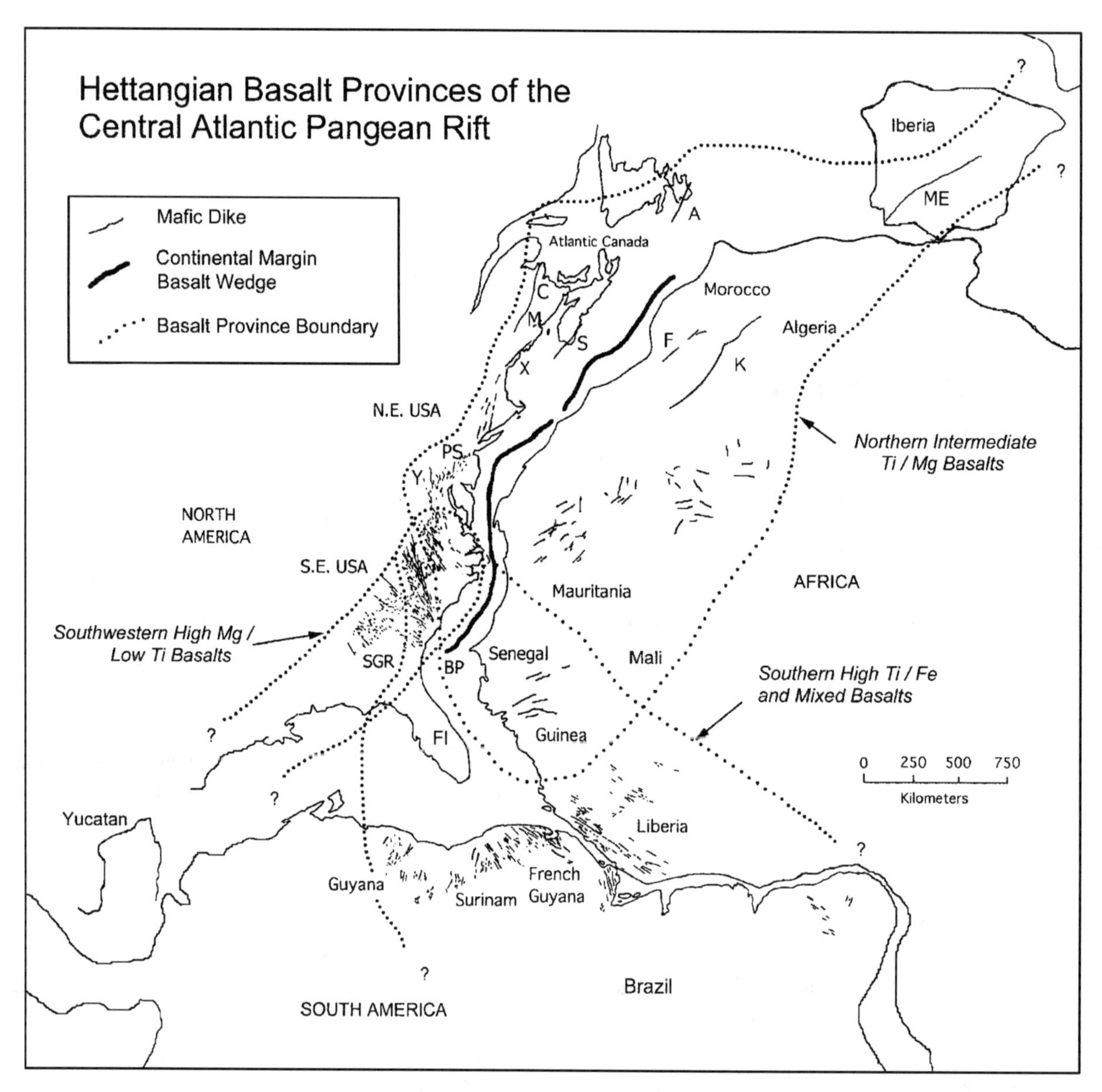

FIGURE 10.1 Areas of potential cover by Early Jurassic flood basalts around the central North Atlantic Pangean rift zone. Large northern province dikes: *A*, Avalon dike; *BP*, Blake Plateau; *C*, Caraquet dike; *F*, Foum-Zguid dike; *H*, Higganum dike; *K*, Ksi-Ksu dike; *ME*, Messejana dike; *S*, Shelburne dike; *SGR*, South Georgia rift; *X*, Christmas Cove dike. (Base information adapted from figures in de Boer et al. 1988; Bertrand 1991; and Deckart, Féraud, and Bertrand 1997)

that the basaltic lavas originated from vents within each basin. Localized sources for basalts are also implied by popular "closed basin" models for their interstratified sediments (Klein 1969; Smoot 1985). Before the fundamental work of Philpotts and Martello (1986), little connection was made between the large regional diabase dike swarms and basalts within the Mesozoic basins.

Continuing field and petrologic studies in eastern North America have shown that Mesozoic diabase dikes are individually extensive (some more than 60 m wide and 250 km long), have ages and magma types similar to those of basalt flows, and can be connected physically to the basin lavas (Philpotts and Martello 1986; Sutter 1988; McHone 1992). Because of these observations, the original extent of surface flood basalts across the Mesozoic Pangean rift terranes should be tied to the distribution of dikes rather than to the present geography of sedimentary basins. Former locations of Early Jurassic flood basalts can be estimated from maps of the dikes and modern basalts, from analyses of Late Triassic to Early Jurassic tectonism and topography, and from analogies with other flood basalt provinces. This model is limited mainly by our poor knowledge of the regional Early Jurassic topography that affected the locations, directions, and thickness of flows and by the timing and nature of both synrift and postrift erosion that reduced the basins and lavas to their present geographic pattern. In addition, it is likely that some or many of the smaller dikes did not reach the surface or made only minor contributions to the surface lavas.

Even allowing for a large error in estimating the actual sizes and geographic extent of these dikes, significant volumes of CO_2, SO_2, and particulates must have been added to the air during fissure eruptions of the Pangean rift basalts, with commensurate effects on animal and plant populations. An estimate of such atmospheric aerosols should be proportional to the effects calculated for other, better-known flood basalt events such as the Laki eruption of 1783 and the Columbia River lavas (Sigurdsson 1990).

Mesozoic Basins and Basalts

Early Mesozoic rift basins are exposed in eastern North America, Iberia, and western Africa, but more basins thought to be of this group are covered by water and/or post-Jurassic sediments of the continental shelves on both sides of the North Atlantic Ocean (Hutchinson et al. 1988). Many of the basins display a half-graben geometry, in some places as pairs of basins that are symmetrically opposed, such as the Hartford and Newark basins (Manspeizer 1988). Subbasins within the Durham basin of North Carolina, partly buried by Cretaceous sediments, would have been split into similar symmetrically opposed basins if another 500 m of erosion had occurred (Manspeizer and Gates 1995). The truncated nature of basin strata beneath the onlap of the Coastal Plain indicates that most of the tilting, uplift, and erosion of the rifted terrane was finished before the deposition of Cretaceous sediments.

From Virginia to Nova Scotia and in Morocco, the Mesozoic rift basins that preserve Jurassic strata also contain one or more horizons of quartz-normative tholeiitic basalts that correlate closely in age, stratigraphic position, chemistry, and petrography (Puffer and Philpotts 1988; Bertrand 1991; Puffer 1992; Olsen, Schlische, and Fedosh 1996). One to three different magmatic types are recognized in these rift basins, with the "best" dates ranging between 201 Ma (Dunning and Hodych 1990) and 196 Ma (Sebai et al. 1991). Stratigraphic mapping shows that the flows are only slightly above the Triassic–Jurassic boundary, or earliest Hettangian, so they provide an important marker for dating that boundary (Olsen, Fowell, and Cornet 1990).

The earliest of the basin basalts (i.e., lowest basalt stratum) is of the same magma type that also comprises most of the sills common to the Culpeper, Gettysburg, Newark, and Hartford basins (Woodruff et al. 1995), and it is the only type known within the large Fundy basin (Dostal and Greenough 1992). This magma has been labeled the Initial Pangean Rift (IPR) basalt (Puffer 1994). Along the western margin of the Fundy basin, the North Mountain Basalt is truncated by a major coast-parallel fault, but underlying Triassic sediments still remain in a few places west of the basin (Stringer and Burke 1985). Basalt is present in the Nantucket basin and is suspected to be present in other offshore basins of the Long Island platform (Hutchinson, Klitgord, and Detrick 1986). In onshore and offshore basins in Morocco, IPR basalts like these earliest units in North America also lie at the base of the Jurassic stratigraphic section (Bertrand 1991; Fiechtner, Friedrichsen, and Hammerschmidt 1992).

Basalt–Dike Correlations

Radiometric dates of Mesozoic diabase dikes and basalts have been problematic. The "best" K-Ar dates typically are scattered between 180 and 210 Ma, even for basalts and dikes that should be the same age, and many such dates have been repeated in the literature as the actual igneous cooling ages (de Boer et al. 1988). However, basin stratigraphy in eastern North America indicates that all known basaltic lavas formed within a 580,000-year period immediately after the beginning of the Jurassic period (Olsen, Schlische, and Fedosh 1996). In addition, diabase dikes within the basins are not found to crosscut any strata above the basalts, but many dikes occur in the Triassic sediments in the basins as well as within basement rocks adjacent to the basins, including those basins with no lava remnants.

All exposed basins that preserve Triassic–Jurassic boundary sediments also preserve basalts.

Several U-Pb isotopic measurements of baddeleyite and zircon of the northeastern North American sills and basalts show ages between 200 and 202 Ma (Dunning and Hodych 1990; Hodych and Dunning 1992). The U-Pb dates agree with substantial circa 201 Ma dates obtained by the $^{40}Ar/^{39}Ar$ method on dikes in eastern North America, western Africa, and northern South America, although a few other Ar dates indicate a second group of ages close to 196 Ma (Sutter 1988; Sebai et al. 1991; Fiechtner, Friedrichsen, and Hammerschmidt 1992; Deckart, Féraud, and Bertrand 1997; West and McHone 1997). These newer works supersede the numerous older, more scattered K-Ar dates of eastern North American diabase dikes and basalts, which were affected by loss or gain of Ar in unpredictable ways. An age of 200 ± 2 Ma appears likely for most, and for possibly all, of the tholeiitic Pangean rift dikes and basalts, but more radiometric work is still needed.

At the base of basaltic lavas within the basins, specific dike-to-flow locations are known in the southeastern Hartford basin (Philpotts and Martello 1986) and in the northern Culpeper basin (Woodruff et al. 1995). Basin lava vents are mapped in the Newark basin of New Jersey (Puffer and Student 1992) and in the Hartford basin in Massachusetts (Foose, Rytuba, and Sheriden 1968); in some cases, subsurface feeder dikes are not well exposed, but the vents can be related to dikes found on-trend in the region. Papezik et al. (1988) hypothesized a major fissure eruption somewhere in the southern area of the Fundy basin to account for a massive northward basalt flow, which fits a source from the Christmas Cove dike in coastal Maine (McHone 1996). In Connecticut, elongate vesicles in basalts indicate flow from southern sources in the southern Hartford basin, but, to the north, basin flows have eastern sources (Gray 1982; Ellefson and Rydell 1985). Likely feeder dikes are mapped across the source locations in both areas (Philpotts and Martello 1986).

After erosion, volcanic fissure sources for flood basalts are located by large diabase dikes, as has been demonstrated in the Columbia River Plateau Basalt group (Reidel and Tolan 1992). In eastern North America, source dikes crosscut the Triassic strata of present-day basins beneath portions of their flood basalt products, and the same or similar (comagmatic) dikes continue far outside the basins (Smith, Rose, and Lanning 1975; Philpotts and Martello 1986) (figure 10.1). Diabase dikes of the Mesozoic basins are also members of large dike swarms that are characterized by particular orientations or magma types or by both and that are found across regions widely separated from the modern basins.

The earliest (IPR) basalt flows are especially noted as having a very similar initial major- and trace-element chemistry in every northern basin (Pegram 1990; Bertrand 1991; Puffer 1994). IPR basalts (and some subsequent basalts) can be correlated with specific comagmatic dikes (table 10.1 and figure 10.2). In major-element chemistry, the TiO_2:MgO ratios of IPR magmas define a crystal fractionation line in which the feeder dikes are centered on a basalt fractionation trend (figure 10.2), which is especially well illustrated by the North Mountain Basalt of the Fundy basin. Dikes and lavas that formed soon after the IPR magmas have slightly lower TiO_2:MgO ratio trends as well as some lower large-ion lithophilic elements (Puffer and Philpotts 1988; Pegram 1990). IPR basalt and diabase average somewhat higher in Ti (at ~1.2%) than the other northern basalts, and thus they are labeled "high-Ti quartz-normative" (HTQ) basalt (Weigand and Ragland 1970). However, some dikes in West Africa and Brazil actually have much higher Ti contents (Puffer 1994).

In the southern and central areas of eastern North America, several hundred dikes of moderate to large size occur in two discrete swarms with generally NW–SE and N–S trends (Smith, Rose, and Lanning 1975; de Boer et al. 1988) (figure 10.1). In comparison, the NE–SW-trending diabase dikes of New England and Atlantic Canada are fewer, but they tend to be very large, up to 20 to 60 m wide and 75 to 400 km long (McHone 1992; Pe-Piper, Jansa, and St. J. Lambert 1992; McHone et al. 1995) (figure 10.1). Likewise, the NE-trending IPR-type Foum-Zguid and Messejana dikes in Morocco and Spain each extend more than 400 km, and the Ksi-Ksou dike in southwestern Algeria may exceed 800 km (Bertrand 1991) (figure 10.1). Portions of major lava flows from these very large dikes remain in exposed and subsurface basins of Morocco and Portugal (Fiechtner, Friedrichsen, and Hammerschmidt 1992).

The petrologic correlation among dikes and be-

TABLE 10.1 Compositions of Initial Pangean Rift Dikes and Basalts

	Source Dike								Related Basalt				
	York Haven	*Palisades Sill*	*Higganum*	*Christmas Cove*	*Ministers Island*	*Foum-Zguid*	*Messejana*		*Mt. Zion Church*	*Orange Mountain*	*Talcott*	*North Mountain*	*High Atlas*
Oxide								*Oxide*					
SiO_2	51.84	51.98	52.62	52.58	52.94	51.99	52.07	SiO_2	51.37	51.45	51.86	52.16	51.06
TiO_2	1.09	1.22	1.17	1.15	1.14	1.10	1.04	TiO_2	1.18	1.02	1.07	1.06	1.06
Al_2O_3	14.34	14.48	14.96	14.47	14.13	15.05	15.90	Al_2O_3	14.24	14.34	14.27	14.29	14.84
FeO*	9.93	10.19	10.20	10.53	10.05	9.52	9.28	FeO*	10.86	10.36	10.86	10.35	10.33
MnO	0.20	0.18	0.17	0.18	0.20	0.18	0.16	MnO	0.17	0.15	0.16	0.16	0.19
MgO	7.72	7.59	7.59	7.81	7.92	7.69	7.16	MgO	7.58	8.19	7.98	7.05	8.17
CaO	10.73	10.33	10.94	10.57	10.94	11.75	11.51	CaO	10.78	10.86	11.24	10.35	12.25
Na_2O	1.96	2.04	2.33	1.98	1.97	1.97	2.16	Na_2O	2.05	2.10	2.06	2.39	1.69
K_2O	0.60	0.84	0.56	0.60	0.55	0.60	0.58	K_2O	0.21	0.54	0.50	0.60	0.49
P_2O_5	0.12		0.20	0.13	0.14	0.15	0.14	P_2O_5	0.13	0.13		0.16	0.16

Note: Each of the numbers in the table is an average of four to 10 analyses.
Sources: Smith, Rose, and Lanning 1975; Philpotts and Martello 1986; Dunn and Stringer 1990; Bertrand 1991; Puffer 1992; McHone, unpublished data.

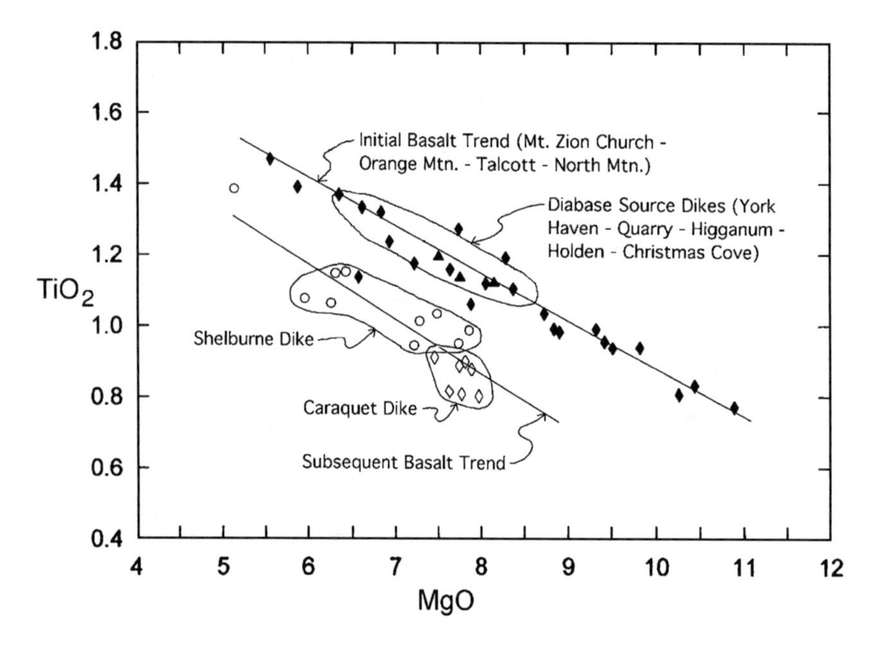

FIGURE 10.2 TiO_2:MgO plot for selected Early Jurassic diabase dikes and basalts in eastern North America. Dikes and basalts of the third flow, which generally plot in positions of high TiO_2 and low MgO, and other basalts of the final rift margin are not shown. Solid triangles represent analyses of the Christmas Cove dike of Maine (McHone, unpublished data). Solid diamonds represent North Mountain Basalt analyses from Dostal and Greenough (1992). (Other data from Smith, Rose, and Lanning 1975; Papezik and Barr 1981; Greenough and Papezik 1986; Philpotts and Martello 1986; Puffer and Philpotts 1988; and McEnroe 1989)

tween dikes and basalts is strong in the northern circum-Atlantic basalt province, but the extrapolation of flood basalts over feeder dikes is more conjectural in the southern region. Limited chemical analyses from drill cores into the subcoastal plain flood basalt near Charleston, South Carolina, indicate a magmatic source from the large quartz tholeiite dikes that extend northward through the Carolinas and Virginia (Ragland 1991), and the western sections of the buried flood basalt are physically close to and probably overlie NW–SE-trending olivine diabase dikes in Georgia. Recent work indicates a similar circa 201 Ma age for dikes in the southeastern United States (W. Hames, personal communication 2000).

Dike swarms and very large sills in western Africa and northern South America need additional correlation work with southeastern U.S. dikes, especially between swarms with similar trends (Ragland, Salters, and Parker, chapter 12 in this volume). Nevertheless, radiometric and petrologic studies (Deckart, Féraud, and Bertrand 1997) show that the magmatic history of the southern province is analogous to that of the northern province, where dikes were feeders to flood basalts. Marzoli et al. (1999) recently showed by new radiometric and chemical analyses that sills and dikes in northern and central Brazil are also mainly quartz tholeiites of circa 200 Ma age. Their major expansion beyond what is shown in figure 10.1 adds at least 2 million km^2 to the province, which Marzoli et al. (1999) have named the Central Atlantic Magmatic Province (CAMP). (An additional description of the CAMP and its geodynamic problems is given in McHone 2000.)

Mesozoic Basaltic Landforms

Massive basalt eruptions occurred along the incipient central North Atlantic Ocean rift, during or soon after the creation of the first ocean crust between 185 and 175 Ma or perhaps earlier (Holbrook and Keleman 1993). Approximately 200,000 km^2 of basalt (or possibly much more) presently underlies Cretaceous sediments across sections of the southeastern United

States and continental shelf, where it merges with a portion of the thick basalt wedge that settled into the new continental margin (Austin et al. 1990; Oh et al. 1995). The final rift basalt occurs as a wedge along most of the eastern margin of the continent, as indicated by the East Coast Magnetic Anomaly (Holbrook and Keleman 1993). Just inland from the hinge zone, ?Middle Jurassic flood basalts presently cover portions of the Georges Bank basin (Hurtubise, Puffer, and Cousminer 1987; Hutchinson et al. 1988). The petrology of these final rift basalts is poorly known relative to that of the older (?) inland basin tholeiites, but if new work can link them by chemistry and stratigraphy, their age and the age of the initial ocean crust may be closer to 201 Ma than to 175 Ma.

Inland from the Cretaceous Coastal Plain of eastern North America, no examples of Early Jurassic basalts or sills are known to be preserved in surface exposures outside the early Mesozoic basins, probably due to uplift and erosion into the surrounding metamorphic basement. However, there are areas of western Africa and northern South America in which sills and surface flows intrude and cover older basement rocks (Bertrand 1991; Deckart, Féraud, and Bertrand 1997). Figure 10.1 includes an estimate of the potential area covered by the Early Jurassic Pangean flood basalts, based on several assumptions of low to moderate relief and of flow lengths in proportion to dike sizes (illustrated mainly by northeastern North America):

1. Early Mesozoic basins originally were covered completely by basalts derived from dikes in the same regions, including basins now covered by younger sediments and offshore basins.
2. Basalts within the basins flowed from fissure dike sources that also fed basalts flowing outside the basins, as controlled by local topography.
3. Basalts flowed from 50 to 400 km or more from dike fissures across regions that are far from modern Mesozoic basins. Such distances are demonstrated by single flows mapped from source dike vents of the eastern Columbia River Plateau Basalt (Reidel and Tolan 1992).
4. Late Triassic topographic relief may have been moderate to low across large sections of the early rift zone, punctuated by particular differential uplift zones known from tectonic studies, fission track studies, and Ar isotopic studies. Low relief is indicated by large regions of meandering fluvial Triassic sediment (Hubert, Feshbach-Meriney, and Smith 1992). In addition, remnants of similar basal Triassic sediments are found in eastern Massachusetts (Kay 1983) and eastern New Brunswick (Stringer and Burke 1985). Wintsch et al. (1996) reported young mineral dates that require sediments of the Hartford basin to have been transported from source terranes that are well east of the Bronson Hill margin of the basin.
5. Major fault activity, related to the development of higher relief within the Pangean rift belt, occurred mainly after the Hettangian age. Evidence is as follows:
   a. Basin activity, shown by sedimentation rates and heat flow, accelerated during the Early Jurassic, starting with the onset of magmatism in the basins (Schlische and Olsen 1990; Roden and Miller 1991).
   b. Direct K-Ar measurements (with tests for Ar corruption) of syntectonic minerals along extensions of the Mesozoic border fault in New Hampshire range around the Middle Jurassic (Lyons and Snellenburg 1971).
   c. Mineral dates in the basins indicate a major thermal event around 175 Ma, perhaps linked to hydrothermal activity that accompanied Middle Jurassic faulting (Roden and Miller 1991).
   d. Elevated zones include the Bronson Hills terrane, which was uplifted as much as 8 km along the eastern border fault of the Hartford basin by 170 Ma (Harrison, Spear, and Heizler 1989). Jurassic uplift of the Bronson Hills is also consistent with the production of fanglomerates along the Hartford eastern border fault and with thinning and absence of strata around the Pelham dome of the Deerfield basin (Stevens and Hubert 1980).
   e. Large phaneritic plutons of the White Mountain Magma Series in central New Hampshire were emplaced mostly during post-Hettangian Early to Middle Jurassic times, and as a thermal zone the White Mountain Magma Series are consistent with uplift ad-

jacent to or connected with the Bronson Hills terrane high (figure 10.1).

f. As shown from organic maturation temperatures and fission track analyses, from 1.5 to 4 km of basin strata were eroded after Early Jurassic sedimentation ceased but before the onlap of Cretaceous Coastal Plain sediments (Pratt, Shaw, and Burruss 1988; Roden and Miller 1991).

g. Pressure-enhanced stability of minerals indicates a much greater depth of crystallization of the Higganum dike, east of the Hartford basin, relative to the adjacent but comagmatic Fairhaven dike within the basin, so that as much as 8 km of postdike differential offset is indicated (Philpotts and Martello 1986).

h. Highway construction temporarily exposed a faulted contact of basin basalt against basement metamorphic rocks on the southeastern border of the Hartford basin (Mikami and Digman 1957). Many other NE-trending high-angle faults offset the basalts within this basin (Rodgers 1985). In southern Maine, Ar mineral data are interpreted to show differential vertical faulting of several kilometers before and into Triassic time (West, Lux, and Hussey 1993), but this tectonism apparently does not coincide with the dike intrusions. NW-trending faults crosscut some of the offshore basins (Hutchinson et al. 1988) and the Minister Island dike (Stringer and Burke 1985). A small northwest fault offsets the Caraquet dike in Maine (J. G. McHone, unpublished data).

The feeder dikes cross all terranes east of the present highlands of the Appalachian Mountains, and the flood basalts would have flowed over all varieties of bedrock surfaces where allowed by topography. As shown by the timing of tectonic activity, interbasin uplift was active during and after magmatism, and basalts would have been eroded quickly from uplifted areas. Basalts and accompanying sediments outside the present-day basins must have been removed during the circa 80 Ma interval before Cretaceous sediments spread unconformably over basins within the Coastal Plain and continental shelf (Klitgord and Hutchinson 1985).

Although basalt clasts are present in the Jurassic border conglomerates of the basins, they are sparse. Basalt erodes rapidly when uplifted, and basalt clasts are not preserved well during stream transportation, as can be observed today in and around basin lavas. Sediments deposited above the basalts within the basins were derived mainly from exposed basement hills and highland source areas that were inland from the flood basalts, and their metamorphic minerals and grain fragments survived fluvial transport better than most basalt clasts.

The total thickness of Hettangian basalts varies widely among basins, but 200 m is a conservative estimate for an average thickness over the rift terrane and adjacent continental dike swarms. For the province area estimate of 6 million km^2, one-half of the surface covered by 200 m of lava requires the extrusion of 0.6 million km^3 of basalt. An equal or greater amount is preserved in the numerous large dikes and sills now exposed within the province, and it remains possible that 1 million km^3 of seaward-dipping reflector basalts are part of this same event.

Atmospheric Emissions

The only historically observed flood basalt of a large dimension was from the Laki fissure eruption in Iceland, which occurred mainly during the summer of 1783 (table 10.2). During the peak of activity, lava flowed several kilometers per day (Thorarinsson 1969). At this and other fissure eruptions, more than half of the gases that were dissolved in the magma escaped at the lava fountains of the fissure, which may have risen to 1 km or more in height (Stothers et al. 1986). Such gases typically are dominated by H_2O, CO_2, and SO_2 in tholeiitic magmas, but halogens such as Cl and F can be considerable components as well. During the summer of the Laki eruption, ash clouds reached across Iceland into northern Scotland, and for much of the summer a strange "cold fog" was observed throughout western Europe. The fog was probably caused by H_2SO_4 droplets formed by reaction of volcanic SO_2 in the atmosphere, and it reduced sunlight enough to cause midsummer frosts and widespread crop failures that led to local famines (Sigurdsson 1982; Bullard 1988). In addition, local fluorine poisoning of Icelandic livestock has been recognized (Sigurdsson 1990).

TABLE 10.2 Comparative Basalt Flows and Atmospheric Emissions

Name	*Fissures (km)*	*Age*	*Area (km^2)*	*Volume (km^3)*	*Rate (km^3/d)*	*CO_2 (m.t.)*	*SO_2 (m.t.)*
Laki	25	A.D. 1783	565	12	0.2	4×10^7	2×10^7
Roza	100	17 Ma	3,000	700	1	4×10^9	2×10^9
H-H-C	400	201 Ma	20,000	2,000	2–4	1×10^{10}	6×10^9
Total est. Pangean	10,000	201 Ma	5×10^6	1.2×10^6	10^2–10^3	6×10^{12}	3×10^{12}

Note: H-H-C, Higganum-Holden–Christmas Cove dikes and estimated lavas; m.t., metric tons.
Sources: Stothers et al. 1986; Bullard 1988; McHone 1996; Martin 1996.

Few published analyses of Pangean rift basalts include CO_2, SO_2, or halides. In table 10.2, we have assumed average emissions of CO_2 and SO_2 at fissure vents to be 0.3 and 0.15 weight percent of the tholeiitic magmas, respectively. Such amounts are roughly comparable to volatile measurements for lavas and feeder dikes at Laki and for the Miocene age Roza flow, an important member of the Columbia River Plateau Basalt group (Stothers at al. 1986). The Roza emissions are estimated to be approximately 100 times that of the Laki eruption, and most of the sulfur in its feeder dikes escaped at fissure eruption vents (Martin 1996). If the Higganum-Holden–Christmas Cove (HHC) dike system of New England had flow volumes and sulfur contents proportional to these younger flood basalts, its atmospheric emissions could have been triple those of the Roza flow, or between 10^9 and 10^{10} metric tons each of CO_2 and SO_2 from this 400 km long vent system (table 10.2). Sigurdsson (1990) calculated that 10^9 metric tons of H_2SO_4 aerosol could cause 3 to 4°C surface cooling, a major global climatic event.

Other preserved IPR basalts that probably were synchronous with the HHC-derived Talcott–North Mountain flood basalts include the Mount Zion Church flows of Virginia, the Orange Mountain (First Watchung) flows of New Jersey, and several High Atlas flows of Morocco (Puffer 1994). A total area covered by IPR basalt of at least 500,000 km^2, or roughly 25 times that of the HHC flow, has been estimated across the northern Pangean rift terrane (McHone 1996). These lavas are probably comagmatic and coeval, so their eruptive history may span only a few decades to centuries. Moreover, this lava-dike system comprises less than 10% of the total contemporaneous flood basalt province.

The total Pangean rift volcanic dike vent systems (now exposed as groups of colinear dikes around the entire Atlantic rift) can be estimated roughly as four dikes approximately 500 km long; 10 dikes approximately 200 km long; 20 dikes approximately 100 km long; and perhaps 400 dikes averaging approximately 10 km long, for a conservative estimate of roughly 10,000 km of fissure eruption vents. In addition, the Pangean dike-fissure system of figure 10.1 is a minimal depiction of global Early Jurassic volcanism in that it excludes southern Karoo volcanism (South Africa), which has been dated around 193 ± 5 Ma by K-Ar and $^{40}Ar/^{39}Ar$ methods (Fitch and Miller 1984), and also excludes contemporaneous Mesozoic volcanic events of the North American Cordilleran and Basin and Range Provinces (Gilluly 1965). At discharges proportional to the younger flood basalt eruptions, gaseous aerosols from the Pangean basalts must have totaled much greater than 10^{12} metric tons (table 10.2). Although possibly spread over several eruptive events totaling 10^5 to 10^6 years, these enormous atmospheric emissions are likely to have caused major environmental problems.

Conclusions

The evidence points to a nearly synchronous event of cogenetic fissure eruptions, with rapid production of at least 10^6 km^3 of an initial rift magma along a pre-Atlantic zone from northern South America to the northern coasts of North America and Europe, as originally assembled within Pangea. Very similar IPR dikes and basalts are found from Virginia to eastern Newfoundland, in northwestern Africa, and in Iberia (Puffer 1994). As pointed out recently (McHone 2000), the distribution of these igneous features does not support an origin from a single plume or "hot-spot" source, but their magmatic uniformity requires a huge linear zone of eutectic-like mass melting, under conditions similar to some other major flood basalt events (Anderson 1994).

The limited chemical fractionation observed within various segments of the initial basalt has been explained by ponding at the base of the crust (Pegram 1990) or within the crust (Philpotts and Martello 1986) and by fractional source melting and crystal–liquid segregation (Philpotts 1992). A model that explains some of the chemical variations between the northern and southern basalt was presented by Cummins, Arthur, and Ragland (1992), who proposed that the homogeneous northern basalts originated close to the axis of the mantle "keel" along the rift zone and farther off the axis for heterogeneous southern magmas.

The slightly younger but also voluminous basalts that followed the IPR basalt have their own distinctive characteristics and widespread distribution, adding to what must have become very large plateaulike flood basalt provinces across the rift terranes around what would become the central North Atlantic Ocean. As indicated by sediments as well as by tectonic studies, the basalt plains were interspersed with basement highlands that were eroding into the basins and surrounding areas. Basalts must have flowed around basin fault scarps and basement hills, and—where dikes crossed highlands—downhill from those regions. The model implies that stream-deposited Triassic sediments might also have been present over many areas between basins, so that we are reminded of the original Broad Terrane hypothesis of Russell (1880), in which some strata of modern basins originally were interconnected at least into Hettangian time. Postbasalt faulting and uplift then accelerated the isolation of the basins. Jurassic faulting and uplift were also responsible for the nearly complete erosion of the northern flood basalt province as well as for the preservation of its remnants within today's basins.

Enormous flood basalt eruptions across what were then the equatorial and tropical belts of the Earth could have had a catastrophic effect on terrestrial lifeforms. The Laki eruption showed that H_2SO_4 haze can reduce sunlight significantly, providing evidence that the problem would have been many times more severe for the Pangean fissure eruptions. Based on stratigraphic relationships of lavas that could be far from their fissure sources, Olsen, Fowell, and Cornet (1990) concluded that the IPR basalts are slightly too young to be related to the Triassic–Jurassic mass-extinction event. They instead proposed an asteroid impact as the cause of the Triassic–Jurassic extinction, but their suggested Quebec Manicouagan event is now known to be too old (Hodych and Dunning 1992). Although Rampino and Stothers (1988) proposed that flood basalts can be triggered by large impacts, the simpler assumption is that major basaltic eruptions cause extinctions by themselves (Stothers 1993).

Fossil and stratigraphic data more recently described by McElwain, Beerling, and Woodward (1999) and by Palfy et al. (2000) support the proposal that CAMP eruptions could have caused or contributed to the Triassic–Jurassic mass extinction. The model of Palfy et al. (2000) suggests that a terrestrial extinction preceded a marine extinction, which might be related to different times of CAMP volcanism. Because volcanism in the northern portion of the pre-Atlantic rift terranes occurred so close in time to the terrestrial extinction event, slightly older flood basalt events in other portions of the CAMP remain a potential cause for the end-Triassic ecological catastrophe.

Acknowledgments

We appreciate discussions with and contributions from Hervé Bertrand, Norman Gray, Paul Olsen, Anthony Philpotts, Paul Ragland, and Robert Wintsch. The manuscript for this chapter was improved by editorial suggestions from Peter M. LeTourneau, Paul Ragland, and Richard Tollo.

Literature Cited

Anderson, D. L. 1994. The sublithospheric mantle as a source of continental flood basalts: The case against the continental lithosphere and plume head reservoirs. *Earth and Planetary Science Letters* 123:269–280.

Austin, J. A., Jr., P. L. Stoffa, J. D. Phillips, J. Oh, D. S. Sawyer, G. M. Purdy, E. Reiter, and J. Makris. 1990. Crustal structure of the southeast Georgia embayment–Carolina trough: Preliminary results of a composite seismic image of a continental suture (?) and a volcanic passive margin. *Geology* 18:1023–1027.

Bertrand, H. 1991. The Mesozoic tholeiitic province of northwest Africa: A volcano-tectonic record of the early opening of the central Atlantic. In A. B. Kampunzu and R. T. Lubala, eds., *Magmatism in Extensional Structural Settings: The Phanerozoic African Plate,* pp. 147–191. New York: Springer.

Bullard, F. M. 1988. *Volcanoes of the Earth.* Austin: University of Texas Press.

Cummins, L. D., J. D. Arthur, and P. C. Ragland. 1992. Classification and tectonic implications of Early Mesozoic magma types of the circum-Atlantic. In J. H. Puffer and P. C. Ragland, eds., *Eastern North American Mesozoic Magmatism,* pp. 119–136. Geological Society of America Special Paper, no. 268. Boulder, Colo.: Geological Society of America.

de Boer, J. Z., J. G. McHone, J. H. Puffer, P. C. Ragland, and D. Whittington. 1988. Mesozoic and Cenozoic magmatism. In R. E. Sheridan and J. A. Grow, eds., *The Atlantic Continental Margin,* pp. 217–241. Vol. I-2 of *The Geology of North America.* Boulder, Colo.: Geological Society of America.

Deckart, K., G. Féraud, and H. Bertrand. 1997. Age of Jurassic continental tholeiites of French Guyana/Surinam and Guinea: Implications to the initial opening of the central Atlantic Ocean. *Earth and Planetary Science Letters* 150:205–220.

Dostal, J., and J. D. Greenough. 1992. Geochemistry and petrogenesis of the early Mesozoic North Mountain basalts of Nova Scotia, Canada. In J. H. Puffer and P. C. Ragland, eds., *Eastern North American Mesozoic Magmatism,* pp. 149–159. Geological Society of America Special Paper, no. 268. Boulder, Colo.: Geological Society of America.

Dunn, T., and P. Stringer. 1990. Petrology and petrogenesis of the Ministers Island dike, southwest New Brunswick, Canada. *Contributions to Mineralogy and Petrology* 105:55–65.

Dunning, G. R., and J. P. Hodych. 1990. U/Pb zircon and baddeleyite ages for the Palisades and Gettysburg sills of the northeastern United States: Implications for the age of the Triassic/Jurassic boundary. *Geology* 18:795–798.

Ellefson, K. J., and P. L. Rydel. 1985. Flow directions of the Hampden basalt in the Hartford basin, Connecticut and Massachusetts. *Northeastern Geology* 7:33–36.

Fiechtner, L., H. Friedrichsen, and K. Hammerschmidt. 1992. Geochemistry and geochronology of early Mesozoic tholeiites from central Morocco. *Geologische Rundschau* 81:45–62.

Fitch, F. J., and J. A. Miller. 1984. Dating Karoo igneous rocks by the conventional K-Ar and $^{40}Ar/^{39}Ar$ age spectrum methods. *Geological Society of South Africa Special Paper* 13:247–266.

Foose, R. M., J. J. Rytuba, and M. F. Sheriden. 1968. Volcanic plugs in the Connecticut Valley Triassic near Mt. Tom, Massachusetts. *Geological Society of America Bulletin* 79:1655–1662.

Gilluly, J. 1965. *Volcanism, Tectonism, and Plutonism in the Western United States.* Geological Society of America Special Paper, no. 80. Boulder, Colo.: Geological Society of America.

Gray, N. H. 1982. Mesozoic volcanism in north-central Connecticut. In R. Joesten and S. S. Quarrier, eds., *Guidebook for Fieldtrips in Connecticut and South-Central Massachusetts,* pp. 173–190. State Geological and Natural History Survey of Connecticut Guidebook, no. 5. Hartford: Connecticut Department of Environmental Protection.

Greenough, J. D., and V. S. Papezik. 1986. Petrology and geochemistry of the early Mesozoic Caraquet dyke, New Brunswick, Canada. *Canadian Journal of Earth Sciences* 23:193–201.

Harrison, T. M., F. S. Spear, and M. T. Heizler. 1989. Geochronologic studies in central New England II: Post-Acadian hinged and differential uplift. *Geology* 17:185–189.

Hodych, J. P., and G. R. Dunning. 1992. Did the Manicouagan impact trigger end-of-Triassic mass extinction? *Geology* 20:51–54.

Holbrook, W. S., and P. B. Kelemen. 1993. Large igneous province on the U.S. Atlantic margin and implications for magmatism during continental breakup. *Nature* 364:433–436.

Hubert, J. F., P. E. Feshbach-Meriney, and M. A. Smith. 1992. The Triassic–Jurassic Hartford rift basin, Connecticut and Massachusetts: Evolution, sandstone diagenesis, and hydrocarbon history. *American Association of Petroleum Geologists Bulletin* 76:1710–1734.

Hurtubise, D. O., J. H. Puffer, and H. L. Cousminer. 1987. An offshore Mesozoic igneous sequence, Georges Bank basin, North Atlantic. *Geological Society of America Bulletin* 98:430–438.

Hutchinson, D. R., K. D. Klitgord, and R. S. Detrick. 1986. Rift basins of the Long Island platform. *Geological Society of America Bulletin* 97:688–702.

Hutchinson, D. R., K. D. Klitgord, M. W. Lee, and A. M. Trehu. 1988. U.S. Geological Survey deep seismic reflection profile across the Gulf of Maine. *Geological Society of America Bulletin* 100:172–184.

Kay, C. A. 1983. Discovery of a Late Triassic basin north of Boston and some implications as to post-Paleozoic tectonics in northeastern Massachusetts. *American Journal of Science* 283:1060–1079.

Klein, G. deV. 1969. Deposition of Triassic sedimentary rocks in separate basins, eastern North America. *Geological Society of America Bulletin* 80:1825–1832.

Klitgord, K. D., and D. R. Hutchinson. 1985. Distribution and geophysical signatures of early Mesozoic rift basins beneath the U.S. Atlantic continental margin. In G. R. Robinson Jr. and A. J. Froelich, eds., *Proceedings of the Second U.S. Geological Survey Workshop on the Early Mesozoic Basins of the Eastern United States,* pp. 45–53. U.S. Geological Survey Circular, no. 946. Washington, D.C.: Government Printing Office.

Lyons, J. B., and J. Snellenburg. 1971. Dating faults. *Geological Society of America Bulletin* 82:1749–1752.

Manspeizer, W. 1988. Triassic–Jurassic rifting and opening of the Atlantic: An overview. In W. Manspeizer, ed., *Triassic–Jurassic Rifting: Continental Breakup and the Opening of the Atlantic Ocean and the Passive Margins,* pp. 41–79. Developments in Geotectonics, no. 22. Amsterdam: Elsevier.

Manspeizer, W., and A. E. Gates. 1995. Episodic rifting and basin inversion: Permo-Triassic events. *Geological Society of America, Abstracts with Programs* 27:66–67.

Martin, B. S. 1996. Sulfur in flows of the Wampum basalt formation, Columbia River basalt group: Implications for volatile emissions accompanying the emplacement of large igneous provinces. *Geological Society of America, Abstracts with Programs* 28:A419

Marzoli, A., P. R. Renne, E. M. Piccirillo, M. Ernesto, G. Bellieni, and A. De Min. 1999. Extensive 200-million-year-old continental flood basalts of the Central Atlantic Magmatic Province. *Science* 284:616–618.

McElwain, J. C., D. J. Beerling, and F. I. Woodward. 1999. Fossil plants and global warming at the Triassic–Jurassic boundary. *Science* 285:1386–1390.

McEnroe, S. A. 1989. *Paleomagnetism and Beochemistry of Mesozoic Dikes and Sills in West-Central Massachusetts.* Department of Geology and Geography Contribution, no. 64. Amherst: University of Massachusetts.

McHone, J. G. 1992. Mafic dike suites within Mesozoic igneous provinces of New England and Atlantic Canada. In J. H. Puffer and P. C. Ragland, eds., *Eastern North American Mesozoic Magmatism,* pp. 1–11. Geological Society of America Special Paper, no. 268. Boulder, Colo.: Geological Society of America.

McHone, J. G. 1996. Broad-terrane Jurassic flood basalts across northeastern North America. *Geology* 24:319–322.

McHone, J. G. 2000. Non-plume magmatism and tectonics during the opening of the central Atlantic Ocean. *Tectonophysics* 316:287–296.

McHone, J. G., D. P. West Jr., A. M. Hussey II, and N. W. McHone. 1995. The Christmas Cove dike, coastal Maine: Petrology and regional significance. *Geological Society of America, Abstracts with Programs* 27:67–68.

Mikami, H. M., and R. E. Digman. 1957. *The Bedrock Geology of the Guilford 15-Minute Quadrangle and a Portion of the New Haven Quadrangle.* State Geological and Natural History Survey of Connecticut Bulletin, no. 86. Hartford: State of Connecticut.

Oh, J., J. A. Austin Jr., J. D. Phillips, M. F. Coffin, and P. L. Stoffa. 1995. Seaward-dipping reflectors offshore the southeastern United States: Seismic evidence for extensive volcanism accompanying sequential formation of the Carolina trough and Blake Plateau basin. *Geology* 23:9–12.

Olsen, P. E., S. J. Fowell, and B. Cornet. 1990. The Triassic/Jurassic boundary in continental rocks of eastern North America: A progress report. In V. L. Sharpton and P. D. Ward, eds., *Global Catastrophes in Earth History,* pp. 585–593. Geological Society of America Special Paper, no. 247. Boulder, Colo.: Geological Society of America.

Olsen, P. E., R. W. Schlische, and M. S. Fedosh. 1996. 580 ky duration of the Early Jurassic flood basalt event in eastern North America estimated using Milankovich cyclostratigraphy. In M. Morales, ed., *The Continental Jurassic,* pp. 11–22. Museum of Northern Arizona Bulletin, no. 60. Flagstaff: Museum of Northern Arizona.

Palfy, J., J. K. Mortensen, E. S. Carter, P. L. Smith, R. M. Friedman, and H. W. Tipper. 2000. Timing the end-Triassic mass extinction: First on land, then in the sea? *Geology* 28:39–42.

Papezik, V. S., and S. M. Barr. 1981. The Shelburne dike, an early Mesozoic diabase dike in Nova Scotia: Mineralogy, petrology, and regional significance. *Canadian Journal of Earth Sciences* 18:1346–1355.

Papezik, V. S., J. D. Greenough, J. A. Colwell, and T. J. Mallinson. 1988. North Mountain Basalt from Digby, Nova Scotia: Models for a fissure eruption from stratigraphy and petrochemistry. *Canadian Journal of Earth Sciences* 25:74–83.

Pegram, W. J. 1990. Development of continental lithospheric mantle as reflected in the chemistry of the Mesozoic Appalachian tholeiites, U.S.A. *Earth and Planetary Science Letters* 97:316–331.

Pe-Piper, G., L. F. Jansa, and R. St. J. Lambert. 1992. Early Mesozoic magmatism on the eastern Canada margin: Petrogenetic and tectonic significance. In J. H. Puffer and P. C. Ragland, eds., *Eastern North American Mesozoic Magmatism,* pp. 13–36. Geological Society of America Special Paper, no. 268. Boulder, Colo.: Geological Society of America.

Philpotts, A. R. 1992. A model for emplacement of magma in the Mesozoic Hartford basin. In J. H. Puffer and P. C. Ragland, eds., *Eastern North American Mesozoic Magmatism,* pp. 137–148. Geological Society of America Special Paper, no. 268. Boulder, Colo.: Geological Society of America.

Philpotts, A. R., and A. Martello. 1986. Diabase feeder dikes for the Mesozoic basalts in southern New England. *American Journal of Science* 286:105–126.

Pratt, L. M., C. A. Shaw, and R. C. Burruss. 1988. Thermal histories of the Hartford and Newark basins inferred from maturation indices of organic matter. In A. J. Froelich and G. R. Robinson Jr., eds., *Studies of the Early Mesozoic Basins of the Eastern United States,* pp. 58–63. U.S. Geological Survey Bulletin, no. 1776. Washington, D.C.: Government Printing Office.

Puffer, J. H. 1992. Eastern North American flood basalts in the context of the incipient breakup of Pangaea. In J. H. Puffer and P. C. Ragland, eds., *Eastern North American Mesozoic Magmatism,* pp. 95–118. Geological Society of America Special Paper, no. 268. Boulder, Colo.: Geological Society of America.

Puffer, J. H. 1994. Initial and secondary Pangean basalts. In B. Beauchamp, A. F. Embry, and D. Glass, eds., *Pangea: Global Environments and Resources,* pp. 85–95. Canadian Society of Petroleum Geologists Memoir, no. 17. Calgary: Canadian Society of Petroleum Geologists.

Puffer, J. H., and A. R. Philpotts. 1988. Eastern North American quartz tholeiites: Geochemistry and petrology. In W. Manspeizer, ed., *Triassic–Jurassic Rifting: Continental Breakup and the Opening of the Atlantic Ocean and the Passive Margins,* pp. 579–605. Developments in Geotectonics, no. 22. Amsterdam: Elsevier.

Puffer, J. H., and J. J. Student. 1992. Volcanic structures, eruptive style, and post-eruptive deformation and chemical alteration of the Watchung flood basalts, New Jersey. In J. H. Puffer and P. C. Ragland, eds., *Eastern North American Mesozoic Magmatism,* pp. 261–278. Geological Society of America Special Paper, no. 268. Boulder, Colo.: Geological Society of America.

Ragland, P. C. 1991. Mesozoic igneous rocks. In J. W. Horton Jr. and V. A. Zullo, eds., *Geology of the Carolinas: Carolina Geological Society 50th Anniversary Volume,* pp. 171–190. Knoxville: University of Tennessee Press.

Rampino, M. R., and R. B. Stothers. 1988. Flood basalt volcanism during the past 250 million years. *Science* 241:663–668.

Reidel, S. P., and T. L. Tolan. 1992. Eruption and emplacement of flood basalt: An example from the large-volume Teepee Butte Member, Columbia River Basalt group. *Geological Society of America Bulletin* 104:1650–1671.

Roden, M. K., and D. S. Miller. 1991. Tectono-thermal history of Hartford, Deerfield, Newark, and Taylorsville basins, eastern United States, using fission-track analysis. *Schweizische Mineralogische und Petrographische Mitteillangen* 71:187–203.

Rodgers, J. 1985. *Bedrock Geological Map of Connecticut* [scale 1:125,000]. Hartford: State Geological and Natural History Survey of Connecticut, Connecticut Department of Environmental Protection.

Russell, I. C. 1880. On the former extent of the Triassic formation of the Atlantic states. *American Naturalist* 14:703–712.

Schlische, R. W., and P. E. Olsen. 1990. Quantitative filling model for continental extensional basins with applications to early Mesozoic rifts of eastern North America. *Journal of Geology* 98:135–155.

Sebai, A., G. Féraud, H. Bertrand, and J. Hanes. 1991. $^{40}Ar/^{39}Ar$ dating and geochemistry of tholeiitic magmatism related to the early opening of the central Atlantic rift. *Earth and Planetary Science Letters* 104:455–472.

Sigurdsson, H. 1982. Volcanic pollution and climate: The 1783 Laki eruption. *Eos: Transactions of the American Geophysical Union* 63:601–602.

Sigurdsson, H. 1990. Assessment of the atmospheric impact of volcanic eruptions. In V. L. Sharpton and P. D. Ward, eds., *Global Catastrophes in Earth History,* pp. 99–122. Geological Society of America Special Paper, no. 247. Boulder, Colo.: Geological Society of America.

Smith, R. C., II, A. W. Rose, and R. M. Lanning. 1975. Geology and geochemistry of Triassic diabase in Pennsylvania. *Geological Society of America Bulletin* 86:943–955.

Smoot, J. P. 1985. The closed-basin hypothesis and its use in facies analysis of the Newark Supergroup. In G. R.

Robinson Jr. and A. J. Froelich, eds., *Proceedings of the Second U.S. Geological Survey Workshop on the Early Mesozoic Basins of the Eastern United States,* pp. 4–10. U.S. Geological Survey Circular, no. 946. Washington, D.C.: Government Printing Office.

Stevens, R. L., and J. F. Hubert. 1980. Alluvial fans, braided rivers, and lakes in a fault-bounded semiarid rift valley: Sugarloaf Arkose (Late Triassic–Early Jurassic), Newark Supergroup, Deerfield basin, Massachusetts. *Northeastern Geology* 2:100–117.

Stothers, R. B. 1993. Flood basalts and extinction events. *Geophysical Research Letters* 20:1399–1402.

Stothers, R. B., J. A. Wolff, S. Self, and M. R. Rampino. 1986. Basaltic fissure eruptions, plume heights, and atmospheric aerosols. *Geophysical Research Letters* 13:725–728.

Stringer, P., and K. B. S. Burke. 1985. *Structure in Southwest New Brunswick: Excursion 9.* Field Guide for the Annual Meeting. Ottawa: Geological Association of Canada and Mineralogy Association of Canada.

Sutter, J. F. 1988. Innovative approaches to the dating of igneous events in the early Mesozoic basins of the eastern United States. In A. J. Froelich and G. R. Robinson Jr., eds., *Studies of the Early Mesozoic Basins of the Eastern United States,* pp. 194–200. U.S. Geological Survey Bulletin, no. 1776. Washington, D.C.: Government Printing Office.

Thorarinsson, S. 1969. The Lakagigar eruption of 1783. *Bulletin of Volcanology* 33:910–929.

Weigand, P. W., and P. C. Ragland. 1970. Geochemistry of Mesozoic dolerite dikes from eastern North America. *Contributions to Mineralogy and Petrology* 29:195–214.

West, D. P., Jr., D. R. Lux, and A. M. Hussey II. 1993. Contrasting thermal histories across the Flying Point fault, southwestern Maine: Evidence for Mesozoic displacement. *Geological Society of America Bulletin* 105:1478–1490.

West, D. P., Jr., and J. G. McHone. 1997. Timing of Early Jurassic "feeder" dike emplacement, northern Appalachians: Evidence for synchronicity with rift basalts. *Geological Society of America, Abstracts with Programs* 29:88.

Wintsch, R. P., M. J. Kunk, J. L. Boyd, and B. P. Hellickson. 1996. Permian and early Mesozoic exhumation of Bronson Hill terrane rocks: Significance for the tectonics of the Hartford basin. In P. M. LeTourneau and P. E. Olsen, eds., *Aspects of Triassic–Jurassic Rift Basin Geoscience: Abstracts,* p. 57. State Geological and Natural History Survey of Connecticut Miscellaneous Reports, no. 1. Hartford: Connecticut Department of Environmental Protection.

Woodruff, L. G., A. J. Froelich, H. E. Belkin, and D. Gottfried. 1995. Evolution of tholeiitic diabase sheet systems in the eastern United States: Examples from the Culpeper basin, Virginia–Maryland, and the Gettysburg basin, Pennsylvania. *Journal of Volcanological and Geothermal Research* 64:143–169.

11

Geochemistry and Origin of Pangean and Rodinian Continental Flood Basalts

John H. Puffer

The rifting of the Pangean supercontinent during the Mesozoic was a rare geologic event. It was preceded, however, by the rifting of the Rodinian supercontinent during the late Proterozoic. In both cases, continental flood basalts (CFBs) were extruded along major portions of intracratonic rifts. With the quantity of geochemical analytical data presently available on a worldwide basis, it now is possible to compare Pangean and Rodinian basalts with each other and with the flood basalts that are not associated with the fragmentation of supercontinents. Such a comparison shows that there are major differences in the composition of most Pangean CFBs compared with most Rodinian CFBs. Most Pangean CFB provinces, particularly those associated with the early rifting stage of fragmentation, are characterized by uniform chemistry, with average TiO_2 contents ranging from 0.9 to 1.2 weight percent and MgO contents ranging from 7.1 to 8.1 weight percent. Most Rodinian CFB provinces are characterized by more diverse chemistry, with average TiO_2 contents ranging from 1.1 to 3.2 weight percent and MgO contents ranging from 5.2 to 9.2 weight percent. Most Rodinian CFBs resemble Cenozoic CFBs that display good evidence of genetic relationships to hot spots and share several of the characteristics of ocean island basalts. Apparently, Rodinia was disassembled by new upwelling mantle hot spots. In contrast, most initial Pangean CFBs share many of the geochemical characteristics of arc volcanics and may have been derived from a similar source, although under different tectonic conditions. Pangean CFB magmatism probably was initiated by global plate-reorganization processes largely along previously developed sutures.

Several major continental flood basalt (CFB) provinces, including the Eastern North American Province (McHone 1996; McHone and Puffer, chapter 10 in this volume), were extruded during the rifting phase of Pangean disassembly. In other work (Puffer 1992, 1994), I showed that most early rift-related Pangean CFBs share a set of geochemical characteristics that differ from those of most late Pangean and hot-spot–related CFBs. I also suggested (Puffer 1992, 1994) that these characteristics may be related to the rare magnitude of rifting that took place along major new plate boundaries as the plates started to separate. Rogers (1996), however, has shown that such processes are not unique in the geologic record because Pangea was preceded by the disassembly of Rodinia during the end of the Proterozoic. The purpose of this chapter, therefore, is to compare Pangean with Rodinian CFB magmatism and to determine the extent to which rift-related Pangean-type CFBs may have also extruded onto Rodinia.

Global Database

The size of the geochemical database pertaining to CFBs on a global basis has reached the point where its presentation on geochemical variation diagrams such as those presented in figures 11.1 through 11.4 generates an indecipherable blur of data points. The use of average compositions pertaining to individual basalt provinces is more manageable, although some averages tend to be skewed by minor or unrepresentative flows. In some cases, more than one distinctive magma type is found in the same province. Therefore, for purposes of comparison, unless a CFB province is compositionally uniform throughout or does not display any systematic top-to-bottom variation, the average values presented in tables 11.1 to 11.3 and in figure 11.1 represent either the initial outpourings of magma or the first thick unit of uniform composition. The first thick flows are more likely to display the geochemical characteristics of rift-related CFB volcanism than later flows that may have extruded during the drifting stage of continental separation.

The database (tables 11.1–11.3) is confined to continental extensional provinces and excludes provinces unrelated either to rifting that opened into new ocean basins or to failed rifting within plate interiors. For example, the Jurassic Kirkpatrick Basalt of Antarctica (Mensing et al. 1984) and the Triassic Wrangellia Basalt of Vancouver Island (Lassiter, DePaolo, and Mahoney 1995) have been described as flood basalts or oceanic flood basalts but were extruded onto leading plate edges and may be related to compressional tectonic settings. Therefore, both were excluded from the database considered here. Tables 11.1 to 11.3 and figure 11.1 include most well-known Pangean and Rodinian CFBs as well as a few of the most thoroughly analyzed pre-Rodinian and post-Pangean CFBs for comparison. In addition, a few representative top-to-bottom CFB geochemical distributions are presented in figures 11.2 and 11.3.

North America

Pangean CFBs

Eastern North America. McHone and Puffer (chapter 10 in this volume) describe two widespread Early Jurassic flood basalt populations in eastern North America. The initial extrusions include the North Mountain Basalt of Nova Scotia, the Talcott Basalt of Connecticut, the Orange Mountain Basalt of New Jersey, and the Mount Zion Church Basalt of Virginia (table 11.1; figure 11.1). These initial extrusions are high-Ti (TiO_2) quartz-normative tholeiites (HTQ type of Weigand and Ragland 1970). Secondary extrusions include the Holyoke Basalt of Connecticut, the Preakness Basalt of New Jersey, and the Sander Basalt of Virginia. These secondary basalts are low-TiO_2 quartz-normative tholeiites (LTQ type of Weigand and Ragland 1970) characterized by some depletion of incompatible elements, particularly high field-strength elements (HFSE), compared with the HTQ type. The TiO_2 and MgO range of the entire Watchung district of New Jersey, which includes both types of flood basalt, is illustrated in figure 11.2a. The TiO_2 and MgO of the Watchung Basalts are distributed as parallel fractionation patterns controlled by pyroxene and plagioclase (Puffer and Philpotts 1988; Husch 1992).

Rodinian CFBs

Appalachia. Widespread flood basalt and rift basalt occurrences of Late Proterozoic to early Paleozoic age are found in the Appalachian Mountains of North America along the margins of the Mesozoic basins. These occurrences include the Belle Isle Basalts of Labrador and the Cloud Mountain flood basalts of Newfoundland (Strong 1974), the Tibbit Hill Basalts of Vermont (Coish et al. 1985), the Rensselaer Basalts of New York (Pestana 1985), and the Catoctin (Appalachian) Basalts of Virginia (Reed and Morgan 1971) (figure 11.3a; table 11.2). The chemical composition of these Appalachian basalts is more diverse than that of the Pangean basalts that structurally overly them and, in particular, contain higher concentrations of HFSEs such as TiO_2. None of the Appalachian districts show any systematic top-to-bottom variation in chemistry. For example, the basal part of the 500 m thick Interstate 64 section through the Catoctin Basalt of

TABLE 11.1 Pangean Flood Basalts

	HTQ Large Igneous Province						Other Pangean		
Name	*Orange Mt.*[a]	*Talcott*[b]	*Mt. Zion*[c]	*North Mt.*[d]	*High Atlas*[e]	*Algarve*[f]	*South Paraná*[g]	*Norilsk*[h]	*Lesotho*[i]
Location	*N.J.*	*Conn.*	*Va.*	*Nova Scotia*	*Morocco*	*Portugal*	*Brazil*	*Siberia*	*South Africa*
Age (Ma ± 10)	*200*	*200*	*200*	*200*	*200*	*200*	*140*	*250*	*200*
Samples	*7*	*7*	*18*	*6*	*14*	*2*	*12*	*58*	*49*
Oxides*									
SiO_2	52.09	51.79	52.48	53.72	52.8	51.9	51.99	51.85	52.08
TiO_2	1.18	1.07	1.16	1.08	1.19	1.01	0.94	1.15	0.96
Al_2O_3	14.58	14.25	14.58	13.77	14.38	14.44	15.84	15.64	15.87
FeO†	10.02	10.84	10.14	8.98	9.57	10.36	9.82	10.1	9.97
MnO	0.16	0.16	0.18	0.16	0.16	0.16	0.16	0.18	0.16
MgO	7.93	7.97	7.79	8.08	8.13	7.91	7.4	7.42	7.09
CaO	10.92	11.23	10.46	10.36	10.98	11.23	10.76	10.44	10.81
Na_2O	2.49	2.06	2.81	2.06	2.1	2.31	2.18	2.24	2.19
K_2O	0.49	0.5	0.27	0.81	0.52	0.54	0.76	0.92	0.71
P_2O_5	0.14	0.13	0.13	0.09	0.17	0.14	0.15	0.06	0.16
ppm									
Elements									
Cr	337	322	282	225	240	241	—	130	283
Ni	85	86	79	64	87	87	—	120	94
Sr	192	186	191	169	187	235	236	240	192
Zr	101	87	99	96	103	98	96	—	94
La	10.9	11.1	10.8	—	10.6	14.6	14	16.5	9.9
Ce	25.1	23.9	24.1	—	22.5	29.7	—	36	23.7

* Weight percent normalized to 100% anhydrous.
† Total Fe as FeO.
[a] Puffer 1992.
[b] Puffer et al. 1981.
[c] Tollo 1988.
[d] Papezik et al. 1988.
[e] Bertrand, Dostal, and Dupuy 1982.
[f] Puffer, unpublished.
[g] Bellieni et al. 1984.
[h] Nesterenko, Avilova, and Smithova 1964.
[i] Cox and Hornung 1966.

Virginia is chemically about the same as the upper flow, although there are random variations through the middle (Badger and Sinha 1988). Some Appalachian districts display fewer flood basalt characteristics than the Catoctin, such as the Rensselaer, which probably flowed into a rift-bounded early Iapetan marine basin (Ratcliffe 1987a), but age, tectonic, and compositional similarities (figure 11.3a) suggest derivation from similar sources.

Most Rodinian Appalachian diabase dike swarms also share these geochemical characteristics, including the Huntington dolerites of Vermont (Coish et al. 1985), the Hudson Highlands dolerites of New York (Ratcliffe 1987b), the diabase dikes of the New Jersey Highlands (Volkert and Puffer 1995) (table 11.2), and the Bakersville dike swarm of North Carolina, Tennessee, and Virginia (Goldberg, Butler, and Fullagar 1986). Each of these occurrences has been interpreted as being related to rifting associated with the opening of the Iapetus Ocean. The timing of the opening of the Iapetus, however, is controversial. For example, the Bakersville dike swarm generally is accepted as representing early stages of rifting and is dated at 734 ± 26 Ma (Goldberg, Butler, and Fullagar 1986). The Catoc-

TABLE 11.2 Rodinian Flood Basalts

Location	*Belle Isle*[a] *Labrador*	*Cloud Mt.*[a] *Newfound*	*Tibbit Hill*[b] *Vermont*	*Rensselaer*[c] *New York*	*Highlands*[d] *New Jersey*	*Catoctin*[e] *Virginia*	*South Grenville*[f] *Canada*	*Bolshaya Kugnamka River*[g] *Siberia*	*Cuddapan*[h] *India*	*Antrim Plateau*[i] *Australia*	*Bømlo*[j] *Norway*
Age (Ma ± 20)	*600*	*600*	*600?*	*600?*	*600+*	*570*	*590*	*Late Proterozoic*	*Late Proterozoic*	*600*	*550?*
Samples	*21*	*11*	*9*	*10*	*35*	*7*	*50*	*3*	*4*	*18*	*1*
Oxides*											
SiO_2	50.96	49.93	49.72	49.28	50.27	50.31	50.44	50.69	48.97	54.33	49.39
TiO_2	2.08	2.14	3.18	2.66	3.04	2.68	2	2.61	1.72	1.08	2.19
$Al2O_3$	14.64	14.86	15.7	14.22	13.48	14.94	15	13.36	12.83	15.07	17.96
FeO†	12.27	13.38	12.85	13.66	13.86	13.51	12.52	13.21	15.62	9.82	10.93
MnO	0.22	0.23	0.11	0.22	0.23	0.26	—	0.05	0.53	0.16	0.2
MgO	6.05	6.37	5.97	6.92	5.19	6.34	6.28	5.88	9.24	6.58	5.29
CaO	9.21	8.31	6.82	9.61	8.75	7.12	10.52	9.44	7.74	9.02	9.03
Na_2O	3.17	3.21	4.56	2.73	3.02	4.08	2.43	3.02	2.67	2.59	4.15
K_2O	1.13	1.35	0.7	0.41	1.46	0.44	0.57	1.33	0.4	1.24	0.7
P_2O_5	0.27	0.22	0.39	0.29	0.7	0.32	0.24	0.41	0.28	0.11	0.16
Ppm											
Elements											
Cr	119	69	95	123	58	77	129	—	181	—	146
Ni	60	66	80	121	26	37	88	—	105	—	55
Sr	227	204	138	365	320	150	230	—	175	—	294
Zr	143	137	218	164	238	120	137	—	90	—	220
La	—	—	23	—	29	24	—	—	—	—	19.16
Ce	—	—	49	—	55.5	—	—	—	—	—	—

* Weight percent normalized to 100% anhydrous.
† Total Fe as FeO.
[a] String 1974.
[b] Cuish et al. 1985.
[c] Pestana 1985.
[d] Volkert and Puffer 1995.
[e] Reed and Morgan 1971.
[f] St. Seymour and Kumarapeli 1995.
[g] Belyakov et al. 1970.
[h] Puffer and Pamganamamula, in progress.
[i] BuHitude 1976.
[j] Gale and Pearce 1982.

TABLE 11.3 Pre-Rodinian and Post-Pangean Flood and Rift Basalts and Calc-alkaline Basalt

	Servilleta[a]	*Gr Ronde*[b]	*East Rift*[c]	*Antrim*[d]	*Skye*[e]	*Deccan*[f]	*Zig Zag Dal*[g]	*Keweenawan*[h]	*Calc-alkaline*[i]
Location	*New Mexico*	*Washington*	*Ethiopa*	*Ireland*	*Scotland*	*India*	*Greenland*	*Lake Superior*	*Southern Andes*
Age (Ma ± 30)	*2 4.5*	*15*	*20*	*60*	*60*	*65*	*1230*	*1100*	*<10?*
Samples	*6*	*72*	*81*	*9*	*12*	*44*	*1*	*5*	*1*
Oxides*									
SiO_2	51.44	54.83	50.81	52.29	48.7	51.03	52.14	48.85	50.43
TiO_2	1.22	1.96	3.02	1	1.81	2.73	1.05	1.7	0.85
Al_2O_3	16.42	14.52	14.29	14.72	13.87	13.06	14.64	14.74	18.92
FeO†	10.51	11.81	11.96	10.58	12.15	13.63	10.28	11.67	9.58
MnO	0.16	—	0.2	0.18	0.17	0.22	0.23	0.15	0.15
MgO	7.1	4.15	5.53	6.5	10.46	5.46	7.81	9.84	5.92
CaO	9.05	7.82	9.66	10.8	9.52	10.47	10.01	9.8	10.61
Na_2O	3.21	3.04	3.02	2.71	2.55	2.42	3.06	2.01	2.96
K_2O	0.67	1.51	1.01	0.74	0.55	0.66	0.66	1.08	0.44
P_2O_5	0.22	0.36	0.5	0.48	0.22	0.32	0.12	0.16	0.14
ppm									
Elements									
Cr	176	20	107	—	466	44	—	586	112
Ni	127	9	31	—	271	44	111	376	50
Sr	352	—	521	—	396	219	239	220	437
Zr	97	186	264	—	125	203	75	122	59
La	12	32	28	—	—	19.3	10	—	6
Ce	27.6	46	—	—	—	43	19	—	15.3

* Weight percent normalized to 100% auhydrous.
† Total Fe as FeO.
[a] BVSP 1981.
[b] Wright, Mangan, and Swanson 1989.
[c] Zanettin et al. 1976.
[d] Patterson and Swain 1955.
[e] Scarrow and Coy 1995.
[f] Ghose 1976.
[g] Kalsbeck and Jepson 1984.
[h] Annells 1973.
[i] Hickey, Frey, and Gerlach 1986.

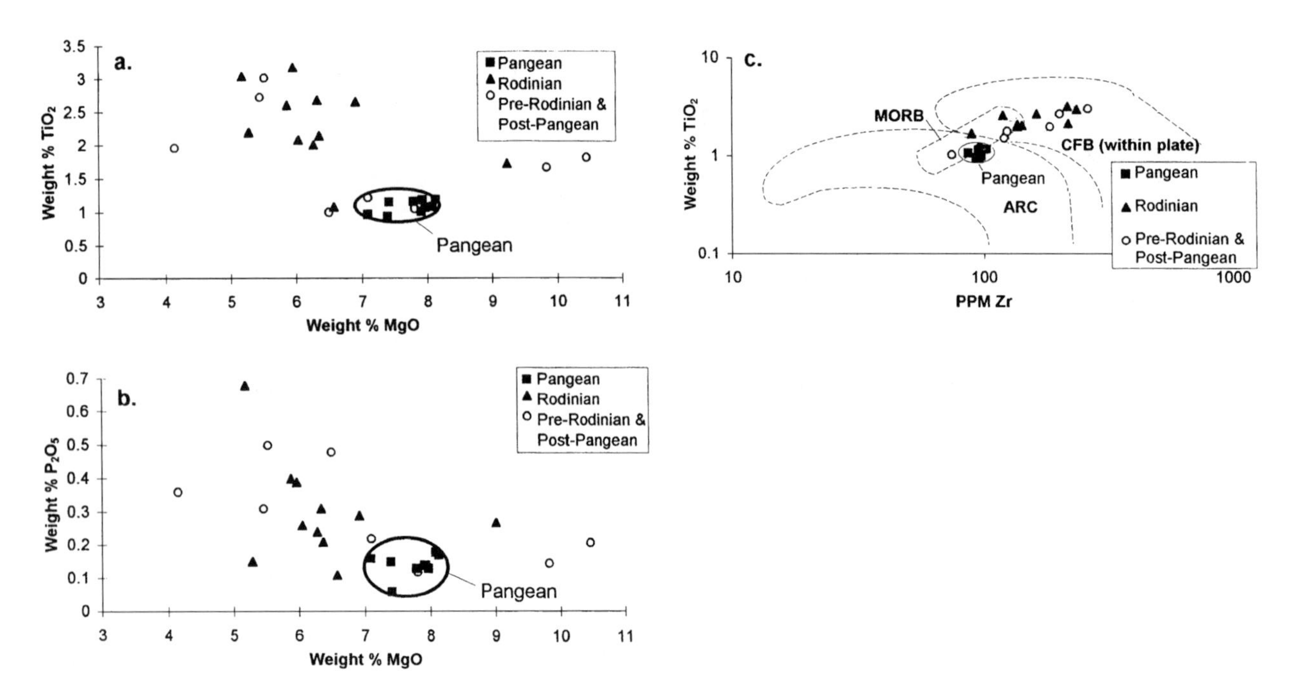

FIGURE 11.1 Average (*a*) TiO_2:MgO, (*b*) P_2O_5:MgO, and (*c*) TiO_2:Zr compositions of CFB provinces. Note the uniform composition of the rift-related Pangean CFBs (*circled*) compared with the diverse composition of non-Pangean CFBs. Note that each of the nine rift-related Pangean CFBs plot within the arc field, whereas most other CFBs plot within the "within-plate" field. (*a, b,* and *c,* data from tables 11.1, 11.2, and 11.3; *c,* labeled dashed field boundaries after Gale and Pearce 1982)

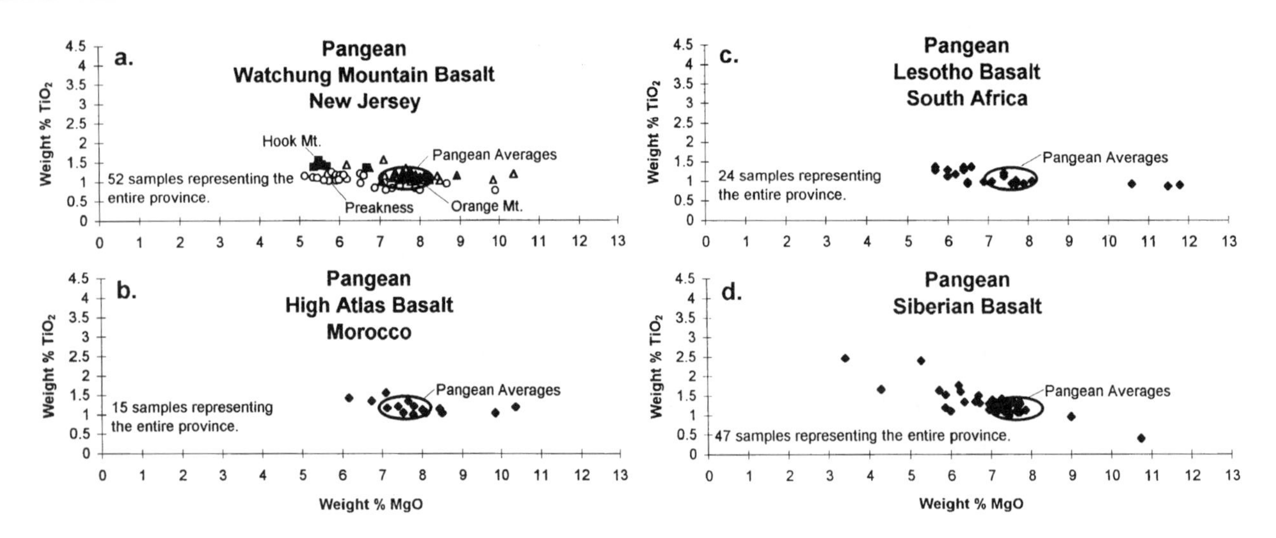

FIGURE 11.2 Complete rift-related Pangean CFB province TiO_2:MgO data sets. (*a*) Watchung Mountain Basalts, New Jersey (Puffer 1992; Puffer and Student 1992). (*b*) High Atlas Basalt, Morocco (Bertrand, Dostal, and Dupuy 1982). (*c*) Lesotho Basalt, South Africa (Cox and Hornung 1966). (*d*) Putorana and Norilsk Basalts, Siberia (Sharma, Basu, and Nesterenko 1991). Note the uniform composition of each province and the clustering of data close to the Pangean average field from figure 11.1a.

tin also generally is regarded as Iapetuan rift-related magmatism, but Badger and Sinha (1988) dated it at 570 ± 36 Ma and suggested that the older ages of other Iapetuan rift basalts indicate either a two-stage or a very long, protracted episode of continental rifting.

Southeastern Canadian Shield. St. Seymour and Kumarapeli (1995) conclude that the approximately 590 Ma Grenville dike swarm is chemically similar to typical CFB (table 11.2) and was intruded into a failed western rift arm of a major triple junction. The north-

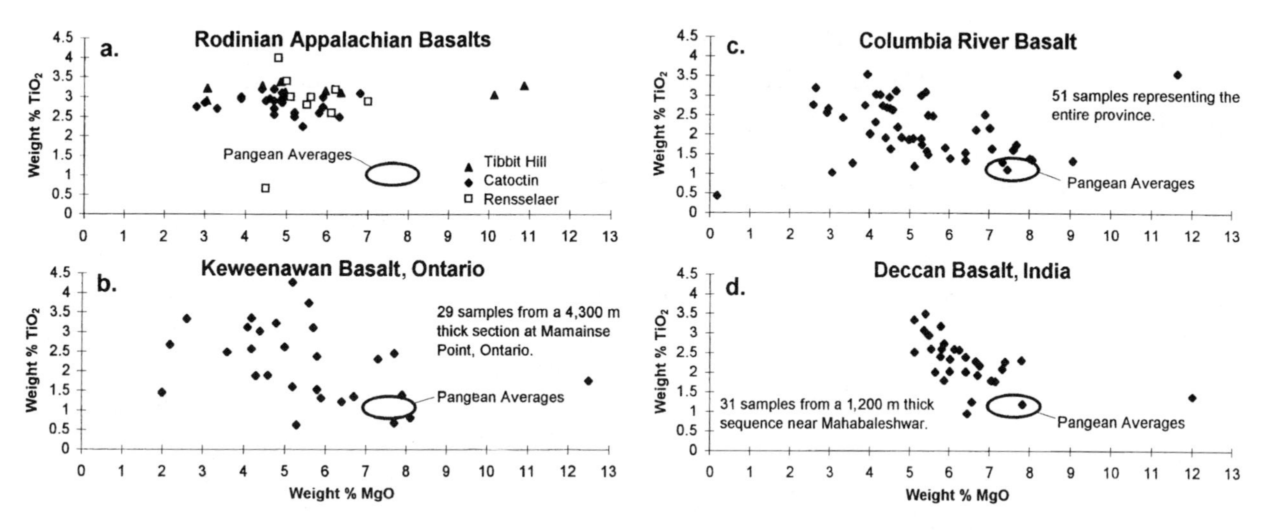

FIGURE 11.3 Complete non-Pangean CFB province TiO_2:MgO data sets. (*a*) Rodinian Tibbit Hill (Coish et al. 1985), Catoctin (Badger and Sinha 1988), and Rensselaer Basalts (Ratcliffe 1987a), Appalachia. (*b*) Keweenawan Basalt, Ontario (Annells 1973). (*c*) Columbia River Plateau Basalt (Hooper 1988). (*d*) Deccan Basalt, India (Cox and Hawkesworth 1985). Note the diverse composition of each province compared with figure 11.2 at the same scale and the rare overlap onto the Pangean average field (*circled*) from figure 11.1a.

ern and southern arms constitute the Appalachian CFB province, which successfully opened into the Iapetus Ocean.

Pre-Rodinian CFBs

Lake Superior Area. The 1.1 Ga Keweenawan CFB of the Lake Superior area is chemically diverse, as represented by 29 core samples through a 4,300 m section at Mamainse Point, Ontario (Annells 1973) (figure 11.3b). The basal 300 m of the section (table 11.3) is relatively uniform but more mafic than upper fractionated portions (Annells 1973). The general consensus of recent research, including that of Nicholson et al. (1997), is that most Keweenawan magmatism is the product of an enriched ocean island–type mantle plume (hot-spot) source.

Post-Pangean CFBs

Columbia River Province. The diverse chemistry of the Tertiary Columbia River Province is illustrated in figure 11.3c, with data presented by Hooper (1988). Few, if any, consistent compositional trends through the lava pile have been reported. The thickest and most homogeneous unit, the Grande Ronde (table 11.3), overlies the relatively heterogeneous Imnaha Basalt at the base of the section.

Taos Plateau. The Quaternary volcanic rocks of the Taos Plateau are rift related, but because they are not dominated by fissure eruptions, they are not CFBs in the restricted sense. The chemical composition of the province is very diverse and includes some highly fractionated andesites (Basaltic Volcanism Study Project 1981). However, the Servilleta Basalt, which aerially is the most extensive volcanic rock in the area, is included in table 11.3 because of its rare chemical resemblance to Pangean CFB. Although the Servilleta Basalt is considerably richer in modal and normative olivine than Pangean CFB, it contains similar HFSE content (table 11.3).

South America

Pangean CFBs

Southern Paraná. Paraná CFB volcanism migrated from south to north at around 140 to 120 Ma (Ernesto 1985). The initial flows of the southern region (table 11.1) are aphyric, low-Ti (0.94 weight percent TiO_2) tholeiites in contrast to younger and diverse but generally high-Ti (3.5 weight percent TiO_2) tholeiites in the northern region (Bellieni et al. 1984).

Africa

Pangean CFBs

Morocco. The Early Jurassic flood basalts of Morocco were extruded as fissure eruptions onto Triassic redbeds in Mesozoic rift basins located in the High Atlas and Meseta Provinces (Manspeizer, Puffer, and Cousminer 1978). The average composition of the massive aphyric flows that dominate the High Atlas section is almost the same as that of the Orange Mountain flows of eastern North America (table 11.1), and the compositional range of the entire Moroccan CFB closely resembles the entire range of the Watchung CFB of eastern North America (figure 11.2b).

South Africa. The Lesotho district, represented by 24 samples of Basutoland Basalt (figure 11.2c), is the first of several flood basalt districts collectively known as the Karoo CFB province. The Lesotho district is described by Cox and Hornung (1966) as "monotonously uniform," unlike later South African CFBs such as the Lebombo and Etendeka districts to the north, which are relatively diverse and enriched in incompatible elements. The Lesotho CFB has been dated at 193 Ma (Cox 1988). Its age, major-element, trace-element, and isotopic characteristics are indistinguishable from HTQ-type eastern North America basalt (Puffer 1992).

Post-Pangean CFBs

Ethiopia. The chemistry of the East African rift basalts is described in detail by Zanettin et al. (1976) and by Kampunzu and Mohr (1991). An average of 81 analyses of Oligocene–Miocene basalt from the Ethiopian plateau (table 11.3) is more alkalic than the tholeiitic Pliocene–Pleistocene flood basalts that followed but is typical of non-Pangean CFB (table 11.3).

Asia

Rodinian CFBs

Siberia. Late Permian and Early Jurassic CFB volcanism in Siberia was proceeded by Late Proterozoic CFB volcanism (Zolotukhin and Al'Mukhamedov 1988). Analytical data are difficult to obtain, but Belyakov and colleagues (1970) presented some Late Proterozoic flow-related diabase analyses, typified by the average of three dikes from the Bolshaya Kuonamka River (table 11.2) that contain 2.55% TiO_2.

Pangean CFBs

Siberia. Just as is the case with North America, the Pangean CFB province of Siberia overlies a Rodinian CFB province. The Siberian province is the most widespread CFB on Earth (Basaltic Volcanism Study Project 1981). Renne and Basu (1991) dated the main extrusive pulse for the province at 248 Ma. Although the altered flows and pyroclastic rocks at the base of the section are chemically diverse, most of the Siberian CFB province is chemically uniform, as typified by Sharma, Basu, and Nesterenko's (1991) data plotted in figure 11.2d. Their samples are a good representation of the entire Siberian province and form a variation pattern very similar to that of the Watchung CFB of New Jersey (figure 11.2a and d). The earliest uniform and thick (200 to 520 m) unit is the Nadezhdinsk Formation, represented by an average of 58 analyses (table 11.1), which overlies the heterogeneous base of the section in the Norilsk region.

Australia

Rodinian CFBs

Antrim Plateau. The Cambrian Antrim basalts of northern Australia, as described by Bulititude (1976), are among the more extensive CFBs on Earth. They were extruded as fissure eruptions separated by thin beds of siltstone. The basalt is uniformly tholeiitic and closely resembles the Pangean CFBs listed in table 11.1 and, coincidentally, the Antrim Plateau basalts of Ireland (table 11.3).

Europe

Rodinian CFBs

Norway. Gale and Pearce (1982) interpreted some of the early Paleozoic greenstone of the Bømlo area of Norway as subaerial, within-plate metabasalt. The metabasalt is located along the northeast projection of the Appalachian system of Newfoundland and chemically resembles Appalachian CFBs (table 11.2).

Pangean CFBs

Portugal. Early Jurassic basalts are exposed as a layer of flows interbedded with redbed sediments and dolomite at the base of the Jurassic section in the Algarve basin of Portugal (Puffer 1993). Most exposures are deeply weathered, but an average of two of the least-altered samples is presented in table 11.1.

Post-Pangean CFBs

Skye, Scotland. The Skye Main Lava Series (table 11.3) is the dominant portion of a 1.5 km thick CFB province exposed on the Isle of Skye. Although these basalts are defined as having MgO greater than 7 weight percent, Scarrow and Cox (1995) interpreted them as having evolved by fractionation from even more mafic picritic parental material generated by melting associated with the Iceland hot spot. The basalts of the Skye Main Lava Series are among the most mafic of all the post-Archean CFBs found on Earth.

Antrim Plateau, Ireland. The Tertiary flood basalts of the Middle Lavas of the Antrim Plateau are described by Patterson and Swaine (1955) as homogenous quartz tholeiites containing 0.7 to 1.2 weight percent TiO_2 (table 11.3), unlike the much more alkalic or mafic flows typical of other portions of the Tertiary British Isles CFB province and the Tertiary Greenland CFB province. The Antrim basalts closely resemble the Pangean CFBs given in table 11.1.

Greenland

Pre-Rodinian CFBs

Eastern North Greenland. The Zig-Zag Dal Basalt of eastern North Greenland is a middle Proterozoic CFB approximately 1,230 Ma in age and 1,350 m thick. Kalsbeek and Jepsen (1984) described these basalts as largely unaffected by deformation or metamorphism. The basalts are divided into three units with chemical variation within each unit governed by fractionation. A representative sample from the main central aphyric unit (AU 438) is presented in table 11.3; it is one of the few published pre-Rodinian basalt analyses that closely resembles the Pangean CFBs of table 11.1.

India

Rodinian CFBs

Cuddapah Basin. The flood basalts of the Late Proterozoic Cuddapah basin of eastern India are a diverse suite, including quartz- and olivine-normative tholeiites and alkalic basalts that are currently being studied by the Geological Survey of India. On the basis of preliminary data (from work in progress by Puffer and Pamganamamula), one of the more common types is an olivine-normative tholeiite that is remarkably unaltered and chemically resembles some of the more mafic Keweenawan flows (table 11.2).

Post-Pangean CFBs

Deccan. The diverse composition of the Late Cretaceous to early Tertiary Deccan Basalt is illustrated in figure 11.3d with 31 basalt analyses from a 1,200 m thick sequence through the Western Ghats escarpment near Mahabaleshwar (Cox and Hawkesworth 1985). Although there is considerable overlap, Ghose (1976) divides the Deccan into the Lower and Upper Traps. The Lower Traps (table 11.3) are uniformly quartz normative and contain less aluminum but more iron than the Upper Traps, which also include alkalic basalts. However, analyses of "low-Ti picrites" from Gujarat State of western India (Melluso et al. 1995) indicate the existence of some HFSE-depleted, Deccan-related basalt that chemically resembles typical Pangean CFB. A few samples are particularly similar, but unlike most Pangean CFB they are interpreted by Melluso et al. (1995) as fractionation products of an abundant, closely related picrite.

Geochemical and Petrographic Characteristics

Pangean CFBs

Most rift-related Pangean CFBs are distinctive and occupy a restricted portion of the general petrographic and geochemical range of CFBs. This is particularly true of the flows near the base of each Pangean CFB province (figure 11.1).

In most cases, Pangean extrusive activity began with fissure eruptions of widespread aerial extent. Most of the basalt in the earliest group of flows is quartz-normative tholeiite, petrographically characterized by low-normative quartz contents and aphyric textures. Thin and altered basal flows or heterogeneous basal pyroclastic layers are found occasionally, but they constitute a minor fraction of the total volume.

Uniform chemistry is a characteristic of several entire Pangean CFB provinces (figure 11.2) and of the initial extrusions of the entire group (figure 11.1). For example, the average TiO_2 contents of the nine rift-related Pangean CFBs in table 11.1 range from only 0.94 to 1.2 weight percent (figure 11.1a) with a P_2O_5

range of only 0.06 to 0.17 weight percent (figure 11.1b) and a Zr range of only 87 to 103 ppm (figure 11.1c). However, a few initial Pangean CFBs, such as the Lesotho and southern Paraná, are followed by CFBs, such as the Lebombo (Sweeney, Duncan, and Erlank 1994) and the northern Paraná (Bellieni et al. 1984), respectively, that are completely unlike the initial flows and are relatively enriched in incompatible elements, approaching ocean island basalt (OIB) chemistry. In addition, Bertrand et al. (1997) determined that several western and central African dike swarms and related CFBs are chemically diverse. The western African dikes are located near the 200 Ma site of the large-scale mantle superplume proposed by Wilson (1997).

Rodinian CFBs

In general, there does not seem to be any CFB criterion—such as mode of emplacement, flow thickness, texture, or chemical composition—that stands out as uniquely Rodinian, and consistent secular trends are elusive. Instead, Rodinian CFBs span the wide range of CFB types (figure 11.1) and in most cases compare in their compositional diversity with Cenozoic CFBs, such as the Deccan and Columbia River (figure 11.3). However, none of the Rodinian CFB averages plot within the compositional range of the rift-related Pangean CFB averages (figure 11.1), with the exception of the Antrim CFB of Australia. Most Rodinian CFBs are distinctly enriched in incompatible elements compared with the rift-related Pangean CFBs (table 11.1), as typified by the representative analyses presented in figure 11.4.

The Origin of Continental Flood Basalts

Deep-Mantle Plume-Related OIB Magma

Considerable evidence has been presented in support of CFB as a product of deep-mantle plume magma that has been fractionated or mixed (or both) with melt derived from continental crustal rocks or subcontinental lithosphere. Several petrologists—including Thompson et al. (1984); Sharma, Basu, and Nesterenko (1991); and Scarrow and Cox (1995)—have developed Morgan's (1981) plume model, showing how initial OIB-like deep-mantle plume magmas can be converted by contamination into CFB magma.

The hot-spot model is consistent with the anorogenic to extensional tectonic setting of CFBs, and a modification by Richards, Duncan, and Courtillot (1989) explains the initial massive outpourings of magma as products of large plume heads that subsequently give way to OIB hot-spot tracks derived from plume tails. The hot-spot model also is consistent with the extent to which several CFBs, including most Rodinian CFBs, share some of their geochemical characteristics with OIBs (figure 11.4). Morgan (1981) presented compelling evidence for a genetic relationship between the Deccan CFB and the Reunion hot spot; between the Columbia River CFB and the Yellowstone hot spot; and between the North Atlantic Tertiary CFBs and the Iceland hot spot. Some other connections, however, are less compelling and require plate-direction reversals and hot-spot occurrences that are not well established.

Any CFBs genetically related to plume heads and OIB magma sources (hereafter referred to as P-CFBs) therefore should be characterized by some geochemical resemblance to the standard OIB of Sun and McDonough (1989). Good candidates for a P-CFB classification include each of the non-Pangean CFB provinces plotted onto figure 11.3 and the representative Rodinian analysis plotted onto figure 11.4.

Enriched Subcontinental Mantle

An alternative to the plume model is CFB derivation from an enriched, ancient (but cooled) subcontinental lithosphere or "keel" (Menzies and Hawkesworth 1987). Gallagher and Hawkesworth (1992) modified this concept and showed that hydration of the cooled lithosphere would enhance partial melting. Anderson (1994a) insisted more recently that the subcontinental lithosphere is actually asthenosphere in a geophysical sense because it is a low-velocity part of the thermal boundary layer. Anderson proposed the term "perisphere" as a substitute and suggested that the perisphere is a viable source for CFB, with adiabatic melting triggered by lithospheric pull-apart during global plate reorganization.

One scenario for the formation of CFB, proposed by Anderson (1994a), calls for the collision of a ridge with a trench followed by pull-apart. A mantle wedge

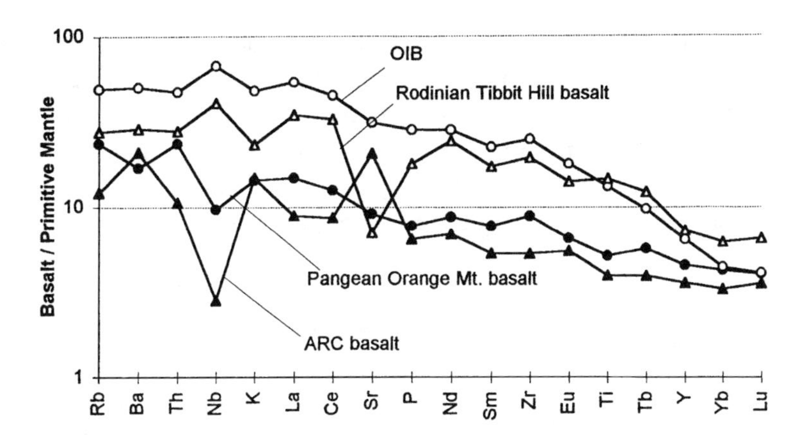

FIGURE 11.4 Spider diagram normalized to primitive mantle values of Sun and McDonough (1989) and arranged according to relative degree of incompatibility in mantle. Note the close resemblance of typical A-CFB from the rift-related Pangean Orange Mountain Basalt, New Jersey (Tollo and Gottfried 1992), and typical arc basalt from the southern zone of the Andes (Hickey, Frey, and Gerlach 1986), and the close resemblance of typical Rodinian P-CFB from Tibbit Hill, Vermont (Coish et al. 1985), and the standard OIB basalt of Sun and McDonough (1989).

fluxed by large ion lithophile (LIL) elements and by H_2O from a subducted oceanic plate melts to form arc volcanics, but only a fraction of the potential LIL element emissions are released. As the stress vectors change upon collision and the accreted terrain lithosphere pulls apart from cratonic lithosphere, CFB is formed from a similar enriched source. Other CFB scenarios offered by Anderson (1994a, 1994b) do not involve arc volcanism but are modeled as short-lived transient events resulting from lithospheric pull-apart at sutures rather than from slower plume heating or lithospheric thinning. CFB is modeled as the result of adiabatic melting from various depths with magma volumes controlled by pull-apart rates, lithospheric thickness, and mantle temperature and fertility.

Any CFBs derived from enriched subcontinental sources following arc magmatism or subduction (hereafter referred to as A-CFB) therefore should include some geochemical resemblance to arc magma. In addition, the best potential candidates for an A-CFB classification should pass three more tests: they should have been extruded through sutures between accreted terrain and cratonic lithosphere that were previously closed by compression and related subduction; they should be short-lived transient events that do not recur in overlying strata; and, finally, they should not be located on a hot-spot track.

Origin of Pangean CFBs

Although Mesozoic CFBs include a wide range of chemical types, each of the CFBs associated with the initial rifting stage of Pangean breakup listed in table 11.1 qualify as A-CFBs.

Test 1: Resemblance to Arc Magma. Each of the rift-related Pangean CFBs listed in table 11.1 plot within the arc field of figure 11.1c. In addition, the uniform chemistry of the group, as typified by the Pangean Orange Mountain Basalt of eastern North America, closely resembles that of typical arc basalt (figure 11.4).

The proposed source of each Pangean CFB listed in table 11.1, therefore, is the same enriched shallow-mantle source generally accepted as the source of arc or calc-alkaline basalt. Compositional similarities between arc basalt and Pangean CFB are obvious (figure 11.4), but they also include overlapping Nd and Sr isotopic values (Pegram 1990; Puffer 1992). For example, Pegram (1990) determined that the source of HTQ basalt magma was isotopically enriched and concluded that the enrichment was probably due to subduction. However, some compositional differences exist between A-CFB magma and arc magma, and at least some of these difference may be due to the contrasting tectonic settings. The compressional tectonic setting of arc volcanism leads to slower upward magmatic movement, resulting in more fractionation and contamination than experienced by aphyric rift-related A-CFB magma that penetrated rapidly through the lithosphere in an extensional setting. In particular, plagioclase fractionation is consistent with the plagioclase-phyric texture of most arc basalt and their elevated levels of Al_2O_3, Na_2O, and Sr.

Test 2: Tectonic Setting. Anderson (1994b) has proposed a model that separates the Earth into areas of hot upwelling mantle characterized by common mantle plumes and areas that have been cooled by heat loss related to subduction processes. The cooled areas are geoid lows shaped like wide bands, have few plumes, and include an oceanic crustal layer subducted into the mantle. Continents tend to drift toward geoid lows as they are displaced by hot upwelling mantle. Pangea, therefore, probably was assembled in a major geoid low over perisphere enriched by Paleozoic subduction of oceanic crust, and then remained intact until it was pulled apart by tectonic stresses that resulted in adiabatic melting of the enriched source.

Wintsch et al. (1996) concluded that a central portion of Pangea (Hartford basin area) was assembled over a rapidly closing convergent margin that remained active until Late Permian time. The mantle under the Hartford basin, therefore, was enriched by the same subduction process that can result in arc basalt.

It is probably not a random coincidence that most Pangean CFBs are located over continental-plate sutures. The closure of continental plates involved subduction of arc source material under the sutures, but it terminated, or perhaps sealed off, arc magmatism. As compression gave way to extensional tectonism, CFBs were extruded along the same sutures without regard to random hot-spot locations.

Test 3: Transient Volcanism. Although most of the rift-related Pangean CFBs were widespread and include some of the most extensive CFBs on record, none of them persist into the Mesozoic or reappear with similar chemistry in overlying strata. The rift-related CFBs of the Karoo and Paraná districts are followed by prolonged episodes of CFBs that are each chemically unlike the initial flows. In at least one case, there is evidence that the Pangean CFB activity was short-lived. Olsen, Schlische, and Fedosh (1996) determined on the basis of Milankovitch cyclostratigraphy that all the Early Jurassic CFBs of the Newark Supergroup of eastern North America were extruded within 580,000 years.

Test 4: Absence of Hot-Spot Association. None of the rift-related Pangean CFBs of table 11.1 are clearly associated with hot-spot tracks or with any current or previous hot-spot position. They are chemically uniform (figures 11.1–11.3) and do not reveal any consistent chemical enrichment or depletion trend toward any geographic location or potential hot spot. For example, the eastern North American HTQ basalts of Virginia are geochemically indistinguishable from the HTQ basalts of New Jersey, Connecticut, Nova Scotia, and Morocco (table 11.1). This chemical uniformity and the absence of most OIB or plume-related geochemical characteristics are not consistent with any genetic relationship to a hot spot in the Carolinas (de Boer and Snider 1979) or to a large 200 Ma superplume beneath the West African craton near Guinea (Wilson 1997).

Pangean Drift-Stage CFBs

Secondary Pangean CFBs that extruded during the drifting stage of continental breakup, such as the LTQ CFB and the olivine-normative dike swarm of southeastern North America (Withjack, Schlische, and Olsen 1998), are geochemically unlike the initial rift-related CFBs and probably had independent origins. In addition, the secondary Pangean CFBs of Brazil and South Africa resemble OIB more closely than the initial rift-related CFB population. Other Pangean CFB and related dikes that are unlike the initial rift-related population listed in table 11.1 include some of the western African dikes described by Bertrand et al. (1997).

It is proposed that the primary eutectic melting of Pangean CFB magma was a short-lived event triggered by extensional thinning and decompression associated with the initial breakup of the supercontinent. However, as breakup continued, A-CFB magmatism gave way to drift-related magmatism, perhaps including some E-MORB magmatism. Wherever continents drifted over existing hot spots, P-CFBs were generated, such as the northern Paraná, Lebombo, Etendeka, and Deccan CFBs. Hot spots apparently were particularly common during Cretaceous time, or plate motion was particularly active.

Why Are Most A-CFBs Pangean?

One Rodinian CFB (Antrim Plateau of Australia), one pre-Rodinian CFB (Zig-Zag Dal of Greenland), and two post-Pangean CFBs (Antrim Plateau of Ireland and Servilleta Basalt of New Mexico) closely resemble initial Pangean CFBs (tables 11.1–11.3). Their occurrence negates the possibility that the processes that developed A-CFBs are uniquely Mesozoic, but the high concentration of A-CFBs associated with Pangean rifting reinforces the likelihood that unusually powerful rifting events may be a genetic requirement.

Origin of Rodinian CFBs

Although Rodinian CFBs are chemically diverse (table 11.1b), most are chemically enriched in incompatible elements to degrees comparable to the standard OIB of Sun and McDonough (1989), as typified by the Tibbit Hill Basalt (figure 11.4). Only one Rodinian CFB (Antrim of Australia) plots in the arc field of figure 11.1c and qualifies as an A-CFB. The reason for the rare occurrence of A-CFBs among the Rodinian group is not clear, but such an occurrence may be caused by a combination of factors. Unlike Pangea, Rodinia probably was disassembled by some new mantle hot spots or perhaps a superplume (Puffer 2002).

However, there is clearly a disparity in the quality of the Pangean and Rodinian stratigraphic records. It is possible to identify both initial rift-related CFBs and subsequent plume-related CFBs in the Pangean record, but less is known about the geometry and timing of Rodinian breakup. Although the huge HTQ flood basalt province of the Atlantic margin originally covered approximately 2.3 million km^2 (McHone 1996), most of the basalt has been eroded away, leaving widely scattered dike swarms as the best evidence of its former extent. In addition, the thickness of most A-CFBs is less than that of most P-CFBs because of their brief duration. Perhaps several unknown Rodinian rift-related basalts, related dike swarms, and metabasalts that resemble Pangean A-CFBs are buried somewhere in the subsurface.

Conclusions

On a global scale, there are two geochemically distinct CFB populations. One population geochemically resembles OIB basalt (P-CFB), and the other shares many of the geochemical characteristics of arc basalt (A-CFB). Most initial rift-related Pangean CFBs belong to the A-CFB group, whereas most Rodinian CFBs belong to the P-CFB group.

A-CFB is not common in the geologic record except for its abundant association with Pangean rifting. In addition to a geochemical resemblance to arc basalt, rift-related Pangean basalts are characterized by uniform chemistry and by extrusion onto continental plate sutures unrelated to hot-spot activity, and they were extruded consistently during brief episodes that do not reappear in the stratigraphic record. Their typically great aerial extent is consistent with abundant adiabatic melting of a wet, enriched source triggered by powerful rifting associated with the breakup of supercontinents. The rift-related Pangean CFBs commonly are followed by P-CFB volcanism during the drifting stage of Pangean breakup.

The uniform chemistry of initial Pangean CFBs on a global basis is of profound importance as a constraint on several petrogenetic factors. It indicates that the CFBs were not influenced significantly by mantle heterogeneity or by proximity to hot spots. Chemical uniformity also is inconsistent with various degrees of fractionation or contamination but is consistent with rapid penetration of an approximately eutectic melt from a wet lithospheric source.

In contrast, most Rodinian CFBs are P-CFBs, suggesting that Rodinia was disassembled by new upwelling mantle plumes or a superplume. However, the Rodinian stratigraphic record is less complete than the Pangean record, and some Rodinian A-CFBs may await future recognition.

Acknowledgments

I thank Gregory McHone for helpful discussions and encouragement, as well as Georgia Pe-Piper, Jonathan Husch, and Peter M. LeTourneau for their thorough review of the manuscript for this chapter.

Literature Cited

Anderson, D. L. 1994a. The sublithospheric mantle as the source of continental flood basalts: The case against the continental lithosphere and plume head reservoirs. *Earth and Planetary Science Letters* 123:269–280.

Anderson, D. L. 1994b. Superplumes or supercontinents? *Geology* 22:39–42.

Annells, R. N. 1973. *Proterozoic Flood Basalts of Eastern Lake Superior: The Keeweenawan Volcanic Rocks of the Mamainse Point Area, Ontario.* Geological Survey of Canada Paper, no. 72–10. Ottawa: Geological Survey of Canada.

Badger, R. L., and A. K. Sinha. 1988. Age and Sr isotopic signature of the Catoctin volcanic province: Implications for subcrustal mantle evolution. *Geology* 16:692–695.

Basaltic Volcanism Study Project. 1981. *Basaltic Volcanism on the Terrestrial Planets.* New York: Pergamon.

Bellieni, G., P. Comin-Chiaramonti, L. S. Marques, A. J. Melfin, A. J. R. Nardy, E. M. Piccirillo, and A. Roisenberg. 1984. High- and low-TiO_2 flood basalts from the Paraná plateau (Brazil): Petrology and geochemical aspects bearing on their mantle origin. *Neues Jahrbuch für Mineralogie Abhandlungen* 150:273–306.

Belyakov, L. P., B. V. Gusev, E. S. Kuteinikov, and L. V. Firskov. 1970. Late Proterozoic trappean intrusion of the western dike of the Anabar anteclise. In A. M. Vilensky, ed., *Geology and Petrology of Intrusive Traps of the Siberian Platform,* pp. 67–79. Moscow: Nauka.

Bertrand, H., J. Dostal, and C. Dupuy. 1982. Geochemistry of early Mesozoic dolerites from Morocco. *Earth and Planetary Science Letters* 58:225–239.

Bertrand, H., G. Féraud, K. Deckart, and J. P. Liegeois. 1997. Timing and genesis of Mesozoic continental tholeiites (MCT) along the African margin of the central Atlantic rift and Guiana: New $^{40}Ar/^{39}Ar$ ages and geochemical constraints. *Geological Society of America, Abstracts with Programs* 29:A32.

Bulititude, R. J. 1976. Flood basalts of probable early Cambrian age in northern Australia. In R. W. Johnson, ed., *Volcanism in Australasia,* pp. 1–17. Amsterdam: Elsevier.

Coish, R. A., F. S. Fleming, M. Larsen, R. Poyner, and J. Seibert. 1985. Early rift history of the Proto-Atlantic Ocean: Geochemical evidence from metavolcanic rocks in Vermont. *American Journal of Science* 285:351–378.

Cox, K. G. 1988. The Karoo province. In J. D. Macdougall, ed., *Continental Flood Basalts,* pp. 239–271. Dordrecht: Kluwer.

Cox, K. G., and C. J. Hawkesworth. 1985. Geochemical stratigraphy of the Deccan Traps at Mahabaleshwar, western Ghats, India, with implications for open system magmatic processes. *Journal of Petrology* 26:355–377.

Cox, K. G., and G. Hornung. 1966. The petrology of the Karroo Basalts of Basutoland. *American Mineralogist* 51:1414–1432.

de Boer, J., and F. G. Snider. 1979. Magnetic and chemical variations of Mesozoic diabase dikes from eastern North America: Evidence for a hotspot in the Carolinas? *Geological Society of America Bulletin* 90:I185–I198.

Ernesto, M. 1985. Paleomagnetismo da formação Serra Geral: Contribução ao estudo do processo de abertura do Atlântico Sul. Ph.D. diss., University of São Paulo, Brazil.

Gale, G. H., and J. A. Pearce. 1982. Geochemical patterns in Norwegian greenstones. *Canadian Journal of Earth Sciences* 19:385–396.

Gallagher, K., and C. J. Hawkesworth. 1992. Dehydration melting and the generation of continental flood basalts. *Nature* 358:57–59.

Ghose, M. C. 1976. Composition and origin of Deccan basalts. *Lithos* 9:65–73.

Goldberg, S. A., J. R. Butler, and P. D. Fullagar. 1986. The Bakersville dike swarm: Geochronology and petrogenesis of Late Proterozoic basaltic magmatism in the southern Appalachian Blue Ridge. *American Journal of Science* 286:403–430.

Hickey, R. L., F. A. Frey, and D. C. Gerlach. 1986. Multiple sources for basaltic arc rocks from the southern zone of the Andes: Trace element and isotopic evidence for contributions from subducted oceanic crust, mantle, and continental crust. *Journal of Geophysical Research* 91:5963–5983.

Hooper, P. R. 1988. The Columbia River Basalt. In J. D. Macdougall, ed., *Continental Flood Basalts,* pp. 1–34. Dordrecht: Kluwer.

Husch, J. M. 1992. Geochemistry and petrogenesis of Early Jurassic diabase from the central Newark basin of New Jersey and Pennsylvania. In J. H. Puffer and P. C. Ragland, eds., *Eastern North American Mesozoic Magmatism,* pp. 169–192. Geological Society of America Special Paper, no. 268. Boulder, Colo.: Geological Society of America.

Kalsbeek, F., and H. F. Jepsen. 1984. The Late Proterozoic Zig-Zag Dal Basalt Formation of eastern North Greenland. *Journal of Petrology* 25:644–664.

Kampunzu, A. B., and P. Mohr. 1991. Magmatic evolution and petrogenesis in the East African rift system. In A. B. Kampunzu and R. T. Lubala, eds., *Magmatism in Extensional Structural Settings: The Phanerozoic African Plate,* pp. 85–136. New York: Springer.

Lassiter, J. G., D. J. DePaolo, and J. J. Mahoney. 1995. Geochemistry of the Wrangellia flood basalt province: Implications for the role of continental and oceanic lithosphere in flood basalt genesis. *Journal of Petrology* 36:983–1009.

Manspeizer, W., J. H. Puffer, and H. L. Cousminer. 1978. Separation of Morocco and eastern North America: A Triassic–Liassic stratigraphic record. *Geological Society of America Bulletin* 89:901–920.

McHone, J. G. 1996. Broad-terrane Jurassic flood basalts across northeastern North America. *Geology* 24:319–322.

Melluso, L., L. Beccaluva, P. Brotzu, A. Gregnanin, A. K. Gupta, L. Morbidelli, and G. Traversa. 1995. Constraints on the mantle sources of the Deccan Traps from the petrology and geochemistry of the basalts of Gujarat State (western India). *Journal of Petrology* 36:1393–1432.

Mensing, T. M., G. Faure, L. M. Jones, J. R. Bowman, and J. Hoefs. 1984. Petrogenesis of the Kirkpatrick Basalt, Solo Nunatak, northern Victoria Land, Antarctica. *Contributions to Mineralogy and Petrology* 87:101–108.

Menzies, M. A., and C. J. Hawkesworth. 1987. Upper mantle processes and composition. In P. H. Nixon ed., *Mantle Xenoliths,* pp. 725–738. Chichester: Wiley.

Morgan, W. J. 1981. Hotspot tracks and the opening of the Atlantic and Indian Oceans. In E. Emiliani, ed., *The Sea,* vol. 7, pp. 443–487. New York: Wiley.

Nesterenko, G. V., N. S. Avilova, and N. P. Smimova. 1964. Rare elements in traps of the Siberian platform. *Geokhimija* 10:1015–1021.

Nicholson, S. W., S. B. Shirey, K. J. Schulz, and J. C. Green. 1997. Rift-wide correlation of 1.1 Ga midcontinent rift system basalt: Implications for multiple mantle sources during rift development. *Canadian Journal of Earth Sciences* 34:504–520.

Olsen P. E., R. W. Schlische, and M. S. Fedosh. 1996. 580 ky duration of the Early Jurassic flood basalt event in eastern North America estimated using Milankovitch cyclostratigraphy. In M. Morales, ed., *The Continental Jurassic,* pp. 11–22. Museum of Northern Arizona Bulletin, no. 60. Flagstaff: Museum of Northern Arizona.

Papezik, V. S., J. D. Greenough, J. A. Colwell, and T. J. Mallinson. 1988. North Mountain Basalt from Digby, Nova Scotia: Models for a fissure eruption from stratigraphy and petrochemistry *Canadian Journal of Earth Sciences* 25:74–83.

Patterson, E. M., and D. J. Swaine. 1955. A petrochemical study of Tertiary tholeiitic basalts: The middle lavas of the Antrim Plateau. *Geochimica et Cosmochimica Acta* 8:173–181.

Pegram, W. J. 1990. Development of continental lithospheric mantle as reflected in the chemistry of the Mesozoic Appalachian tholeiites, U.S.A. *Earth and Planetary Science Letters* 97:316–331.

Pestana, E. M. 1985. Geochemistry and tectonic significance of the volcanic rocks associated with the Rensselaer graywacke of east-central New York. M.S. thesis, Rutgers University, Newark, N.J.

Puffer, J. H. 1992. Eastern North American flood basalts in the context of the incipient breakup of Pangea. In J. H. Puffer and P. C. Ragland, eds., *Eastern North American Mesozoic Magmatism,* pp. 95–119. Geological Society of America Special Paper, no. 268. Boulder, Colo.: Geological Society of America.

Puffer, J. H. 1993. Early Jurassic basalts of the Algarv Basin, Portugal. In *Program and Abstracts: Canadian Society of Petroleum Geologists,* p. 254. Calgary: Canadian Society of Petroleum Geologists.

Puffer, J. H. 1994. Initial and secondary Pangean basalts. In B. Beauchamp, A. F. Embry, and D. Glass, eds., *Pangea: Global Environments and Resources,* pp. 85–95. Canadian Society of Petroleum Geologists Memoir, no. 17. Calgary: Canadian Society of Petroleum Geologists.

Puffer, J. H. 2002. A Late Neoproterozoic eastern North American superplume: Location, size, chemical composition, and environmental impact. *American Journal of Science* 302:1–27.

Puffer, J. H., D. O. Hurtubise, F. J. Geiger, and P. Lechler. 1981. Chemical composition and stratigraphic correlation of Mesozoic basalt units of the Newark basin, New Jersey, and the Hartford basin, Connecticut. *Geological Society of America Bulletin* 92:515–553.

Puffer, J. H., and A. R. Philpotts. 1988. Eastern North American quartz tholeiites: Geochemistry and petrology. In W. Manspeizer, ed., *Triassic–Jurasic Rifting: Continental Breakup and the Origin of the Atlantic Ocean and the Passive Margins,* pp. 579–606. Developments in Geotectonics, no. 22. Amsterdam: Elsevier.

Puffer, J. H., and J. J. Student. 1992. Volcanic structures, eruptive style, and post-eruptive deformation and chemical alteration of the Watchung flood basalts, New Jersey. In J. H. Puffer and P. C. Ragland, eds., *Eastern North American Mesozoic Magmatism*, pp. 261–278. Geological Society of America Special Paper, no. 268. Boulder, Colo.: Geological Society of America.

Ratcliffe, N. M. 1987a. Basaltic rocks in the Rensselaer Plateau and Chatham slices of the Taconic allochthon: Chemistry and tectonic setting. *Geological Society of America Bulletin* 99:511–528.

Ratcliffe, N. M. 1987b. High TiO_2 metadiabase dikes of the Hudson Highlands, New York and New Jersey: Possible Late Proterozoic rift rocks in the New York recess. *American Journal of Science* 287:817–850.

Reed, J. C., and B. A Morgan. 1971. Chemical alteration and spilitization of the Catoctin greenstones, Shenandoah National Park, Virginia. *Journal of Geology* 79:526–548.

Renne, P. R., and A. R. Basu. 1991. Rapid eruption of the Siberian traps flood basalts at the Permo-Triassic boundary. *Science* 253:176–178.

Richards, M. A., R. A. Duncan, and V. E. Courtillot. 1989. Flood basalts and hot-spot tracks: Plume heads and tails. *Science* 246:103–107.

Rogers, J. J. W. 1996. A history of continents in the past three billion years. *Journal of Geology* 104:91–107.

Scarrow, J. H., and K. G. Cox. 1995. Basalts generated by decompressive adiabatic melting of a mantle plume: A case study from the Isle of Skye, NW Scotland. *Journal of Petrology* 36:3–22.

Sharma, M., A. R. Basu, and G. V. Nesterenko. 1991. Nd-Sr isotopes, petrochemistry, and origin of the Siberian flood basalts, USSR. *Geochemica et Cosmochimica Acta* 55:1183–1192.

Strong, D. F. 1974. Plateau lavas and diabase dikes of northwestern Newfoundland. *Geological Magazine* 111:501–514.

St. Seymour, K., and P. S. Kumarapeli. 1995. Geochemistry of the Grenville dyke swarm: Role of plume-source mantle in magma genesis. *Contributions to Mineralogy and Petrology* 120:29–41.

Sun, S. S., and W. F. McDonough. 1989. Chemical and isotopic systematics of oceanic basalts: Implications for composition and processes. In A. D. Saunders and M. J. Norry, eds., *Magmatism in the Ocean Basins*, pp. 313–345. Geological Society Special Publication, no. 42. London: Geological Society.

Sweeney, R. J., A. R. Duncan, and A. J. Erlank. 1994. Geochemistry and petrogenesis of central Lebombo basalts of the Karoo igneous province. *Journal of Petrology* 35:95–125.

Thompson, R. N., M. A. Morrison, G. L. Hendry, and S. J. Parry. 1984. An assessment of the relative roles of a crust and mantle in magma genesis: An elemental approach. *Philosophical Transactions of the Royal Society of London*, part A 310:549–590.

Tollo, R. P. 1988. Petrographic and major-element characteristics of Mesozoic basalts, Culpeper basin, Virginia. In A. J. Froelich and G. R. Robinson Jr., eds., *Studies of the Early Mesozoic Basins of the Eastern United States*, pp. 105–113. U.S. Geological Survey Bulletin, no. 1776. Washington, D.C.: Government Printing Office.

Tollo, R. P., and D. Gottfried. 1992. Petrochemistry of Jurassic basalt from eight cores, Newark basin, New Jersey: Implications for the volcanic petrogenesis of the Newark Supergroup. In J. H. Puffer and P. C. Ragland, eds., *Eastern North American Mesozoic Magmatism*, pp. 233–259. Geological Society of America Special Paper, no. 268. Boulder, Colo.: Geological Society of America.

Volkert, R. A., and J. H. Puffer. 1995. Late Proterozoic diabase dikes of the New Jersey Highlands: A remnant of Iapetan rifting in the north-central Appalachians. *U.S. Geological Survey Professional Paper* 1565-A:1–22.

Weigand, P. W., and P. C. Ragland. 1970. Geochemistry of Mesozoic dolerite dikes from eastern North America. *Contributions to Mineralogy and Petrology* 29:195–214.

Wilson, M. 1997. Thermal evolution of the central Atlantic passive margins: Continental break-up above a Mesozoic super-plume. *Journal of the Geological Society of London* 154: 491–495.

Wintsch, R. P., M. L. Kunk, J. L. Boyd, and B. P. Hellickson. 1996. Permian and early Mesozoic exhumation of Bronson Hill terrane rocks: Significance for the tectonics of the Hartford basin. In P. M. LeTourneau and P. E. Olsen, eds., *Aspects of Triassic–Jurassic Rift Basin Geoscience: Abstracts*, p. 57. State Geological and Natural History Survey of Connecticut Miscellaneous Reports, no. 1. Hartford: Connecticut Department of Environmental Protection.

Withjack, M. O., R. W. Schlische, and P. E. Olsen. 1998. Diachronous rifting, drifting, and inversion on the passive margin of eastern North America: An analog for other passive margins. *American Association of Petroleum Geologists Bulletin* 82:817–835.

Wright, T. L., M. Mangan, and D. A. Swanson. 1989. *Chemical Data for Flows and Feeder Dikes of the Yakima Basalt Subgroup, Columbia River Basalt Group, Washington, Oregon, and Idaho, and Their Bearing on a Petrogenetic Model.* U.S. Geological Survey Bulletin, no. 1821. Washington, D.C.: Government Printing Office.

Zanettin, B., A. Gregnanin, E. Justin-Visentin, G. Mezzacasa, and E. M. Piccirillo. 1976. *New Chemical Analyses of the Tertiary Volcanics of the Central-Eastern Ethiopian Plateau.* Padua: Consiglo Naz Ricerche, Instituto di Mineralogia, University of Padua.

Zolotukhin, V. V., and A. I. Al'Mukhamedov. 1988. Traps of the Siberian platform. In J. D. Macdougail, ed., *Continental Flood Basalts,* pp. 273–310. Dordrecht: Kluwer.

12

A Geographic Trend for MgO-Standardized Major Oxides in Lower Mesozoic Olivine Tholeiites of the Southeastern United States

Paul C. Ragland, Vincent J. M. Salters, and William C. Parker

A dataset of 325 screened major-oxide analyses from lower Mesozoic, olivine-tholeiitic diabase dikes of the southeastern United States was compiled. These tholeiites are part of the recently recognized ~200 Ma Central Atlantic Magmatic Province (CAMP), a large igneous province found on four circum-Atlantic continents. The data were examined in the light of petrologic considerations, such as degree and depth of melting of a mantle source, as well as with regard to correlations between geographic location and geochemistry. Most of the samples are from the swarm of NW-trending dikes that extends from central Virginia to Alabama. The analyses were standardized to 8.0 weight percent MgO to remove effects of low-pressure crystal fractionation of olivine ± plagioclase ± clinopyroxene. The standardized oxides (written as $Na_{8.0}$, $Fe_{8.0}$, and so on) yield patterns similar to the "local trends" reported for many Mid-Ocean Ridge Basalts.

These differences in concentrations of MgO-standardized oxides are related to their geographic position; this relationship has been confirmed statistically by the use of principal-components analysis in conjunction with nonparametric Mann-Whitney rank-sum tests. On average, compositions from Virginia through the Carolinas to Georgia and Alabama progressively become enriched in $Na_{8.0}$, $Ti_{8.0}$, and $Fe_{8.0}$, whereas $Si_{8.0}$ and $Ca_{8.0}$ systematically decrease. These systematic variations in chemical compositions can be interpreted to indicate that, toward the southwest, melts on average are generated at deeper levels and represent smaller melt percentages or that, toward the southwest, melts are derived from a mantle source that becomes progressively more enriched in "fertile" components. In the first case, the compositional changes can be the effect of a progressively thicker continental crust to the southwest; in the latter case, the increased "fertility" can be interpreted as a larger influence in the Georgia and Alabama region of a Mesozoic hot spot. A more recent proposal is that the center of the mantle superplume responsible for CAMP is in the southeastern United States, the Caribbean, or the Liberia–Guyana–Surinam region on Pangea.

After an initial effort made in the early 1970s (Weigand and Ragland 1970), a number of classification schemes for the lower Mesozoic diabases (dolerites) of eastern North America, especially the olivine tholeiites, have been proposed (Ragland, Cummins, and Arthur 1992). At least four schemes have been suggested for the olivine-tholeiitic diabases, all of which depend on differences in trace or major incompatible-element content or both. All four schemes, however, are based on different specific criteria. Our purpose here is not to propose another classification for the olivine tholeiites or even to resolve differences among the existing classifications. Rather, we examine a dataset of 325 screened major-element analyses from the southeastern United States as a whole without being concerned about classification; that is, we intend to be "lumpers" rather than "splitters." Phase-equilibria studies (e.g., Jaques and Green 1980; Stolper 1980; Fujii and Scarfe 1985; Kinzler and Grove 1992a, 1992b) are our guides for our interpretation of the variations in major-element compositions of basalts and allow us to investigate mechanisms and physicochemical conditions of mantle-source melting that produces basaltic magma. We examine the dataset in the light of such petrologic considerations as possible relations with degree and depth of melting of the mantle source for the tholeiites, variations in mantle-source composition, and correlations between geographic location and geochemistry.

More than 500 major-element analyses of these rocks are available, but we rejected these samples because they were clearly altered by secondary processes; had oxide totals outside 98 to 102%; contained large percentages of phenocrysts (many of which were likely accumulative); had a reported oxide value that was particularly anomalous; or contained greater than 51 weight percent SiO_2. We included the fifth criterion because, although quartz tholeiites are also present in the southeastern United States (primarily in the north–south swarm; see the discussion of three main swarms later in the chapter), our intent was to study only the olivine tholeiites. Although many workers have suggested that at least some of the olivine and quartz diabases are comagmatic (e.g., Puffer 1992; Ragland, Cummins, and Arthur 1992), the issue is still open to question (e.g., Milla and Ragland 1992). In order to "reduce the number of variables" and deal with compositions that (we hope) are closer to those of primary melts, we consider only the olivine diabases from the northwest swarm in the southeastern United States.

Moreover, a compositional gap of almost 0.5% at 51 weight percent SiO_2 is present between the quartz and olivine tholeiites of the entire southeastern United States (figure 12.1). Given an approximate 2% error in the calculation of either olivine or quartz in a CIPW norm and the use of 51 weight percent SiO_2 as the approximate dividing line between calculating normative quartz or olivine in these rocks, the fifth screening criterion seems appropriate. Three quartz-normative samples had less than 51 weight percent SiO_2 and were included in the 325 samples comprising the database. In addition, to avoid biasing the dataset, we took only a few samples from transverse profiles across dikes at one locality. We included only analyses from dikes. Despite the pervasive alteration in flows from beneath the Coastal Plain of Georgia, South Carolina, and Florida, some of the individual flow samples were probably acceptable but lacked sufficient petrographic descriptions to warrant their inclusion; thus all were excluded. We compiled the dataset from major-element analyses in Ragland, Rogers, and Justus (1968), Weigand (1970), Steele (1971), Bell et al. (1980), Campbell (1985), Warner et al. (1985), Warner et al. (1986), Cummins (1987), Whittington (1988b), and Milla (1990). It is available on disk from the senior author on request.

Researchers (see discussion in de Boer et al. 1988) have recognized three main lower Mesozoic diabase dike swarms in eastern North America: (1) NE-trending dikes from northern Virginia to southeastern Canada, (2) N-trending dikes from underneath the South Carolina Coastal Plain (determined by aeromagnetic surveys) to central Virginia, and (3) NW-trending dikes from Alabama to central Virginia (figure 12.2). Geophysical evidence (e.g., Popenoe and Zietz 1977; Daniels, Zietz, and Popenoe 1983) indicates that dikes from all these swarms continue beneath the post-Jurassic sediments of the Coastal Plain, as shown schematically in figure 12.2. The first two swarms are dominated by quartz tholeiites, and the third, from which most of the samples for this chapter were collected, is the most mafic of the three swarms and contains primarily olivine tholeiites. Recent evidence suggests that rocks in all three swarms are about the same age, circa 200 Ma (Sutter 1985, 1988). These dikes crosscut the strata and border faults of the (dominantly) Triassic basins.

The dikes are part of an enormous (> 7 million km^2) circum-Atlantic large igneous province (LIP) that exists on four continents and is considered to represent the late rifting–early drifting stage of the Pangean

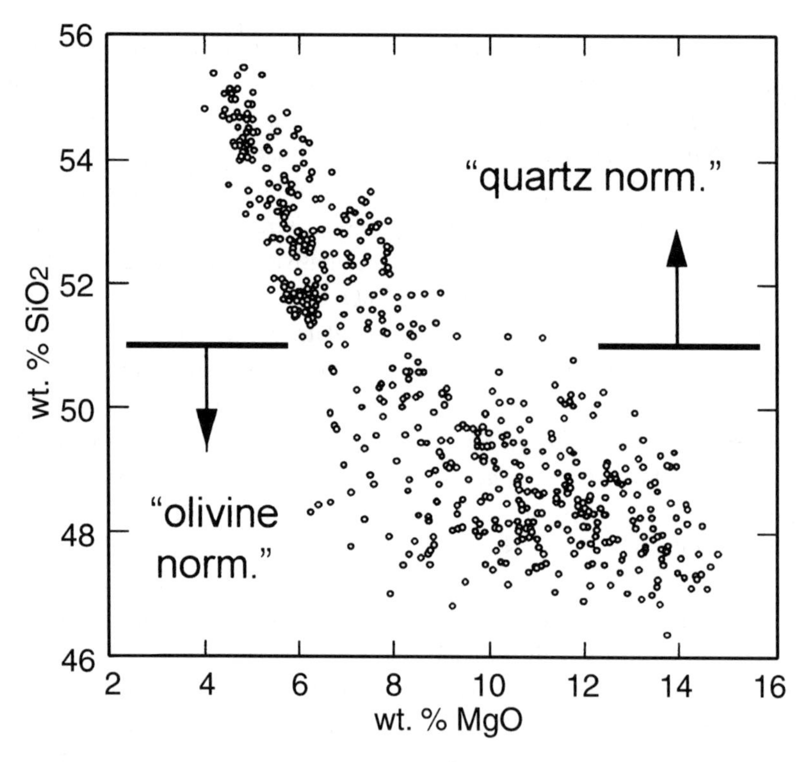

FIGURE 12.1 Plot of MgO versus SiO_2 (all data in weight percent) for 568 analyses of lower Mesozoic diabase dikes from central Virginia to Alabama. The approximate fields of quartz- and olivine-normative diabases can be delineated on the basis of 51 weight percent SiO_2; also note the compositional gap at about 51 weight percent SiO_2. The olivine diabases are primarily from the NW-striking swarm, whereas the quartz diabases are mostly from the N-striking swarm and subsurface basalts from beneath the Coastal Plain. In general, the olivine diabases contain more phenocrysts than do the quartz diabases, so the greater scatter for the olivine diabases may be due to more differential crystal accumulation and fractionation.

breakup (May 1971; Hill 1991; Cummins, Arthur, and Ragland 1992; Holbrook and Kelemen 1993; Wilson 1997; Leitch, Davies, and Wells 1998; Withjack, Schlische, and Olsen 1998; Courtillot et al. 1999; Marzoli et al. 1999). The LIP has been referred to as the Central Atlantic Magmatic Province (CAMP) by Marzoli et al. (1999). Some of these writers (e.g., Hill 1991; Wilson 1997; Leitch, Davies, and Wells 1998; Courtillot et al. 1999) suggest that the LIP and the continental breakup were caused by a superplume originating under Pangea at the core–mantle boundary, whereas others (e.g., Holbrook and Kelemen 1993; as well as McHone 1999; Puffer 1999) disagree with the plume model. McHone and Puffer (chapter 10 in this volume) have suggested that these dikes might be the feeder dikes for a series of dominantly olivine-tholeiitic flood basalts that were once in the southeastern United States and parts of West Africa but presently exist only in the subsurface. A companion flood basalt province composed mainly of high-Ti quartz (HTQ) tholeiites (also referred to as Initial Pangean Rift, or IPR, basalt [Puffer 1994]) reportedly existed northward of this olivine-tholeiite province on Pangea (McHone and Puffer, chapter 10 in this volume).

As mentioned previously, most of these rocks apparently have been affected by low-pressure (dominantly crustal) fractionation and accumulation of olivine and, to a lesser degree, of plagioclase and clinopyroxene (Ragland, Cummins, and Arthur 1992; Ragland, Kish, and Parker 1997). Chemical trends on variation diagrams are affected by these low-pressure processes and commonly obscure more fundamental aspects of their petrology, such as magma formation by partial melting of an upper-mantle peridotitic

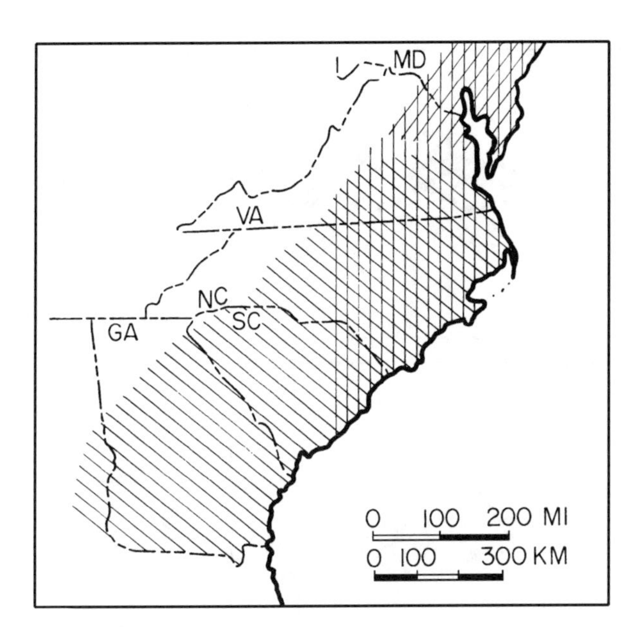

FIGURE 12.2 Schematic map showing the three overlapping lower Mesozoic swarms of diabase dikes in the southeastern United States. Directions of the ruled lines simulate dike attitudes; that is, they represent the NW-striking swarm to the southwest, the N-striking swarm in the center, and the NE-striking swarm to the northeast. Extensions of the swarms under the Coastal Plain are based on aeromagnetic surveys (Popenoe and Zietz 1977; Daniels, Zietz, and Popenoe 1983). (Modified from de Boer et al. 1988)

source. This problem originally was addressed for arc basalts and andesites in the 1960s and 1970s by standardizing the data to a constant value of SiO_2; this oxide was considered to be one of the most sensitive to fractional crystallization (Ragland and Defant 1983 and references therein). Plank and Langmuir (1988) standardized a major-oxide dataset for arc basalts to 6.0% MgO, which is generally considered to be a better measure of crystal fractionation and accumulation than SiO_2 for mafic rocks. A similar technique was employed in Cummins, Arthur, and Ragland (1992) by MgO standardizing TiO_2 for lower Mesozoic basaltic rocks around the circum-Atlantic. Turner and Hawkesworth (1995) also utilized MgO standardization in a study of several continental flood basalt provinces. Standardizing to a constant value of MgO for Mid-Ocean Ridge Basalts (MORBs) (commonly 8.0%) has been a common practice in recent years (e.g., Klein and Langmuir 1987, 1989). Oxides such as Na_2O, FeO* (total Fe as FeO), and SiO_2 standardized to 8.0% MgO (written as $Na_{8.0}$, $Fe_{8.0}$, and $Si_{8.0}$, respectively) have been used widely (Klein and Langmuir 1987, 1989). This standardization to constant MgO is not commonly used for continental rift–related tholeiites, but we attempted to use it in our study.

Thus the main purpose of this chapter is to search for compositional trends in MgO-standardized major oxides that relate to geographic distribution or to conditions of tholeiite magma formation. A multivariate numerical technique, principal-components analysis, augments conventional bi- or trivariate trends analysis. A number of papers from this laboratory have proposed affinities of these olivine tholeiites with MORBs, especially E-MORBs (7 to 9% MgO) (Ragland, Kish, and Parker 1997 and references therein). The oldest seafloor in the Atlantic is in fact Jurassic in age (Bryan, Frey, and Thompson 1977), so these rocks, approximately 200 Ma in age, were apparently the precursors to the Atlantic MORBs. Hence an additional purpose of this chapter is to explore further this possible MORB kinship. We have no conceptual problem with these dikes being feeders for a vast Pangean flood basalt province, on the one hand, but also having E-MORB or arc basalt chemistries, on the other.

Results

Figure 12.3 provides a series of Fenner diagrams for the 325 major-oxide analyses of olivine-tholeiitic diabase dikes from the southeastern United States. Several points are noteworthy:

1. The scatter in K_2O is probably a function of alteration or perhaps minor deep crustal assimilation.
2. The picritic (highest MgO) samples have Mg# values (molar (Mg × 100) / (Mg + Fe_2)) approaching 70, which indicates that their magmas were mantle equilibrated, and those that are aphyric represent primary melts.
3. The overall trend is apparently due to fractionation and accumulation dominated by olivine and to a lesser degree by plagioclase and clinopyroxene ("main trend" [Ragland, Kish, and Parker 1997]).
4. The scatter in FeO* and CaO are not a function of alteration but are some fundamental petrologic processes involving mantle sources (Ragland, Kish, and Parker 1997 and references therein).

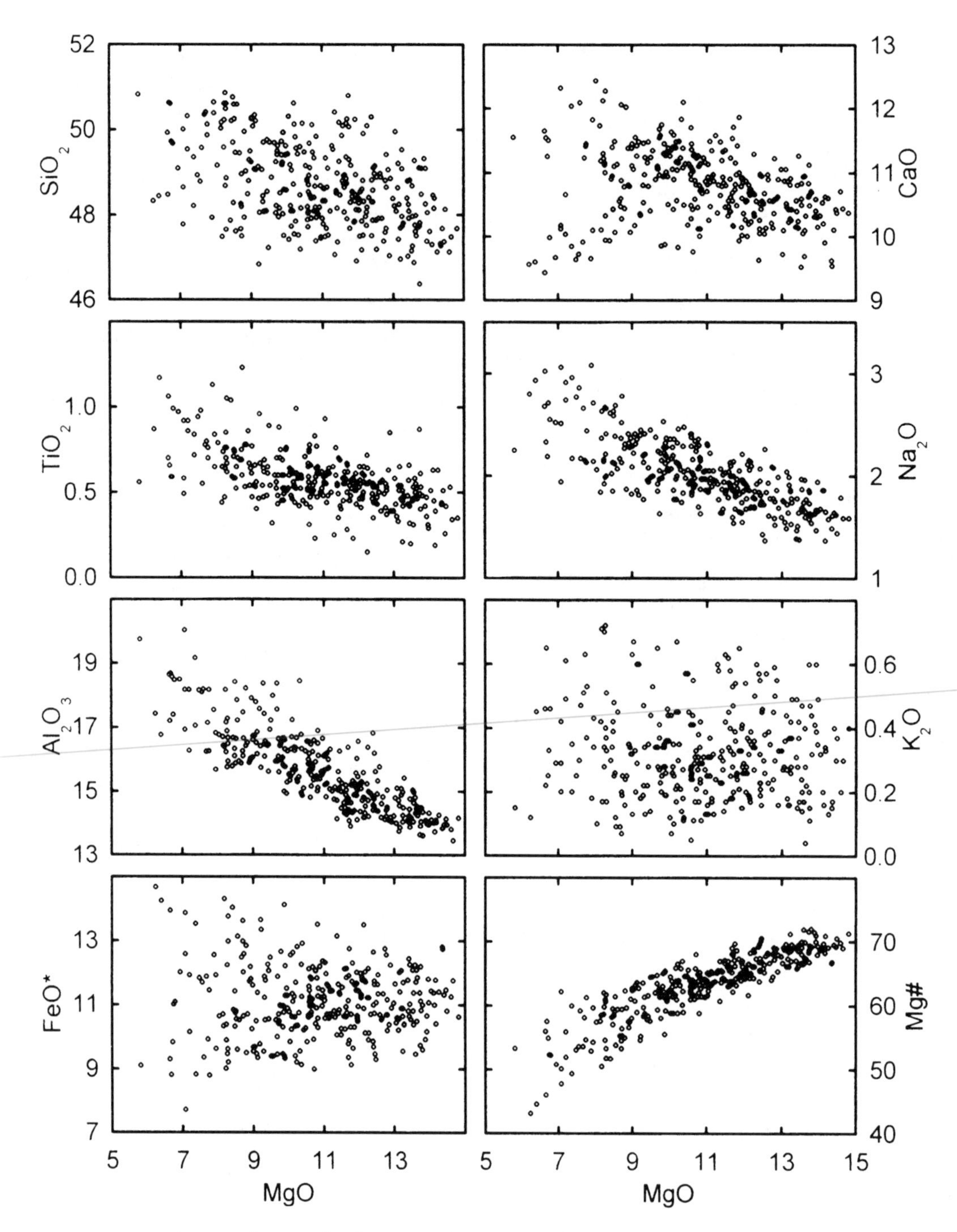

FIGURE 12.3 Fenner diagrams for the 325 analyses used in this study (all values in weight percent). Mg# is defined in the footnote of table 12.1; FeO* is total Fe as FeO. For sources, see the text.

The fourth point is explored further later in the chapter. With regard to the third point, a solution of about 37% fractionation of an extract assemblage containing 58% olivine, 31% augite, and 11% plagioclase from the most picritic sample can produce the most evolved sample using the least-squares mixing program of Bryan, Finger, and Chayes (1969). The solution to this approximation is quite good because the sum of squares for its solution (0.079) is very low. These rocks are either phyric with phenocrysts of olivine ± plagioclase ± clinopyroxene (augite and pigeonite) or aphyric. In addition, it should be pointed out here, however, that the samples with less than approximately 9% MgO and anomalously low CaO (< 10%) are

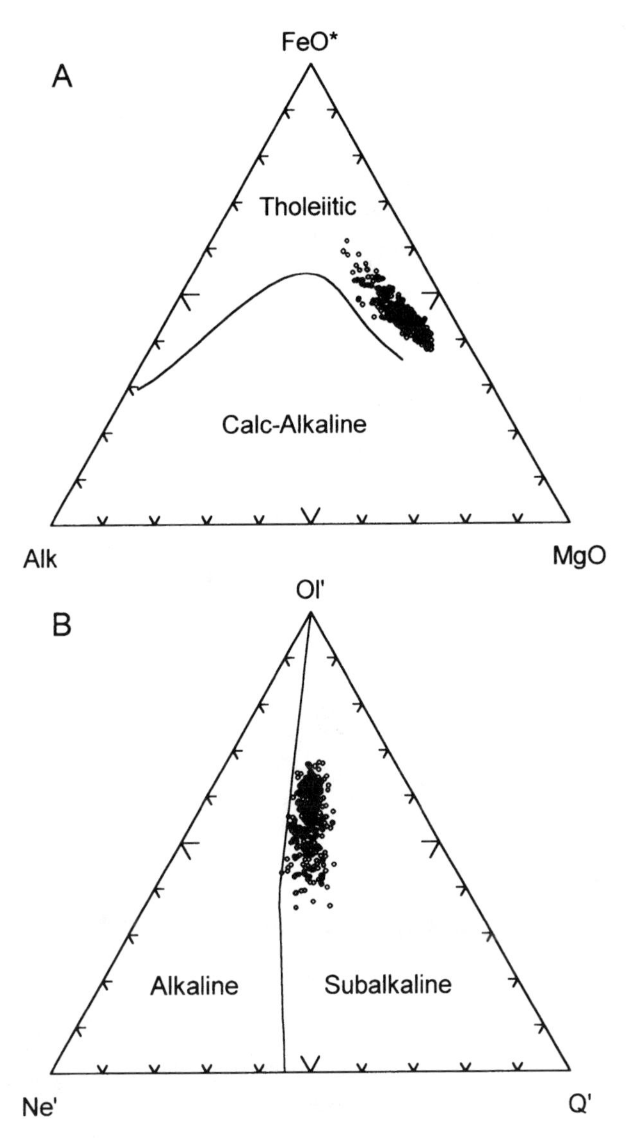

FIGURE 12.4 Classification diagrams showing that the rocks are subalkaline and tholeiitic. (*A*) Alk represents $Na_2O + K_2O$. Note the classic Fe-enrichment trend, but in this case it is not explained primarily by crystal fractionation. (*B*) The symbols (Ne′, Ol′, and Q′) are based on molecular norms for nepheline-olivine-quartz and are defined by Irvine and Baragar (1971). For details, see the text. (After Irvine and Baragar 1971)

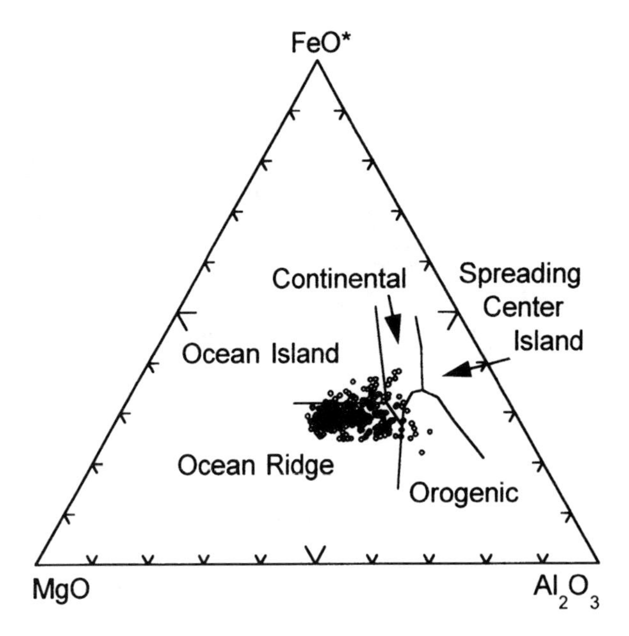

FIGURE 12.5 Major-element tectonic discrimination diagram (all data in weight percent). FeO* is total Fe as FeO. Most of the diabase compositions plot in the MORB field. (After Pearce, Gorman, and Birkett 1977)

among the highest in FeO*. Whittington (1988a, 1988b) referred to these high-Fe rocks as the "high-Fe olivine" (HFO) diabases; Ragland, Kish, and Parker (1997) simply called them the "high-Fe group" (in contrast to the "main group").

Figures 12.4A and 12.4B are conventional classification plots from Irvine and Baragar (1971). Figure 12.4B demonstrates that the rocks are clearly subalkaline, whereas the classic Fe-enrichment trend for tholeiitic rocks is apparent in figure 12.4A. Neither of these plots is a surprise, but they clearly show that the olivine diabases in the southeastern United States compositionally are "as advertised": subalkaline tholeiites. Figure 12.5 shows that the composition of the basalts analyzed in this study fall within the field of midocean ridge basalts on a $MgO–FeO–Al_2O_3$ plot. Although some of the data fall within the fields for basalts from different tectonic settings, the large majority of the basalts from this study have a major-element chemistry similar to that of MORBs. As a consequence, the phase relations that govern the major-element compositions are similar to those of MORBs, and we can use the experimental studies on MORBs to guide the interpretation of the major-element variations. Moreover, the sample compositions plot in the MORB field on several tectonic discrimination diagrams that are based on trace elements (Ragland, Kish, and Parker 1997 and references therein).

The MgO-standardized oxides plotted versus $Na_{8.0}$ are shown in figure 12.6. The equation used in this chapter to standardize the individual oxide analyses to 8.0 weight percent MgO (modified after Ragland and Defant 1983) is

$$C_s = C_0 + m\,(8 - C_{MgO})$$

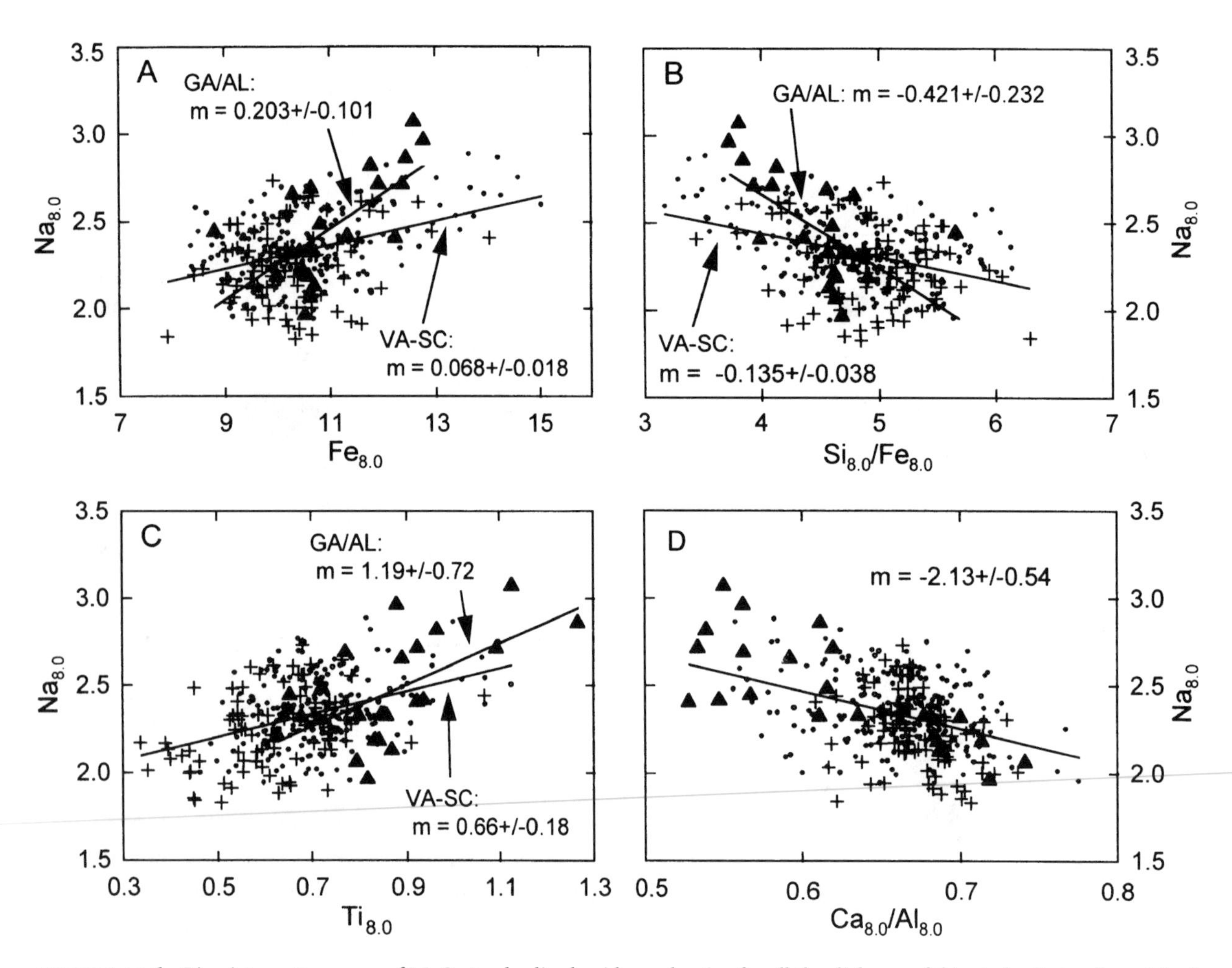

FIGURE 12.6 Bivariate scattergrams of MgO-standardized oxides and ratios for all the diabases of this study. Dots refer to North Carolina and South Carolina (NC/SC); triangles, to Georgia and Alabama (GA/AL); and plus signs, to Virginia (VA). (*A–C*) Best-fit linear regression lines are for GA/AL and for the remainder of the data (VA-SC). (*D*) The linear regression line is for all the data. Errors on slopes are for two standard deviations and represent approximate 95% confidence limits. Given their lower limits, all slopes in *A* and *C* are positive at confidence ≥95%. Likewise, all slopes in *B* and *D* are negative at confidence ≥95%.

where C_s is the standardized concentration of the unknown oxide ($Na_{8.0}$, $Fe_{8.0}$, and so on); m is the slope of the MgO-oxide linear-regression line (with MgO as the *X* variable and the unknown oxide as the *Y* variable); C_{MgO} is the MgO concentration for that sample; and C_0 is the original composition of the unknown oxide. We used 8.0 weight percent MgO as this value at the location where the HFO group is centered. As a consequence, the possible effect of slightly different modal proportions of crystallization (resulting in a different liquid line of decent) for the high-Fe group is minimized. We calculated regression lines based on the complete dataset for all oxides but FeO* and CaO; for these two oxides, the lines were based on the "main-trend" samples (i.e., the trend that includes all high-MgO samples and the low-FeO* and high-CaO samples at < 9% MgO) (figure 12.3). In addition, K_2O was not included in the standardization because of the scatter and lack of correlation with MgO (figure 12.3). Separate regression of the high-Fe group does not lead to a significant change in the variations among the groups and is therefore not presented here. Slopes used for the MgO standardization are: $Si_{8.0}$: − 0.221; $Ti_{8.0}$: − 0.0462; $Al_{8.0}$: − 0.522; $Fe_{8.0}$: + 0.300; $Ca_{8.0}$: − 0.250; $Na_{8.0}$: − 0.125.

The data standardized to 8.0 weight percent MgO is now assumed to be corrected for fractionation, and the standardized values can be used to infer the conditions during melting. Klein and Langmuir (1987, 1989) first introduced this parameterization for

MORBs and argued that basalts with relatively high $Na_{8.0}$ represent low extents of melting, whereas basalts with high $Fe_{8.0}$ represent melts that are generated at relatively deep levels in the mantle. Niu and Batiza (1991, 1993) added $Ca_{8.0}/Al_{8.0}$ as a good indicator of the extent of melting and $Si_{8.0}/Fe_{8.0}$ as a good indicator for depth of melting. These interpretations assume that the mantle source for the basalts is similar in chemical composition and that chemical heterogeneities are not responsible for the overall trends in MORBs.

Our data show a positive correlation of $Na_{8.0}$ with $Fe_{8.0}$ and $Ti_{8.0}$ as well as a negative correlation of $Na_{8.0}$ with $Si_{8.0}/Fe_{8.0}$ and $Ca_{8.0}/Al_{8.0}$ (figure 12.6). Excluding the $Ti_{8.0}$ variations, these variations are similar to the so-called local trends displayed by MORBs (see also Ragland, Kish, and Parker 1997). The reasons for these "local" trends are that

> differences in degrees of partial melt formed and removed during polybaric partial melting of the mantle column beneath the oceanic ridge. MORBs are assumed to represent aggregates of melt generated at different depths in the mantle. At a single ridge segment melting beneath the ridge starts at similar levels irrespective of the distance from the ridge axis. On-axis melting columns, however, will exist to shallower levels as compared to off-axis melting columns, resulting in a triangular-shaped melting regime. Melt aggregates from the off-axis melting columns are generated at comparatively deeper levels and represent smaller degrees of melting, which favor enrichment of Fe and Na relative to Si in the melts. Thus these off-axis melt aggregates have higher $Fe_{8.0}$ and $Na_{8.0}$ but lower $Si_{8.0}$, compared to on-axis melts. (Ragland, Kish, and Parker 1997:14)

These differences result in "local" trends exhibiting positive correlations between $Fe_{8.0}$ and $Na_{8.0}$ but negative trends between $Si_{8.0}/Fe_{8.0}$ and $Ca_{8.0}/Al_{8.0}$ The standardized major-element variations in the southeastern U.S. olivine tholeiites thus can be interpreted to represent a variation in the length of the melting column. Melting in all columns starts at similar depths, but those basalts with high $Fe_{8.0}$ and $Na_{8.0}$ as well as low $Ca_{8.0}/Al_{8.0}$ and $Si_{8.0}/Fe_{8.0}$ are derived from relatively short and deep melting columns (melting is terminated at deeper levels) compared with the low $Fe_{8.0}$ and $Na_{8.0}$ basalts.

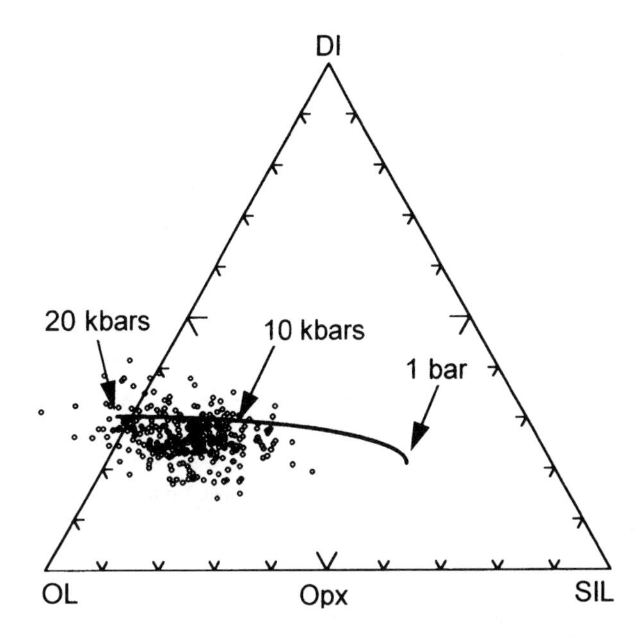

FIGURE 12.7 Pseudoliquidus diagram, plotting compositions for all the diabases of this study in the pseudoternary system olivine-diopside-silica (OL-DI-SIL). The curve is the approximate locus of minimum points from 1 bar to 20 kbars (data from Stolper 1980). (Modified from Walker, Shibata, and DeLong 1979 and Stolper 1980)

Evidence supporting this possibility of different depths on melting for southeastern U.S. diabases can be found by plotting the data on a pseudoliquidus diagram. Many of these diagrams exist, but the diagram used for this study is the projection from plagioclase onto the olivine-diopside-silica "system" developed for MORBs (Walker, Shibata, and DeLong 1979; Stolper 1980). Results are given in figure 12.7, which shows the approximate locus of pseudo-invariant points (the point at 1 bar is a pseudo-eutectic; the remainder are pseudoperitectics) from 1 bar to 20 kbars for mantle-equilibrated basaltic liquids (pseudo-invariant points from Stolper 1980). The 325 points representing the dataset follow the high-pressure ($\sim$ 10 to 20 kbars) half of the locus but are displaced toward the olivine (OL) apex. The fact that most data points plot below the locus of pseudo-invariant points suggests that many of these tholeiitic diabases contain accumulative olivine.

The $Na_{8.0}$ concentrations are related to the trend of data points in figure 12.7. This relationship is shown in figure 12.8, a plot of $Na_{8.0}$ versus normative olivine and quartz (the bottom two variables on the triangular diagrams of figure 12.7). A parenthetical aside is necessary here. In figure 12.7, a number of samples plot

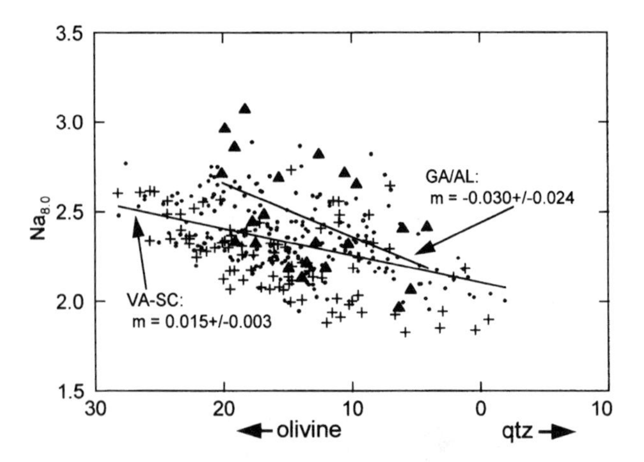

FIGURE 12.8 Bivariate scattergram of $Na_{8.0}$ plotted against CIPW norms for olivine and quartz. Note that only three samples are quartz normative. Both slopes are negative at confidence ≥95%. Dots refer to North Carolina and South Carolina (NC/SC); triangles, to Georgia and Alabama (GA/AL); and plus signs, to Virginia (VA).

to the left of the olivine-diopside (OL-DI) sideline, in the SiO_2 "critically undersaturated" alkali-olivine basalt field. Compositions of all 325 samples in the database, however, are hypersthene normative rather than nepheline normative (i.e., tholeiitic). This graphically illustrates the difference between CIPW norms and at least one set of variables used for pseudoliquidus diagrams, in this case those used by Walker, Shibata, and DeLong (1979). It does not negate, however, the usefulness of figures 12.7 and 12.8. Although the absolute degree of SiO_2 undersaturation may be different between the "Walker projections" and CIPW norms, the trends will be similar.

The negative correlation between $Na_{8.0}$ and the degree of SiO_2 saturation, which is correlative with decreasing pressure of melting, is apparent in figure 12.8. In fact, this relationship bears on a classic tenant of igneous petrology: alkali-olivine basalts represent smaller amounts of partial melting at deeper mantle levels than do tholeiites. Note the negative relationship between $Na_{8.0}$ and $Si_{8.0}/Fe_{8.0}$ in figure 12.6B. The higher $Na_{8.0}$ and the lower $Si_{8.0}/Fe_{8.0}$, the more SiO_2 undersaturated is the magma and the closer the basalt (diabase) composition approaches the compositional field of alkali-olivine basalts.

A positive $Na_{8.0}$ to $Ti_{8.0}$ correlation for the diabases exists as well (figure 12.6C). Langmuir, Klein, and Plank (1992:202, fig. 17) found a positive correlation between $Na_{8.0}$ and $Ti_{8.0}$ for MORBs, although they reported a great deal of variation in slopes of the trends for different ocean basins. They concluded that although $Ti_{8.0}$ and $Na_{8.0}$ exhibit similar global systematics, and both are affected by variations in the degree of melting, $Ti_{8.0}$ also shows large regional differences as well as differences that may be related to such phenomena as hot spots. In this case, we use the term "hot spot" in its simplest context. Stiegler (1976) defines hot spots as "areas on the earth's surface that have a higher than average heat flow . . . thought to develop within the mantle, giving rise to mantle plumes."

Although the standardized major-element variations are entirely consistent with variations in degree and depth of melting of a homogeneous source, it must be kept in mind that this interpretation is not necessarily the correct one. Hirschmann and Stolper (1996) were the latest in a long list of researchers to point out that the MORB source, especially when under the influence of a hot spot, can have a second lithology present in addition to Iherzolite/peridotite. They suggest that the presence of (garnet) pyroxenite (enriched in the "fertile" components Na, Fe, Al, and Ti) can have significant effects on the chemistry of the associated basalts and results in relative enrichment in these basalts by fertile components. The $Ti_{8.0}$ variations can be interpreted as variations in source enrichment as well as variations in extent of melting. The trends of increased $Fe_{8.0}$ and $Ti_{8.0}$ and decreased $Ca_{8.0}/Al_{8.0}$ with increasing $Na_{8.0}$ are consistent at least qualitatively with greater source enrichment in the high $Na_{8.0}$ basalts. The lack of experimental data on pyroxenitic compositions at higher pressure and temperature prevents a more quantitative examination of this possibility.

Discussion and Conclusions

Principal-Components Analysis

Figure 12.6 includes a suggested geographic control for MgO-standardized oxides in the diabases. We are well aware of the fact that the dikes were emplaced around 200 Ma before the state lines were established, but it is convenient to discuss the samples in a geographic context based on the states in which they occur. Samples from Georgia and Alabama tend to plot on one trend and contain relatively high $Na_{8.0}$ and correlative elements, whereas those from Virginia and the Carolinas generally fall on the other (figure 12.6). Moreover, assuming a homogeneous mantle source for the whole southeastern United States, the Virginia samples, on average, represent higher percentages of partial melting at relatively shallow levels, relative to those from the Carolinas (those from Georgia and Alabama would then represent the smallest degrees of partial melting at the deepest levels). In this section, we explore the statistical significance of this apparent NE–SW trends and a possible explanation for the two trends on some of the diagrams.

Principal-components analysis (PCA) is one of several numerical techniques that allow a search for patterns in a large database. Le Maitre (1982) provides an excellent discussion of the basis for PCA and its applications in petrology. It is typically the first step in R-mode factor analysis. PCA searches for these patterns by projecting the data points onto axes of a coordinate system that is different from the original coordinate system of the raw variables; the main intent here is to redistribute the variances. Given a data array with n variables in n-dimensional space, the linear best-fit line through this array is the first principal component (PC1, first eigenvector); the next best-fit line is the second principal component (PC2, second eigenvector); and so on until n principal components (eigenvectors) have been calculated. If the data points are projected onto each of these eigenvectors, the spread (variance) of these projected points measures the length of the eigenvector, which is referred to as the eigenvalue of that eigenvector.

PCA requires that n new variables (the principal components) are calculated as linear combinations of n original variables in such a way that PC1 accounts for the largest possible percentage of the total variance (sum of the eigenvalues). Successive principal components (PC2, PC3, and so on) have increasingly small eigenvalues and thus account for increasingly smaller percentages of total variance. All the n new variables account for the same amount of total variance in the entire dataset, as do the n original variables. The first two or three new variables, however, normally account for much more of the total variance than the first two or three original variables. A tacit assumption here is that, given high-precision data, variance is process driven. Igneous processes such as partial melting, crystal fractionation, assimilation, and so on produce chemical trends that are reflected in increased variances for those variables.

PCA analysis of the Mg-standardized data was carried out using SPSS for Windows, version 6.1.3. Visual inspection of histograms of each standardized variable revealed that $A_{8.0}$, $Fe_{8.0}$, and $Na_{8.0}$ are strongly right skewed and approach log normality. Natural-log transformation of these variables restored them to approximate normality, which is necessary before PCA is run. In addition, this version of PCA recalculates the data as Z-scores. A Z-score is defined as $(x_i - x_m)/s$, where x_i is the numerical value of a variable, and x_m and s are the mean and standard deviation, respectively, of all values measured for that variable. PCA with varimax rotation (to maximize the variance on each eigenvector) was used to extract two components that jointly accounted for 67.8 % of the original variance. The loadings, which represent correlations of the variables with the principal components, are shown in table 12.1 and plotted on figure 12.9. PC1, which accounts for 47.7% of the original variance, shows strong positive correlations with $Fe_{8.0}$, $Na_{8.0}$, and $Ti_{8.0}$; that is, these variables are strongly loaded on PC1. This finding is in agreement with figure 12.4 and suggests a relationship with depth and degree of melting, as discussed earlier in the chapter. In contrast, $Si_{8.0}$ and $Ca_{8.0}$ are loaded negatively on PC1, indicating negative correlations with the three variables given earlier, again in agreement with figure 12.6. $Al_{8.0}$ does not correlate with any of the other variables and is loaded strongly positively on PC2 (figure 12.9).

A geographic trend that may relate to depth and degree of melting becomes evident when the group centroids for sample scores—means for projections of

TABLE 12.1 Variable Loadings on the First Two Principal Components

Variables	*PC1*	*PC2*
$Al_{8.0}$ (ln)	.2527	.9084
$Ca_{8.0}$	.7604	.0100
$Fe_{8.0}$ (ln)	.8623	.0240
$Na_{8.0}$ (ln)	.7300	.0440
$Si_{8.0}$	.6910	.5590
$TI_{8.0}$	.6817	.2577
Percentage of variance	47.7%	20.1%

Note: "8.0" indicates that the variable was standardized to 8% MgO. $Al_{8.0}$, $Fe_{8.0}$, and $Na_{8.0}$ were transformed to natural logs to improve normality.

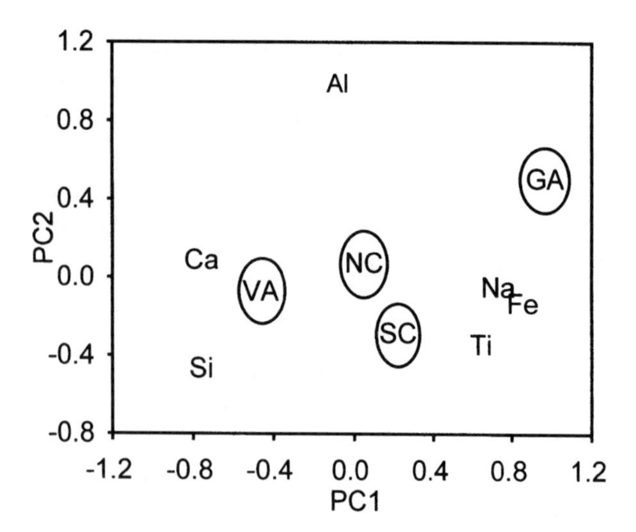

FIGURE 12.9 Plot of variable loadings (element symbols; see also table 12.2) and group-score centroids for each state (*ovals; GA,* Georgia and Alabama) for PC1 versus PC2.

the samples onto the principal components—are separated into their respective geographic subsets. As shown in figure 12.9, the southwesternmost samples, from Georgia and Alabama, have relatively high centroids on PC1 and PC2. The other geographic areas have progressively lower centroids (assuming samples from the Carolinas are grouped together). This trend is especially evident in PC1, as it should be inasmuch as PC1 accounts for more than twice the variance of PC2.

The significance levels for the differences in mean (group centroid) scores for the four areas, shown in table 12.2 (upper-right triangle), are related to their geographic separation. Significance levels for scores from the standardized data display a pattern of progressive increase in probability of difference with increasing geographic separation, so that most of the differences are highly significant. The differences in the standardized data are statistically significant at the 97% confidence level or higher (with the exception of South Carolina and North Carolina). This is based on a nonparametric Mann Whitney rank-sums test, which is more robust than standard parametric procedures such as Student *t*-tests. For comparison, the probabilities for differences in scores for a similar analysis of the unstandardized data (starred values) are shown in the lower-left triangle of table 12.2. Note that none of the differences between geographic groups for the unstandardized data are statistically significant at greater than the 95% confidence level, thus demonstrating the importance of MgO standardization in the observation of this trend. In summary, results of the PCA, in conjunction with the Mann-Whitney test, corroborate and add statistical validity for the geographic trend suggested by more conventional bivariate plots discussed earlier.

One more bivariate plot, however, encapsulates many of the observations we have discussed. The MgO-standardized variables $Na_{8.0}$ and $Ti_{8.0}$ are loaded strongly on PC1 (table 12.1; figure 12.9) and show differences among geographic groups when means and standard deviations are taken into account concurrently (table 12.3). These differences are also apparent in figure 12.10—a plot similar to figure 12.6C, but with much less clutter. Both these variables progressively decrease from Georgia and Alabama to Virginia, in agreement with the observations made earlier. When PCA is used and all MgO-standardized variables are taken into account together, differences among groups are statistically significant (except North Carolina and South Carolina [table 12.2]). We did not test

TABLE 12.2 Statistical Probabilities for Differences in Mean Scores on PC1 for the Four Geographic Groups

	Georgia/ Alabama	*South Carolina*	*North Carolina*	*Virginia*
Georgia/Alabama	—	0.9713	0.9984	0.9999
South Carolina	0.0089*	—	0.7942	0.9999
North Carolina	0.7143*	0.8478*	—	0.9999
Virginia	0.8314*	0.9471*	0.3892*	—

Note: The upper-right triangle is from the analysis of data standardized to 8.0% MgO; the lower-left triangle with starred values is for unstandardized data. Probabilities are based on the nonparametric Mann-Whitney rank-sum test.

TABLE 12.3 Mean Compositions and Standard Deviations for Geographic Groups

	Georgia/Alabama		*South Carolina*		*North Carolina*		*Virginia*			
	Mean	*Stdv*	*Mean*	*Stdv*	*Mean*	*Stdv*	*Mean*	*Stdv*	*Total Mean*	*Total Stdv*
n	22	22	56	56	157	157	90	90	325	325
SiO_2	48.50	0.64	48.58	1.00	48.72	1.02	48.75	0.89	48.69	0.96
TiO_2	0.81	0.20	0.65	0.13	0.58	0.12	0.47	0.14	0.58	0.16
Al_2O_3	16.80	1.22	15.39	0.97	15.79	1.26	15.22	1.12	15.63	1.24
FeO*	11.33	0.81	11.20	1.11	11.12	1.22	10.94	0.98	11.10	1.12
MgO	9.37	1.91	11.15	1.42	10.64	2.05	11.58	1.93	10.90	2.00
CaO	10.60	0.86	10.67	0.52	10.82	0.56	10.80	0.49	10.77	0.57
Na_2O	2.31	0.43	2.01	0.26	2.07	0.32	1.85	0.27	2.01	0.33
K_2O	0.27	0.11	0.35	0.15	0.27	0.12	0.40	0.13	0.32	0.14
Mg#	59.04	5.56	63.77	4.97	62.61	5.12	65.02	4.18	63.23	5.09
$Si_{8.0}$	48.93	0.68	49.61	1.18	49.59	0.83	49.92	0.64	49.64	0.88
$Ti_{8.0}$	0.87	0.15	0.79	0.10	0.69	0.11	0.62	0.12	0.70	0.13
$Al_{8.0}$	17.47	0.63	16.90	0.75	17.05	0.67	16.94	0.58	17.02	0.68
$Fe_{8.0}$	11.04	0.98	10.56	1.31	10.59	1.25	10.22	0.99	10.51	1.19
$Ca_{8.0}$	10.84	0.95	11.16	0.54	11.23	0.55	11.36	0.40	11.23	0.56
$Na_{8.0}$	2.46	0.30	2.37	0.17	2.36	0.20	2.25	0.21	2.34	0.21
or	1.60	0.64	2.05	0.87	1.60	0.73	2.34	0.76	1.88	0.83
ab	19.49	3.65	17.01	2.19	17.45	2.69	15.62	2.26	17.01	2.77
an	34.61	2.25	31.86	2.54	32.93	2.76	31.99	2.67	32.60	2.76
di	14.54	3.65	16.96	2.73	16.74	2.30	17.42	1.41	16.82	2.40
hy	11.43	5.23	12.31	6.06	12.75	5.12	13.50	4.53	12.80	5.18
mt	3.34	0.28	3.11	0.19	3.01	0.18	2.85	0.20	3.00	0.23
il	1.53	0.37	1.23	0.25	1.10	0.23	0.89	0.27	1.09	0.31
ol	13.47	4.83	15.46	5.09	14.37	6.33	15.38	6.38	14.78	6.09

Notes: mean, arithmetic mean; stdv, standard deviation; *n*, number of analyses; Mg#, molar [Mg × 100/(Mg + Fe_2+)]; FeO*, total Fe as FeO; SiO_2, etc., weight percent oxides; $Si_{8.0}$, etc., weight percent oxides standardized to 8.0% MgO; or, ab, etc., CIPW norms.

for significant differences of means for $Na_{8.0}$ and $Ti_{8.0}$; figure 12.10 makes the point sufficiently well.

Possible Tectonic Implications

The observations with regard to the geographic trend in the standardized major-element compositions of the diabases have some tectonic implications. Two possibilities exist. If the major-element variations are interpreted assuming a homogeneous mantle source, then the depth of melting progressively increases and the degree of melting progressively decreases from Virginia to Georgia and Alabama. Plank and Langmuir (1988) have shown that $Na_{6.0}$ (Na_2O standardized to 6.0% MgO) increases and $Ca_{6.0}$ decreases with increasing continental crustal thickness because the arc basalts travel through the crust from their origin in the mantle wedge. This relationship for arc basalts can be extended to these continental rift–related basalts as well. Thickening of the continental lithosphere will result in a shortening of the melting column, resulting in lower degrees of melting that are generated at relatively deep levels in the mantle. The thickest continental crust is then projected to be under Georgia and Alabama. A well-documented late Paleozoic collision event took place in northern Florida and southern Georgia (e.g., McBride and Nelson 1988; Hall 1990), which almost certainly created thicker continental crust in this region at that time. Given increased erosion rates that must have occurred after this event, and a Triassic rifting event that took place in the same region (South

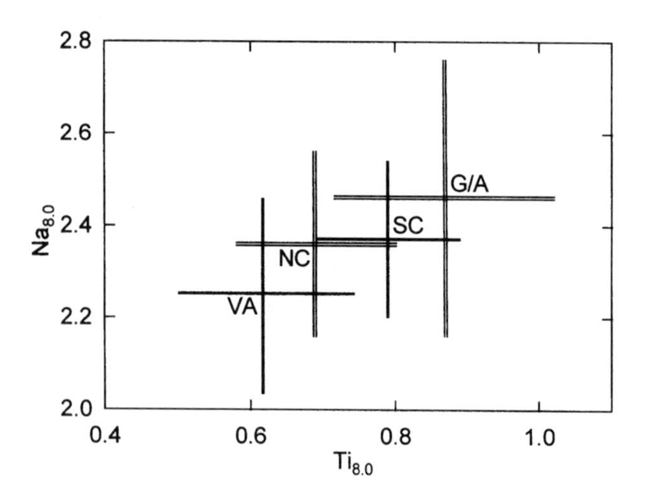

FIGURE 12.10 Bivariate scattergram of $Ti_{8.0}$ versus $Na_{8.0}$, showing limits on the mean ± one standard deviation for the four geographic groups; *G/A*, Georgia/Alabama. The southwestward increase in mean values is apparent.

Georgia rift basin [Daniels, Zietz, and Popenoe 1983; Smith 1983]), the question is whether that thicker crust could have persisted for more than 100 Ma until the generation of these tholeiitic magmas. Another possibility for thickest crust under Georgia and Alabama is based on the assumption that at around 200 Ma the lateral distance from the mantle sources of these rocks to the axis of the main spreading center (to the east) progressively increased from Virginia to Georgia and Alabama. This interpretation seems possible based on some tectonic reconstructions (e.g., de Boer et al. 1988; Withjack, Schlische, and Olsen 1998).

The second possible interpretation is assuming a variation in enrichment of the mantle source. In terms of percentage, the greatest progressive increase in MgO-standardized oxides from northeast to southwest is for $Ti_{8.0}$ (table 12.3; figures 12.6 and 12.10). Langmuir, Klein, and Plank (1992) have noted the possible relationship between $Ti_{8.0}$ and hot spots for MORBs, with highest $Ti_{8.0}$ values for MORBs with the strongest hot-spot signature. De Boer and Snider (1979) have offered evidence for the presence of a lower Mesozoic hot spot in the Carolinas, and it has been suggested (Ragland, Hatcher, and Whittington 1983; Ragland 1991) that the Clubhouse Crossroads region near Charleston, South Carolina (Gottfried, Annell, and Schwarz 1977; Gottfried, Annell, and Byerly 1983), might be a lower Mesozoic center for a mantle plume. According to the model proposed here, a lower Mesozoic hot spot in the Carolinas is too far north. Moreover, it has been demonstrated (Ragland 1991; Ragland, Cummins, and Arthur 1992) that the subsurface basalts in the Clubhouse Crossroads area are compositionally associated with high-Fe quartz (HFQ) diabases that radiate northward from Charleston and are within the N-trending dike swarm (figure 12.2). The mean strike of these dikes is N–S, but they radiate about a maximum angle of approximately 40°. These rocks are not associated with the olivine tholeiites of the NW swarm reported in this study.

Hill (1991) and Leitch, Davies, and Wells (1998) propose the center of the "Newark" (CAMP) plume head, based in part on May's (1971) map, to be near the Georgia coastline (south of Charleston, South Carolina) or in the Caribbean. In contrast, Wilson (1997) places the center in West Africa and refers to it as the center of a superplume. Dupuy et al. (1988) and Bertrand et al. (1999) reported very high-Ti tholeiites in Guyana, Liberia, and Surinam (adjacent to Liberia on Pangea), which may represent the center of the plume. A proposed Mesozoic mantle plume in Florida (Mullins and Lynts 1977; Mueller and Porch 1983) should be considered as well. Chowns and Williams (1983) discovered high-Ti olivine tholeiites (up to approximately 2% TiO_2) under the Georgia Coastal Plain, and Arthur (1988) reported one high-Ti olivine-tholeiite sample from beneath the Florida Coastal Plain. Thus a number of localities in the southeastern United States, the Caribbean, and West Africa have been suggested as centers for a mantle plume (a superplume?) in lower Mesozoic time. It should be pointed out, however, that nonplume models have been proposed to explain the existence of CAMP as well (e.g., Holbrook and Kelemen 1993; McHone 1999; Puffer 1999).

In conclusion, at least two possibilities could explain the geographic variations in major elements: progressively thicker continental crust to the southwest or the center of a mantle plume in the southern part of the southeastern United States, the Caribbean, or, possibly, West Africa. In the southeastern United States, these hypotheses have to be tested with trace elements and isotopes. Some of these data are available, but very few, and virtually none in some key areas, such as Virginia and beneath the Florida–Georgia Coastal Plain. We plan to rectify this problem shortly.

Finally, we should briefly comment on how the conclusions reached in this paper bear on recent developments in research on the CAMP large igneous province. These rocks in the southeastern United States may have been formed during the initial drifting stage of the Pangean breakup (Withjack, Schlische, and Olsen 1998), during a period of only a few million years or less. In addition, they may have been part of an enormous dominantly olivine-tholeiite flood basalt province that existed primarily in the southeastern United States, near the center of CAMP and related to the superplume ultimately responsible for CAMP. The question remains as to whether the rocks' chemical compositions are appropriate for rocks in the center of a superplume. The high-Ti tholeiites in Guyana, Liberia, and Surinam may be more appropriate. Ample aeromagnetic evidence exists for a widespread flood basalt province of presumably lower Mesozoic age beneath some parts of the Atlantic and Gulf Coastal Plains in this region (Withjack, Schlische, and Olsen 1998 and references therein). Exposed dikes of the NW swarm in the southeastern Piedmont and Triassic basins may well have been feeders for the flood basalts, most of which are now eroded away.

If, indeed, the rocks in the Southeast represent the initial drifting stage of the Pangean breakup, then considerably attenuated continental crust undoubtedly existed, and the petrogenetic models proposed here based at least partly on MORBs may have additional validity. Unfortunately, precise and accurate dating of these rocks has proved so difficult that the exact timing of this igneous activity in the southeastern United States is not known (Hames, Ruppel, and Renne 1999), although neither of the two models precludes a synchronous igneous event throughout the entire region. If the rocks are synchronous, a possibility for future research would be the use of chemical data to estimate crustal thicknesses at the time of magma genesis and initial Pangean drifting. At this time, we can only point out with certainty the geographic trend in compositions of these dikes and suggest some possible explanations for those trends.

Acknowledgments

We thank Jon Arthur and Joel Duncan for fruitful discussions and help with the literature, and Rosemarie Raymond for drafting the map (figure 12.2). Paul Ragland is indebted in particular to three former members of the U.S. Geological Survey, who over the years provided leadership in this research area, financial support, and, above all, friendship: Al Froelich (deceased), Henry Bell (deceased), and Dave Gottfried (retired). He also gratefully acknowledges the contributions of several former graduate students—Steve Campbell, Laura Cummins, Katherine Milla, Ken Steele, Pete Weigand, and Dave Whittington—who provided many of the chemical analyses on which the work in this chapter was based. The manuscript for this chapter benefited considerably from Greg McHone's review and Peter M. LeTourneau's suggestions. The graphics and least-squares mixing calculations used here are based on the program Igpet for Windows (Igpetwin) by Michael Carr.

Literature Cited

Arthur, J. D. 1988. *Petrogenesis of Early Mesozoic Tholeiite in the Florida Basement and an Overview of Florida Basement Geology.* Florida Geological Survey Report of Investigations, vol. 97. Tallahassee: Florida Geological Survey.

Bell, H., III, K. G. Books, D. L. Daniels, W. E. Huff, and P. Popenoe. 1980. *Diabase Dikes in the Haile-Brewer Area, South Carolina, and Their Magnetic Properties.* Shorter Contributions to Geophysics, U.S. Geological Survey Professional Paper, no. 1123-C. Washington, D.C.: Government Printing Office.

Bertrand, H., J. P. Liegeois, K. Deckart, and G. Feraud. 1999. High-Ti tholeiites in Guiana and their connection with the central-Atlantic CFB province: Elemental and Nd–Sr–Pb isotopic evidence for a preferential zone of mantle upwelling in the course of rifting. *Eos: Transactions of the American Geophysical Union* 80: S317.

Bryan, W. B., L. W. Finger, and F. Chayes. 1969. Estimating proportions in petrographic mixing equations by least-squares approximation. *Science* 163:926–927.

Bryan, W. B., F. A. Frey, and G. Thompson. 1977. Oldest Atlantic seafloor, Mesozoic basalts from the western North Atlantic margin and eastern North America. *Contributions to Mineralogy and Petrology* 64:223–242.

Campbell, S. K. 1985. Geology, petrography, and geochemistry of the Lemon Springs pluton and associated rocks, Lee County, North Carolina. M.S. thesis, East Carolina University.

Chowns, T. M., and C. T. Williams. 1983. Pre-Cretaceous rocks beneath the Georgia Coastal Plain: Regional implications. In G. S. Gohn, ed., *Studies Related to the*

Charleston, South Carolina, Earthquake of 1886: Tectonics and Seismicity, pp. L1–L42. U.S. Geological Survey Professional Paper, no. 1313. Washington, D.C.: Government Printing Office.

Courtillot, V., C. Jaupart, I. Manigheitti, P. Tapponier, and J. Besse. 1999. On causal links between flood basalts and continental breakup. *Earth and Planetary Science Letters* 166:177–195.

Cummins, L. E. 1987. Geochemistry, mineralogy, and origin of Mesozoic diabase dikes of Virginia. Ph.D. diss., Florida State University.

Cummins, L. E., J. D. Arthur, and P. C. Ragland. 1992. Classification and tectonic implications of early Mesozoic magma types of the circum-Atlantic. In J. H. Puffer and P. C. Ragland, eds., *Eastern North American Mesozoic Magmatism,* pp. 119–136. Geological Society of America Special Paper, no. 268. Boulder, Colo.: Geological Society of America.

Daniels, D. L., L. Zietz, and P. Popenoe. 1983. Distribution of subsurface lower Mesozoic rocks in the southeastern United States as interpreted from regional aeromagnetic and gravity maps. In G. S. Gohn, ed., *Studies Related to the Charleston, South Carolina, Earthquake of 1886: Tectonics and Seismicity,* pp. K1–K24. U.S. Geological Survey Professional Paper, no. 1313. Washington, D.C.: Government Printing Office.

de Boer, J., J. G. McHone, J. H. Puffer, P. C. Ragland, and D. Whittington. 1988. Mesozoic and Cenozoic magmatism. In R. E Sheridan and G. A. Grow, eds., *The Atlantic Continental Margin,* pp. 217–241. Vol. I-2 of *The Geology of North America.* Boulder, Colo.: Geological Society of America.

de Boer, J., and F. G. Snider. 1979. Magnetic and chemical variations of Mesozoic diabase dikes from eastern North America: Evidence for a hot spot in the Carolinas? *Geological Society of America Bulletin* 90:185–198.

Dupuy, C., J. Marsh, J. Dostal, A. Michard, and S. Testa. 1988. Asthenospheric and lithospheric sources for Mesozoic dolerites from Liberia (Africa): Trace element and isotopic evidence. *Earth and Planetary Science Letters* 87:100–110.

Fujii, T., and C. M. Scarfe. 1985. Compositions of liquids coexisting with spinel lherzolite at 10 kbar and the genesis of MORBs. *Contributions to Mineralogy and Petrology* 90:18–28.

Gottfried, D., C. S. Annell, and G. R. Byerly. 1983. Geochemistry and tectonic significance of subsurface basalts near Charleston, South Carolina: Clubhouse Crossroads test holes #2 and #3. In G. S. Gohn, ed., *Studies Related to the Charleston, South Carolina, Earthquake of 1886: Tectonics and Seismicity,* pp. A1–A19. U.S. Geological Survey Professional Paper, no. 1313. Washington, D.C.: Government Printing Office.

Gottfried, D., C. S. Annell, and L. J. Schwarz. 1977. Geochemistry of subsurface basalt from the deep corehole (Clubhouse Crossroads core 1) near Charleston, South Carolina. In D. W. Rankin, ed., *Studies Related to the Charleston, South Carolina, Earthquake of 1886: A Preliminary Report,* pp. 91–114. U.S. Geological Survey Professional Paper, no. 1028. Washington, D.C.: Government Printing Office.

Hall, D. J. 1990. Gulf coast–east coast magnetic anomaly: Root of the main crustal decollement for the Appalachian-Ouachita orogeny. *Geology* 18:862–865.

Hames, W., C. Ruppel, and P. Renne. 1999. Age of basaltic dikes and flows of the southeastern U.S. in the context of the circum-Atlantic large igneous province. *Eos: Transactions of the American Geophysical Union* 80: S318.

Hill, R. I. 1991. Starting plumes and continental breakup. *Earth and Planetary Science Letters* 104:398–416.

Hirschmann, M. M., and E. M. Stolper. 1996. A possible role for garnet pyroxenite in the origin of the "gamet signature" in MORB. *Contributions to Mineralogy and Petrology* 124:185–208.

Holbrook, W. S., and P. B. Kelemen. 1993. Large igneous province on the US Atlantic margin and implications for magmatism during continental breakup. *Nature* 364:433–436.

Irvine, T. N., and W. R. A. Baragar. 1971. A guide to the chemical classification of the common volcanic rocks. *Canadian Journal of Earth Sciences* 8:523–548.

Jaques, A. L., and D. H. Green. 1980. Anhydrous melting of peridotite at 0–15kb and the genesis of tholeiitic basalt. *Contributions to Mineralogy and Petrology* 73: 287–310.

Kinzler, R. J., and T. L. Grove. 1992a. Primary magmas of mid-ocean ridge basalts. 1. Experiments and methods. *Journal of Geophysical Research* 97:6885–6906.

Kinzler, R. J., and T. L. Grove. 1992b. Primary magmas of mid-ocean ridge basalts. 2. Applications. *Journal of Geophysical Research* 97:6907–6926.

Klein, E. M., and C. H. Langmuir. 1987. Global correlations of ocean ridge basalt chemistry with axial depth and crustal thickness. *Journal of Geophysical Research* 92:8089–8115.

Klein, E. M., and C. H. Langmuir. 1989. Local versus global variations in ocean ridge basalt composition. Reply. *Journal of Geophysical Research* 94:4241–4252.

Langmuir, C. H., E. M. Klein, and T. Plank. 1992. Petrological systematics of mid-ocean ridge basalts: Constraints on melt generation beneath ocean ridges. In J. P. Morgan, D. K. Blackman, and J. M. Sinton, eds., *Mantle Flow and Melt Generation at Mid-ocean Ridges,* pp. 183–280. American Geophysical Union Geophysical Monograph, no. 71. Washington, D.C.: American Geophysical Union.

Leitch, A. M., G. F. Davies, and M. Wells. 1998. A plume head melting under a rifting margin. *Earth and Planetary Science Letters* 161:161–177.

Le Maitre, R. W. 1982. *Numerical Petrology: Statistical Interpretation of Geochemical Data.* Amsterdam: Elsevier.

Marzoli, A., P. R. Renne, E. M. Piccirillo, M. Ernesto, G. Bellieni, and A. De Min. 1999. Extensive 200-million-year-old continental flood basalts of the Central Atlantic Magmatic Province. *Science* 284:616–618.

May, P. O. 1971. Pattern of Triassic–Jurassic diabase dikes around the North Atlantic in the context of pre-drift positions of the continents. *Geological Society of America Bulletin* 82:1285–1292.

McBride, J. H., and K. D. Nelson. 1988. Integration of COCORP deep reflection and magnetic anomaly analysis in the southeastern United States: Implications for the origin of the Brunswick and East Coast Magnetic Anomalies. *Geological Society of America Bulletin* 100: 436–445.

McHone, J. G. 1999. Non-plume magmatism and rifting during the opening of the central north Atlantic Ocean. *Eos: Transactions of the American Geophysical Union* 80:S317.

Milla, K. A. 1990. Geochemistry of the lower Mesozoic Talbotton diabase dikes of west-central Georgia. M.S. thesis, Florida State University.

Milla, K. A., and P. C. Ragland. 1992. Early Mesozoic Talbotton diabase dikes in west-central Georgia: Compositionally homogeneous high-Fe quark tholeiites. In J. H. Puffer and P. C. Ragland, eds., *Eastern North American Mesozoic Magmatism,* pp. 347–360. Geological Society of America Special Paper, no. 268. Boulder, Colo.: Geological Society of America.

Mueller, P. A., and J. W. Porch. 1983. Tectonic implications of Paleozoic and Mesozoic igneous rocks in the subsurface of peninsular Florida. *Gulf Coast Association Geological Society Transcripts* 33:169–183.

Mullins, H. T., and G. W. Lynts. 1977. Origin of the northwestern Bahamas platforms: Review and reinterpretation. *Geological Society of America Bulletin* 88:1447–1461.

Niu, Y., and R. Batiza. 1991. An empirical method for calculating melt compositions produced beneath mid-ocean ridges: Application for axis and off-axis (sea mounts) melting. *Journal of Geophysical Research* 96:21753–21777.

Niu, Y., and R. Batiza. 1993. Chemical variation trends at fast and slow spreading mid-ocean ridges. *Journal of Geophysical Research* 98:7887–7902.

Olsen, P. E., R. W. Schlische, and M. S. Fedosh. 1996. 580 ky duration of the Early Jurassic flood basalt event in eastern North America estimated using Milankovitch cyclostratigraphy. In M. Morales, ed., *The Continental Jurassic,* pp. 11–22. Museum of Northern Arizona Bulletin, no. 60. Flagstaff: Museum of Northern Arizona.

Pearce, T. H., B. E. Gorman, and T. C. Birkett. 1977. The relationship between major element chemistry and tectonic environment of basic and intermediate volcanic rock. *Earth and Planetary Science Letters* 36:121–132.

Plank, T., and C. H. Langmuir. 1988. An evaluation of the global variations in the major element chemistry of arc basalts. *Earth and Planetary Science Letters* 90: 349–370.

Popenoe, P., and L. Zietz. 1977. The nature of the geophysical basement beneath the Coastal Plain of South Carolina and northeastern Georgia. In D. W. Rankin, ed., *Studies Related to the Charleston, South Carolina, Earthquake of 1886: A Preliminary Report,* pp. 119–138. U.S. Geological Survey Professional Paper, no. 1028. Washington, D.C.: Government Printing Office.

Puffer, J. H. 1992. Eastern North American flood basalts in the context of the incipient breakup of Pangaea. In J. H. Puffer and P. C. Ragland, eds., *Eastern North American Mesozoic Magmatism,* pp. 95–118. Geological Society of America Special Paper, no. 268. Boulder, Colo.: Geological Society of America.

Puffer, J. H. 1994. Initial and secondary Pangean basalts. In B. Beauchamp, A. F. Embry, and D. Glass, eds., *Pangea: Global Environments and Resources,* pp. 85–95. Canadian Society of Petroleum Geologists Memoir, no. 17. Calgary: Canadian Society of Petroleum Geologists.

Puffer, J. H. 1999. Tectonic controls on geochemistry of continental flood basalts. *Eos: Transactions of the American Geophysical Union* 80:S317.

Ragland, P. C. 1991. Mesozoic magmatism. In J. W. Horton Jr. and V. A. Zullo, eds., *Geology of the Carolinas: Carolina Geological Society 50th Anniversary Volume,* pp. 171–190. Knoxville: University of Tennessee Press.

Ragland, P. C., L. E. Cummins, and J. D. Arthur. 1992. Compositional patterns for early Mesozoic diabases from South Carolina to central Virginia. In J. H. Puffer and P. C. Ragland, eds., *Eastern North American Mesozoic Magmatism,* pp. 309–331. Geological Society of America Special Paper, no. 268. Boulder, Colo.: Geological Society of America.

Ragland, P. C., and M. J. Defant. 1983. Silica standardization: A discriminant technique applied to a volcanic arc system. *Earth and Planetary Science Letters* 64:387–395.

Ragland, P. C., R. D. Hatcher Jr., and D. Whittington. 1983. Juxtaposed Mesozoic diabase dike sets from the Carolinas. *Geology* 11:394–399.

Ragland, P. C., S. A. Kish, and W. C. Parker. 1997. Compositional patterns for lower Mesozoic olivine tholeiitic diabase dikes in the Deep River basin, North Carolina. In *Proceedings from TRIBI Workshop on Triassic Basins,* p. 14. Southeastern Geology. Durham, N.C.: Department of Geology, Duke University.

Ragland, P. C., J. J. W. Rogers, and P. S. Justus. 1968. Origin and differentiation of Triassic dolerite magmas, North Carolina, U.S.A. *Contributions to Mineralogy and Petrology* 20:56–80.

Smith, D. W. 1983. Basement model for the panhandle of Florida. *Gulf Coast Association Geological Society Transactions* 33:203–208.

Steele, K. F. 1971. Chemical variation parallel and perpendicular to strike in two Mesozoic dolerite dikes, North Carolina and South Carolina. Ph.D. diss., University of North Carolina.

Stiegler, S. E. 1976. *A Dictionary of Earth Sciences.* London: Macmillan.

Stolper, E. M. 1980. A phase diagram for mid-ocean basalts: Preliminary results and implications for petrogenesis. *Contributions to Mineralogy and Petrology* 74:13–27.

Sutter, J. F. 1985. Progress on geochronology of Mesozoic diabases and basalts. In G. P. Robinson Jr. and A. J. Froelich, eds., *Proceedings of the Second U.S. Geological Survey Workshop on the Early Mesozoic Basins of the Eastern United States,* pp. 110–114. U.S. Geological Survey Circular, no. 946. Washington, D.C.: Government Printing Office.

Sutter, J. F. 1988. Innovative approaches to the dating of igneous events in the early Mesozoic basins of the eastern United States. In A. J. Froelich and G. P. Robinson Jr., eds., *Studies of the Early Mesozoic Basins of the Eastern United States,* pp. 194–200. U.S. Geological Survey Bulletin, no. 1776. Washington, D.C.: Government Printing Office.

Turner, S., and C. Hawkesworth. 1995. The nature of the sub-continental mantle: Constraints from the major-element composition of continental flood basalts. *Chemical Geology* 120:295–314.

Walker, D., T. Shibata, and S. E. DeLong. 1979. Abyssal tholeiites from the Oceanographer fracture zone. II. Phase equilibria and mixing. *Contributions to Mineralogy and Petrology* 70:111–125.

Warner, R. D., D. S. Snipes, L. L. Burnett, J. A. Wylie, L. A. Sacks, and J. C. Steiner. 1986. Shoals Junction and Due West dolerites, South Carolina. *Southeastern Geology* 26:141–152.

Warner, R. D., D. S. Snipes, S. S. Hughes, J. C. Steiner, M. W. Davis, P. R. Manoogian, and R. A. Schmitt. 1985. Olivine-normative dolerite dikes from western South Carolina: Mineralogy, chemical composition, and petrogenesis. *Contributions to Mineralogy and Petrology* 90:386–400.

Weigand, P. W. 1970. Major and trace element geochemistry of Mesozoic dolerite dikes from eastern North America. Ph.D. diss., University of North Carolina.

Weigand, P. W., and P. C. Ragland. 1970. Geochemistry of Mesozoic dolerite dikes from eastern North America. *Contributions to Mineralogy and Petrology* 29:195–214.

Whittington, D. 1988a. Chemical and physical constraints on petrogenesis and emplacement of ENA olivine diabase magmas. In W. Manspeizer, ed., *Triassic–Jurassic Rifting: Continental Breakup and the Origin of the Atlantic Ocean and the Passive Margins,* pp. 557–577. Developments in Geotectonics, no. 22. Amsterdam: Elsevier.

Whittington, D. 1988b. Mesozoic diabase dikes of North Carolina. Ph.D. diss., Florida State University.

Wilson, M. 1997. Thermal evolution of the central Atlantic passive margins: Continental breakup above a Mesozoic super-plume. *Journal of the Geological Society of London* 154:491–495.

Withjack, M. O., R. W. Schlische, and P. E. Olsen. 1998. Diachronous rifting, drifting, and inversion on the passive margin of eastern North America: An analog for other passive margins. *American Association of Petroleum Geologists Bulletin* 82:817–835.

13

Evidence for Predominant Lateral Magma Flow Along Major Feeder-Dike Segments of the Eastern North America Swarm Based on Magnetic Fabric

Jelle Zeilinga de Boer, Richard E. Ernst, and Andrew G. Lindsey

Studies of magma flow can be used to test source model hypotheses for giant radiating diabase dike swarms. Long "master" dikes in the eastern North American segment of the 200 Ma circum-Atlantic dike system were sampled, and flow directions were determined using the anisotropy of magnetic susceptibility (AMS) technique. Measurements were made on 215 cores collected at 25 sites distributed along the Higganum-Holden and Christmas Cove dikes of New England. These dikes can be traced for distances of approximately 250 km and 100 km, respectively.

Chemical and paleomagnetic data suggest that emplacement of these dikes was contemporaneous and that they represent different segments of a single intrusion. Magnetic susceptibility varied from 10 to 31×10^{-3} SI. The data are characterized by AMS ratios typically of 1 to 8%. A few samples exhibited anisotropy values of 9 to 12%. The higher values (> 10%) usually occurred in narrow dike offshoots. Three principal magnetic fabrics were identified. Most sites provide dike-parallel subhorizontal susceptibility maxima (K_{max}) axes, and susceptibility minima (K_{min}) axes at right angles to the dike plane, suggesting lateral subhorizontal flow. Evidence for subvertical flow in the form of subvertical K_{max} is found only for sites in or near "doglegs" (abrupt changes in dike trend). Some sites show a magnetic fabric with K_{min} in the dike plane, which might be due to late-stage compaction of magma in the conduit.

The AMS data for the Christmas Cove dike provided relatively tight groupings with subhorizontal K_{max}. For the Higganum-Holden dike, K_{max} directions are also predominantly subhorizontal, but show more scatter, perhaps because the southernmost segment of the dike underwent a minor thermochemical overprint, identified through paleomagnetism.

The rather consistent evidence for lateral flow in dikes from New England and the Canadian Maritime Provinces (determined previously by Canadian researchers) supports a plume model for the magma source. This evidence is consistent with previous models based on dike trend convergence, suggesting that this plume may have developed near the present-day Blake Plateau during the embryonic stage of the Atlantic opening.

The 200 Ma Eastern North America dike swarm forms part of one of the largest intrusive systems in the world, the Central Atlantic Magmatic Province (CAMP)

(Marzoli et al. 1999). The dikes extend over a distance of approximately 3,000 km along the eastern coasts of the United States and Canada, from Georgia to Newfoundland (de Boer et al. 1988). Other swarms belonging to this igneous province occur in western Africa (Sebai et al. 1991) and northern South America (Marzoli et al. 1999). Overall the dikes show a convergent pattern, with a focal point near the present-day Blake Plateau (May 1971; Bertrand 1991; Ernst et al. 1995) (figure 13.1).

The dikes were emplaced before the breakup of Pangea, during the embryonic phase in the opening of the Atlantic. Flows and sills fed by some of the dikes were emplaced in the Newark group of rift basins, which generally parallel the Appalachian tectonic grain (Philpotts 1992). McHone (1996, chapter 9 in this volume) believes that the flows extended beyond the rift basins and formed a huge flood basaslt province. So far, however, no traces have been found of flows emplaced outside the rift basins.

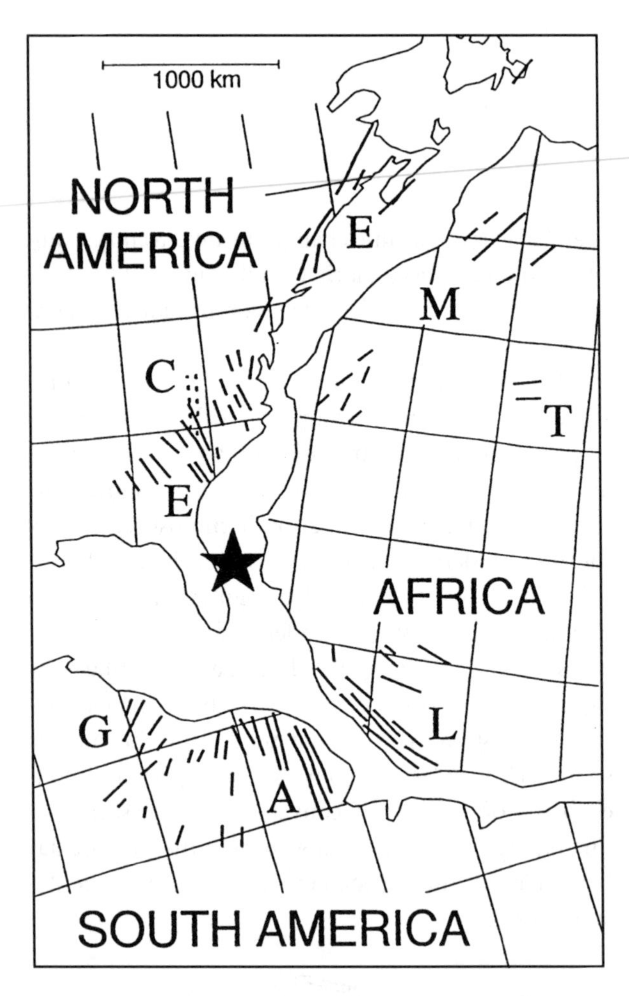

FIGURE 13.1 Circum-Atlantic dike swarms of Early Jurassic age. The star indicates the region of dike convergence near Blake Plateau; *A*, Amapa; *C*, Charleston; *E*, eastern North America; *G*, Guiana; *L*, Liberia; *M*, Morocco; *T*, Taoudenni dike swarms. (After Ernst and Buchan 1997)

Subsidence in the basins was initiated prior to the Eastern North America event, around 230 Ma (de Boer et al. 1988; Manspeizer and Cousminer 1988; Hill 1991; de Boer 1992). A magmatic event associated with this earliest tectonic activity was restricted to the coastal belt of New England, where the dikes (termed Coastal New England dikes) have trends similar to those of the Eastern North America Province but are compositionally distinct (generally alkalic) (McHone and Butler 1984; Pe-Piper, Jansa, and Lambert 1992; Swanson 1992).

The Eastern North America dikes and their extrusive equivalents in the rift valleys have been classified into two major geochemical groups (Ragland, Cummins, and Arthur 1992). In the southern Appalachians, the dikes are predominantly olivine-normative tholeiites, whereas those in the northern Appalachians are exclusively quartz-normative tholeiites with two distinct affinities, Ti enriched and Ti depleted (Cummins, Arthur, and Ragland 1992; Ragland, Cummins, and Arthur 1992).

The Higganum-Holden and Christmas Cove dikes in New England (figure 13.2) possess high-Ti quartz-normative (HTQ) chemistry affinities and are characterized by an unusual monoclinic morphology of their orthopyroxene phenocrysts (McHone 1996). Furthermore, chemical analyses (specifically the ratio of titanium to magnesium) indicate that their magmas are probably derived from the same source area (figure 13.3).

Paleomagnetic evidence suggests that dike emplacement was contemporaneous. McEnroe (1989, 1993) obtained a paleopole with latitude 59.7° north and longitude 77.9° east (n = 55; dp 2.0°, dm 3.6°) for the Holden segment of the Higganum-Holden dike (dp and dm represent the semiminor and semimajor axes of the ellipse at 95% confidence limit around the pole). New paleomagnetic data (J. Z. de Boer, unpublished) for the Higganum segment of the Higganum-Holden dike at its type locality, however, have provided a pole with latitude 57.2° north and longitude 92.8° west (n = 19; dp 2.6°, dm 3.5°). The Christmas Cove dike at its type locality at Christmas Cove has provided a paleopole with latitude 56.9° north and longitude 94.5° west (n = 7; dp 1.8°, dm 2.4°). The latter two

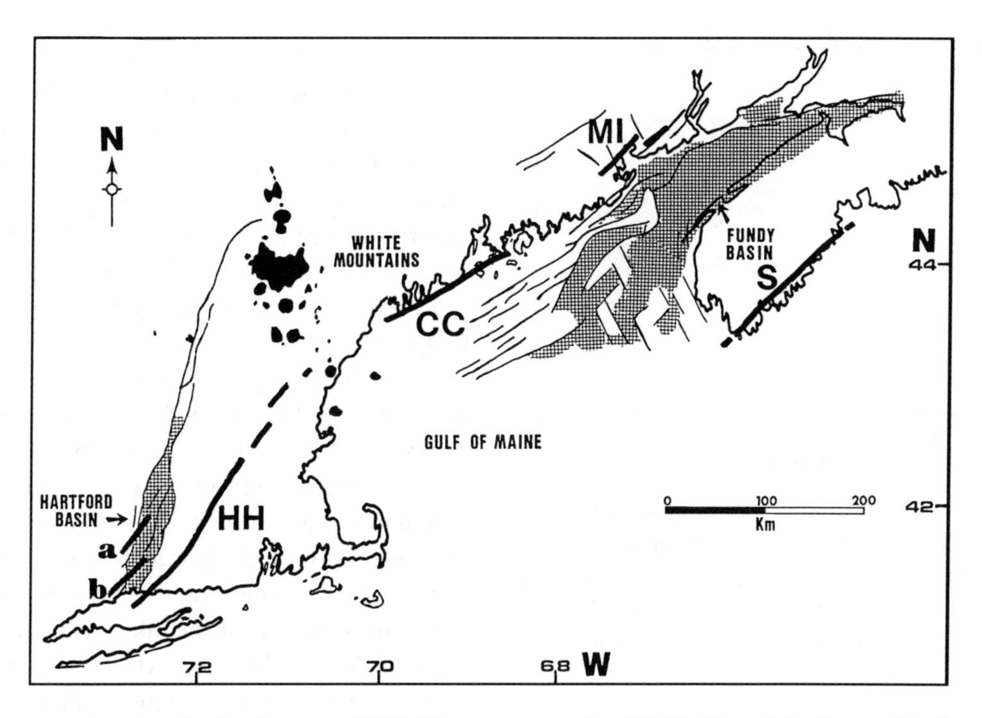

FIGURE 13.2 Regional extent of high-Ti quartz-normative (HTQ) dikes of the Eastern North America swarm in New England and the Maritime Provinces: *HH*, Higganum-Holden; *CC*, Christmas Cove; *MI*, Ministers Island; *S*, Shelburne; *a* and *b*, other dikes of the Eastern North America swarm in Connecticut. The crosshatched pattern represents rift basins; thin lines are faults.

poles plot within each other's cone of confidence and indicate that emplacement of magma along the Higganum-Holden and Christmas Cove dikes was most probably contemporaneous. Chemical similarities, petrographic peculiarity, and paleomagnetic data thus suggest that the Higganum-Holden and Christmas Cove dikes represent a single intrusion.

It is generally believed that an age range from 205 to 195 Ma is the most representative for all Eastern North America dikes and flows. U-Pb ages on the Palisades sill, which also belongs to the HTQ group, suggest emplacement around 201 ± 1 Ma (Dunning and Hodych 1990). This age agrees well with ^{40}Ar-^{39}Ar plateau ages of 200 ± 1, 201 ± 1, and 202 ± 1 Ma obtained by Sutter (1988) for various diabase sheets and 202 ± 4 for the Shelburne dike of the Canadian Maritimes (Dunn et al. 1998). Similar ^{40}Ar-^{39}Ar ages are found in Iberia (Messejana dike [Dunn et al. 1998]), Africa (Sebai et al. 1991), and South America, indicating an overall age 200 ± 4 Ma for the event (Marzoli et al. 1999).

Some authors have attributed the origin of the Early Jurassic Circum-Atlantic Dike System ("Central Atlantic Reconstructed Swarm" [Ernst et al. 1995]) to the arrival of a mantle plume (e.g.,White and McKenzie 1989; Oliveira, Tarney, and João 1990; Hill 1991). Alternative models link the Eastern North America and other swarms of the Circum-Atlantic Dike System to magmatic processes related to proto-Atlantic rifting. Thus, on one end of the spectrum, the magma source is modeled as one major mantle plume located near Florida below the Blake Plateau (Greenough and Hodych 1990) or two hot spots (de Boer 1992), one off Florida and the other in the Gulf of Maine. At the other end of the spectrum are the models that suggest that magmas were derived from multiple, rather localized sources below the rift valleys (Bertrand 1991; Philpotts 1992). Cummins, Arthur, and Ragland (1992) proposed the concept of a mantle "keel" that developed below the zone of maximal crustal extension and in which a low degree of partial melting provided the magmas that rose subvertically to the surface.

To distinguish between these models, analyses of flow directions should be carried out in dikes proximal to the site of the hypothetical plume head(s) and also in dikes far removed from that site. A mantle plume model would imply subvertical flow in proximity to the plume center and lateral subhorizontal flow in dis-

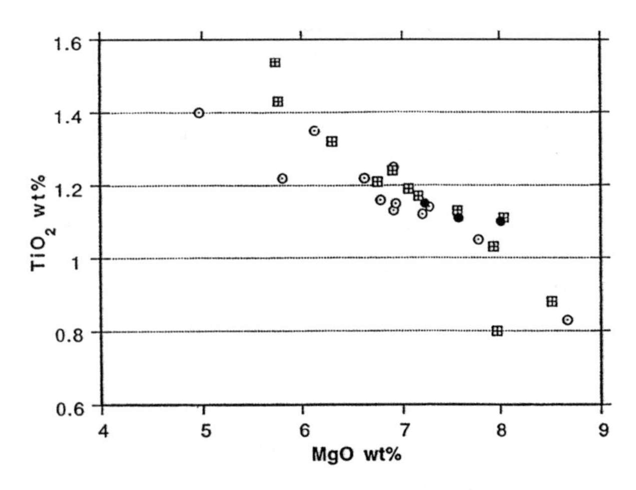

FIGURE 13.3 TiO_2:MgO plot for the Higganum-Holden and Christmas Cove dikes. Squares represent data from the Higganum-Holden dike (Weigand and Ragland 1970; Koza 1976); open circles, data from the Christmas Cove dike (McEnroe 1989, 1993); and closed circles, unpublished data for the Christmas Cove dike.

tal zones (Ernst and Baragar 1992). The multiple (or distributed) source model would predict mainly vertical flow in the distal areas as well.

This study reports new flow direction data from the New England dikes that are distal from the Blake Plateau and, hence, would be predicted to exhibit predominantly lateral flow if a mantle plume beneath the plateau were the source. Alternative models implying that the magmas originate from partial melting in a underlying mantle "keel" would predict predominant vertical flow in the sampled dikes.

Previous Evidence for Magma Flow in Distal Eastern North America Dikes

Based on measurements of the anisotropy of anhysterictic magnetic susceptibility, Greenough and Hodych (1990) determined subhorizontal flow in the Shelburne dike of Nova Scotia and the Ministers Island dike (also referred to as Passamaquoddy Bay dike) of New Brunswick and Maine (figure 13.2). Furthermore, these dikes contain ramping structures tens of centimeters long that indicate flow toward the northeast.

Papezik et al. (1988) noticed systematic chemical changes along strike in the North Mountain lava flows of Nova Scotia and concluded that the magma in their feeder dike(s), which are not exposed, flowed northeast. A comparison of North Mountain Basalt chemistry from the southwesternmost (Digby) and northeasternmost (Cape Split) exposures showed the former to have higher Ba:La and La:Yb ratios and lower Ti content, all characteristics that would be expected closer to the center of a plume (Dostal and Greenough 1992; Pe-Piper, Jansa, and Lambert 1992).

In contrast, detailed petrographic observations by Philpotts and Asher (1994) on three exposures (approximately 5 km apart) in a single en echelon segment of the Higganum dike in Connecticut show a more complicated flow history. Using imbrication of tabular feldspar laths, sheared and/or rotated phenocrysts and granophyre wisps near its upper contact, they found evidence that suggests upward flow in the central (Higganum) site and subhorizontal flow outward from this center to both ends of the dike segment. In addition, their work provided evidence for a late-stage backflow—that is, down-dip motion of the magma—a phenomenon that may be due either to drainback of the magma caused by widening or lengthening of the dike or to emplacement of sills at depth.

Higganum-Holden and Christmas Cove Dikes

The Higganum-Holden and Christmas Cove dikes were selected for this anisotropy of magnetic susceptibility (AMS) study because they are distal to the region of the hypothetical plume head near the Blake Plateau and because they provide good exposures of fresh rock on a regional scale. Sampling details are given in Lindsey (1995).

The Higganum-Holden dike trends generally NE–SW and extends from East Haven, Connecticut, to Manchester, New Hampshire, over a distance of approximately 250 km (figure 13.4A). Farther north in southeastern New Hampshire, it probably continues as the Onway dike (Sundeen and Huff 1992). The attitude of the dike varies locally. The intrusive occasionally "doglegs" from NE to N–NE and is frequently offset en echelon, consistently stepping north in a northeast direction. Figure 13.5A shows two maxima for faults that controlled magma emplacement: NE-trending extensional fractures and N–NE transform faults. Dike segments are predominantly subvertical, but in a few places the dip is approximately 45°, and

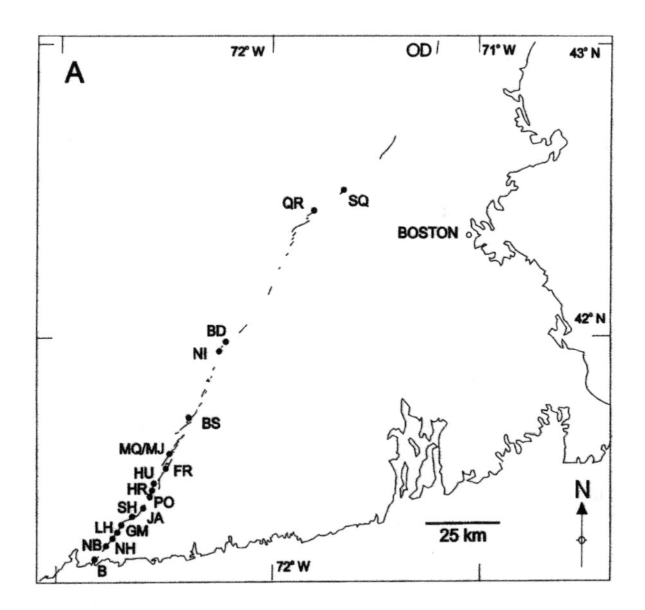

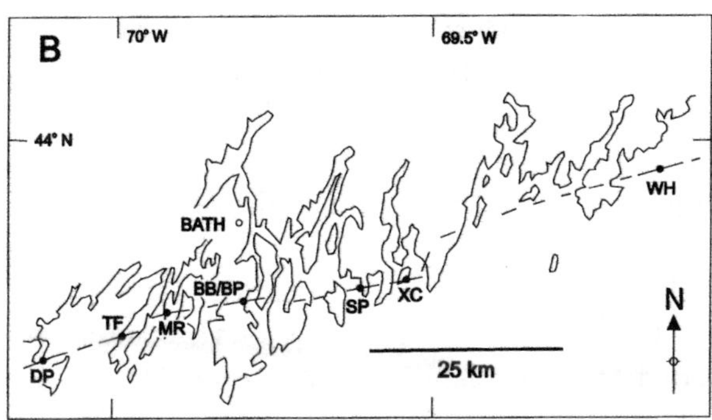

FIGURE 13.4 (*A*) Higganum-Holden dike with sampling sites: *B*, Branford; *BD*, Bearden; *BS*, Bear Swamp; *FR*, Flatbrook Road; *GM*, Guilford Monastery; *HR*, Higganum Railroad; *HU*, Hurd State Park; *JA*, Jackson Road; *LH*, Long Hill Road; *MJ*, Marlborough Johnson Road; *MQ*, Marlborough Quarry; *NB*, North Branford; *NH*, Notch Hill; *NI*, Nipmuck State Forest; *PO*, Ponset Road; *QR*, Quanapoxet Reservoir; *SH*, Summer Hill Road; *SQ*, Sterling Quarry. The Onway dike (*OD*) is a probable extension of the Higganum-Holden dike. (*B*) Christmas Cove dike with sampling sites: *BB*, Below Bijhouwer's; *BP*, Bijhouwer's Property; *DP*, Doyle Point; *MR*, Mountain Road; *SP*, Spruce Point; *TF*, Tank Farm; *WH*, Whitehead Island; *XC*, Christmas Cove.

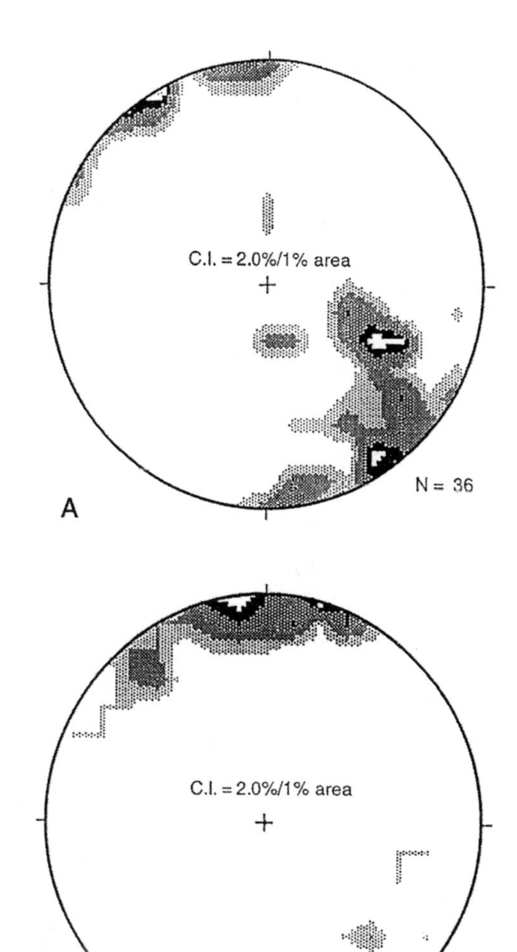

FIGURE 13.5 (*A*) Equal-area stereonet (lower hemisphere) of normals to contacts of the Higganum-Holden dike (2% contour interval). (*B*) Equal-area stereonet (lower hemisphere) of normals to contacts of the Christmas Cove dike (2% contour interval).

in at least one site (JA) the intrusion formed as a subhorizontal sill. Cooling columns, which commonly have diameters up to 1 m, provide an accurate measure of local dike attitude in most outcrops.

Evidence exists for at least two magmatic pulses in the Higganum-Holden dikes. The younger intrusion is generally finer-grained and exhibits clear chill contacts against the older dike. The older dike is usually more weathered and frequently exhibits anomalous magnetizations. Samples for this study were collected from only the younger dike. Dike width varies from 10 to 50 m; thin (1 to 3 m) offshoots occur but are relatively rare.

The Christmas Cove dike (Hussey 1971; McHone et al. 1995) trends generally E–NE to W–SW (figure 13.4B). It extends from Portland to White Head Island, Maine, a distance of approximately 100 km. The dike may extend farther east to Schoodic Point and may connect with the Ministers (Passamaquoddy) dike in New Brunswick (Dunn and Stringer 1990). If so, its cumulative length is approximately 400 km. This dike, too, varies considerably in attitude and width on a local scale. "Doglegs" are common and caused the dike to step north in an E–NE direction. One of the best exposures occurs on White Head Island, where both contacts are exposed over distances of more than 100 m. The magma here used E–W extensional fractures and N–NE transform faults, resulting in a general ENE–WSW trend (figure 13.5B). Dike attitude varies from subvertical to steeply dipping. A major subhorizontal sill, which appears to be associated with the dike, crops out on the shores of Pemaquid Bay (near site XC). Dike width varies from 10 to 30 m.

Anisotropy of Magnetic Susceptibility

Apparent directions for magma flow in the Higganum-Holden and Christmas Cove dikes were determined using the magnetic fabric technique. Magnetic susceptibility is a second-rank tensor quantity that relates the strength of the Earth's magnetic field to the induced magnetization in rocks. The AMS fabric is represented graphically as a triaxial ellipsoid, and data typically are presented in terms of the orientation (declination and inclination) and magnitude of the three principal axes of this ellipsoid (Rochette, Jackson, and Aubourg 1992; Tarling and Hrouda 1993). In diabase dikes, the AMS technique records the anisotropic distribution of the titano-magnetite grains.

The rock fabric is commonly dominated by lath- and tabular-shaped plagioclase crystals, and the magnetite distribution mimics this fabric regardless of the timing of magnetite growth. Early-crystallizing magnetite grains are redistributed by flow during plagioclase growth, and late-crystallizing magnetite infills between plagioclase grains (Hargraves, Johnson, and Chan 1991). The result in both cases is that the distribution of the magnetite grains mimics that of plagioclase and hence that of the overall rock texture. The anisotropy in the distribution of the isotropic magnetite crystals is the key control on the magnetic fabric (Hargraves, Johnson, and Chan 1991; Stephenson 1994; Cañón-Tapia 1996).

The AMS technique is sensitive to a few percent anisotropy, making it useful for revealing very subtle fabrics in diabase dikes. When the AMS fabric is caused by magma flow, it typically is aligned such that the minimum susceptibility axis (K_{min}) is perpendicular to the dike wall, and the maximum axis (K_{max}) is in the direction of flow (Knight and Walker 1988).

Methodology

From 25 sites along the Higganum-Holden and Christmas Cove dikes, 215 samples were collected (figure 13.4A and B; table 13.1). All samples were drilled and oriented using a magnetic compass held 40 cm from the rock surface. Orientation was checked in several sites using the sun compass. Samples were collected as near to the dike margin as possible, and most were obtained at more or less regular intervals toward the dike center. The low-field AMS fabric in the second specimen from each core was measured using an SI-2 Magnetic Susceptibility and Anisotropy Instrument manufactured by Sapphire Instruments, which is designed to measure magnetic susceptibility in different specimen orientations.

Mean results from each site were determined according to the tensor-averaging procedure proposed by Jelinek (1978), based on the work of Hext (1963) and discussed by Ernst and Pearce (1989) and by Lienert (1991). Calculations and plots were generated using the computer program discussed by Lienert (1991).

The data are displayed in figure 13.6 and summarized in table 13.1. Figure 13.7 shows site means converted to a common vertical dike orientation. The data are interpreted in terms of several fabric types. Those in which the maxima are parallel to the dike plane and the minima are aligned perpendicular to the dike plane are most easily interpreted in terms of magma flow. In this case, the inclination of the maximum susceptibility axis (in the dike plane) provides the direction of magma flow. Thus shallow-dipping K_{max} axes indicate subhorizontal flow (e.g., figure 13.6D, site WH), whereas steep inclinations give a subvertical direction of flow (e.g., figure 13.6A, site GM). In near-margin samples, it is also possible to observe an imbrication wherein the susceptibility axes are inclined at low angles to the margin. Such imbrication has been used to determine the polarity of flow (Knight and Walker 1988; Ernst 1994) and is under investigation for this dataset. Another fabric type is one in which the minimum susceptibility axis is aligned steeply within the dike plane. This fabric is interpreted as being due to vertical compaction (and horizontal extension through dike widening or dike lengthening) of partially crystallized magma (Park, Tanczyk, and Desbarats 1988; Ernst and Baragar 1992).

TABLE 13.1 Summary of AMS Data for the Higganum-Holden and Christmas Cove Dikes

Site	N	D_{max}	I_{max}	A_{max}	D_{int}	I_{int}	A_{int}	D_{min}	I_{min}	A_{min}	K_{mean} ($\pm$ 1σ)	$K_{n\text{-}max}$	$K_{n\text{-}int}$	$K_{n\text{-}min}$	P_J	T	$A\%$	$B\%$	$H\%$
B	14	57.6	28.6	26:10	277.8	54.4	27:16	158.6	19.2	19:10	2.24 (0.75)	1.008	1.000	0.992	1.016	0.004	1.18	0.00	1.6
NB	13	27.9	31.7	10:6	250.4	50.1	12:10	132.1	21.6	12:6	2.62 (0.57)	1.011	1.000	0.990	1.021	−0.042	1.57	0.10	2.1
NH	9	50.9	40.7	18:12	252.3	47.2	33:14	150.4	10.8	33:12	1.59 (0.37)	1.010	0.999	0.992	1.018	−0.218	1.42	0.39	1.8
LH	3	81.9	34.9	79:62	329.4	28.8	88:60	210.1	41.6	88:50	1.68 (0.78)	1.008	0.998	0.995	1.014	−0.536	1.14	0.69	1.3
GM	8	358.7	84.9	25:11	92.2	0.3	62:13	182.3	5.0	62:13	1.87 (0.67)	1.010	0.996	0.994	1.017	−0.748	1.48	1.18	1.6
SH	6	67.1	74.5	34:19	289.0	11.7	43:27	196.9	10.1	45:11	1.20 (0.44)	1.010	0.999	0.992	1.018	−0.218	1.42	0.39	1.8
JA	5	85.6	18.6	38:26	343.4	32.1	79:26	200.8	51.7	79:34	1.07 (0.41)	1.005	0.998	0.996	1.009	−0.554	0.79	0.50	0.9
PO	28	241.7	19.6	11:5	341.7	26.2	15:8	119.4	56.3	13:5	2.35 (1.21)	1.012	1.002	0.986	1.026	0.237	1.75	−0.58	2.6
HR	13	238.5	15.8	15:6	339.6	34.2	15:4	127.7	51.4	7:4	2.04 (0.78)	1.014	1.002	0.984	1.031	0.169	2.13	−0.48	3.1
HU	7	298.3	67.0	25:12	144.1	20.9	48:9	50.5	9.2	48:24	3.16 (1.78)	1.008	0.998	0.994	1.014	−0.426	1.18	0.59	1.4
FR	3	292.2	12.3	54:44	72.6	74.2	78:25	200.0	9.7	78:40	1.76 (0.28)	1.038	0.997	0.965	1.076	−0.087	5.48	0.74	7.3
MQ-MJ	10	55.1	2.9	13:9	320.1	59.9	37:13	146.8	29.9	37:9	1.57 (0.32)	1.015	0.995	0.990	1.026	−0.596	2.20	1.46	2.5
BS	5	2.6	20.1	65:19	269.7	7.8	65:21	159.5	68.3	25:7	1.95 (0.37)	1.008	1.003	0.990	1.019	0.448	1.13	−0.79	1.8
NI	4	322.2	14.8	43:18	53.8	6.0	43:7	165.3	73.9	49:17	2.66 (0.19)	1.012	0.997	0.990	1.022	−0.359	1.81	0.78	2.2
BD	3	284.8	48.6	68:37	57.3	30.8	74:9	163.3	24.8	69:36	1.93 (0.08)	1.016	1.005	0.979	1.039	0.376	2.41	−1.35	3.8
QR	12	82.8	19.8	19:3	260.2	70.2	19:3	352.5	0.8	4:3	1.81 (0.68)	1.023	1.013	0.964	1.066	0.654	3.44	−3.77	5.9
SQ	6	79.2	25.8	26:19	199.4	46.1	28:25	331.1	32.6	28:20	1.57 (0.61)	1.013	1.002	0.985	1.029	0.179	1.99	−0.49	2.9
DP	7	65.4	14.1	20:7	283.1	72.4	25:8	158.0	10.3	18:7	1.72 (0.12)	1.009	1.001	0.989	1.020	0.205	1.37	−0.39	2.0

(continued)

TABLE 13.1 Continued.

Site	*N*	D_{max}	I_{max}	A_{max}	D_{int}	I_{int}	A_{int}	D_{min}	I_{min}	A_{min}	K_{mean} ($\pm 1\sigma$)	$K_{n\text{-}max}$	$K_{n\text{-}int}$	$K_{n\text{-}min}$	P_J	*T*	*A%*	*B%*	*H%*
TF	9	268.3	38.9	25:9	74.3	50.3	55:13	172.7	6.9	55:22	1.75 (0.17)	1.009	0.997	0.995	1.015	−0.713	1.28	0.99	1.4
MR	8	97.1	4.6	5:4	0.6	54.7	4:4	190.3	34.9	5:4	1.71 (0.12)	1.029	0.998	0.973	1.057	−0.074	4.16	0.47	5.5
BP	6	241.1	40.3	12:10	46.5	48.8	14:12	144.9	7.3	15:7	1.96 (0.12)	1.017	0.998	0.985	1.033	−0.204	2.57	0.68	3.2
BB	5	102.3	1.4	75:26	348.5	86.6	75:12	192.4	3.1	32:10	1.58 (0.27)	1.014	1.010	0.976	1.043	0.798	2.07	−2.98	3.8
SP	6	289.5	1.3	17:4	25.3	77.3	17:7	199.2	12.6	8:5	2.08 (0.10)	1.034	1.015	0.952	1.090	0.549	4.88	−4.24	8.1
XC	14	100.5	35.8	12:4	246.2	48.9	12:8	357.3	17.5	9:4	1.80 (0.12)	1.008	1.001	0.990	1.018	0.226	1.23	−0.39	1.8
WH	11	262.8	35.2	14:4	69.9	54.1	14:5	168.5	6.2	5:4	1.84 (0.23)	1.012	1.004	0.984	1.029	0.434	1.75	−1.17	2.8
Means*																			
Higganum-Holden dike mean†																			
		45.2	6.7	27:10	253.6	82.4	28:11	135.6	3.6	14:11	—	1.010	1.002	0.988	0.278 1.023	1.49−0.59	2.2		
Christmas Cove dike mean‡																			
		256.7	5.4	30:10	33.8	82.6	30:9	176.2	5.0	11:9	—	1.014	1.003	0.983	0.297 1.032	2.07−0.89	3.1		

Notes: Site abbreviations are B, Branford; NB, North Branford; NH, Notch Hill; LH, Long Hill Road; GM, Guilford Monastery; SH, Summer Hill Road; JA, Jackson Road; PO, Ponset Road; HR, Higganum Railroad; HU, Hurd State Park; FR, Flatbrook Road; MQ, Marlborough Quarry; MJ, Marlborough Johnson Road; BS, Bear Swamp; NI, Nipmuck State Forest; BD, Bearden; QR, Quanapoxet Reservoir; SQ, Sterling Quarry; DP, Doyle Point; TF, Tank Farm; MR, Mountain Road; BP, Bijhouwer's Property; BB, Below Bijouwer's; SP, Spruce Point; XC, Christmas Cove; and WH, Whitehead Island.

N is the number of samples used for computation. Subscripts max, int, and min denote the maximum, intermediate, and minimum principal AMS axes, respectively. D and I are the declination and inclination of the axes. A consists of two numbers that equal the long and short radii of the 95% uncertainty ellipse. K_{mean} is the magnetic susceptibility in 10^{-2} SI units. $K_{n\text{-}max}$, $K_{n\text{-}int}$, $K_{n\text{-}min}$ are the susceptibilities of the principal axes normalized such that their sum equals 3. T and P_J are the ellipsoid shape and degree of anisotropy (Jelinek, 1981). $T = (2LnK_{int} - LnK_{max} - LnK_{min})/(LnK_{max} - LnK_{min})$ and varies between +1 and −1. (Ln is natural logarithm.) $P_J = \exp$ [sqrt $\{2[(LnK_{max} - LnK_{mean})^2 + (LnK_{int} - LnK_{mean})^2 + (LnK_{min} - LnK_{mean})^2]\}$] where $LnK_{mean} = (LnK_{max} + LnK_{int} + LnK_{min})/3$. P_J 1. The observation of both negative and positive values of T reflects the presence of both prolate and oblate ellipsoid fabrics, respectively. The percentage of anisotropy, P_J, of sites varies from 1% to 9% (1.01–1.09). A% is degree of anisotropy $\{= 100[1-(K_{min} + K_{int})/2K_{max}]$, range 0, 100}, and B% is magnetic fabric, $\{= 100[1 + (K_{min} - 2K_{int})/K_{max}]$, range−100, 100} after Cañón-Tapia, Walker, and Herrero-Bervera (1996). H(%) is $100^*(K_{max} - K_{min})/K_{mean}$ (Owens 1974; Tarling and Hrouda 1993).

* Higganum-Holden and Christmas Cove dike means calculated after site means rotated to a common dike strike and dip (s/d) and also calculated on the basis of equal weighting for each site.

† 11 sites: B, NB, NH, LH, JA, PO, HR, FR, MQ-MJ, QR, and SQ. Reference orientation; s/d = 45°/90°.

‡ 8 sites: DP, TF, MR, BP, BB, SP, XC, and WH. Reference orientation: s/d = 75°/90°.

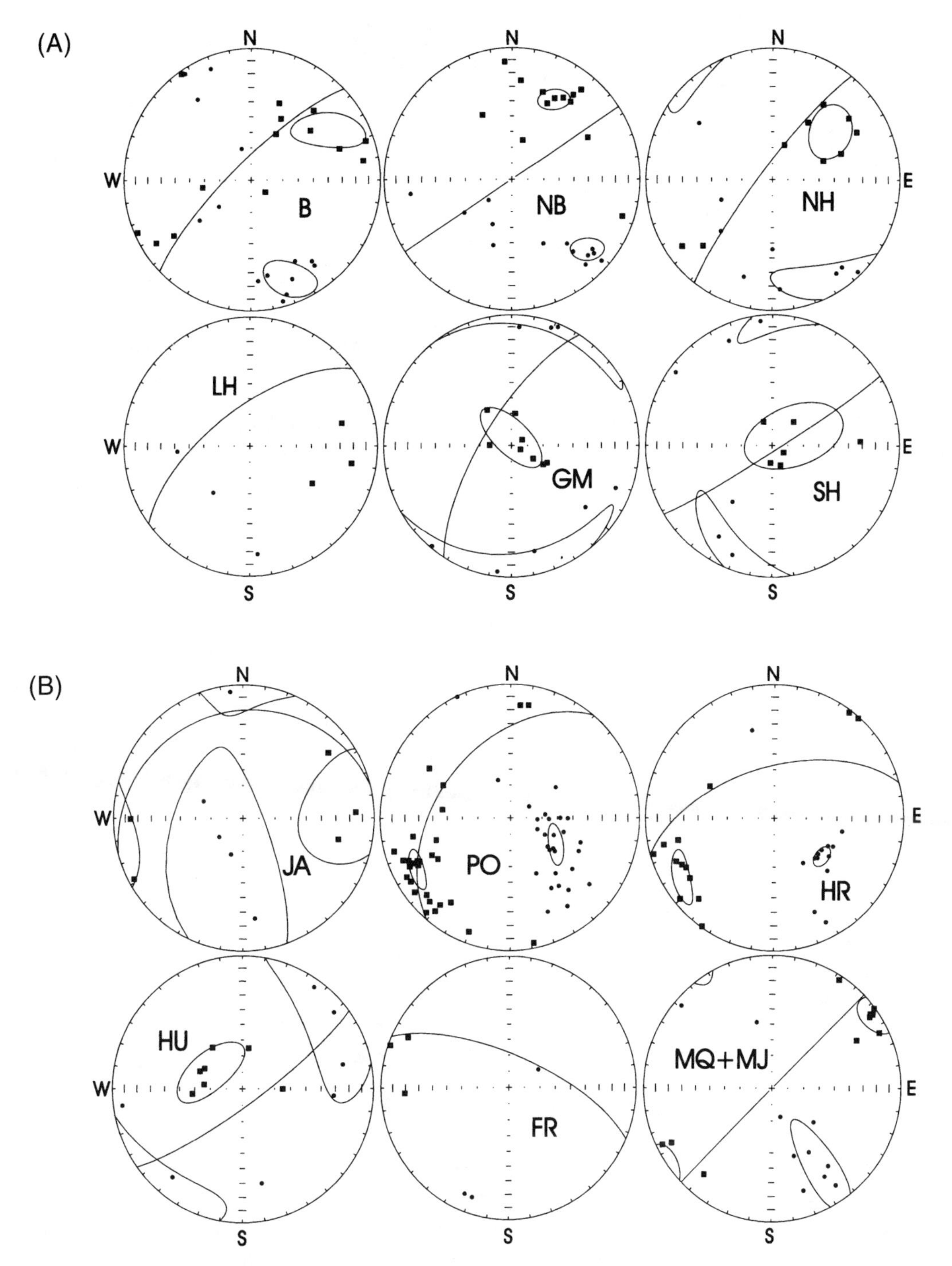

FIGURE 13.6 Stereonet plots (lower hemisphere) of AMS data for individual sites. AMS maxima are indicated by squares; AMS minima, by dots. Also plotted on the lower hemisphere are 95% confidence ellipses (*thin lines*) and the average attitude of dike segment (*thick lines*). Trace of dike plane (on lower hemisphere) is shown in solid line. Plots were generated using a computer program, which is discussed in Lienert (1991). (*A*) *B*, Branford; *GM*, Guilford Monastery; *LH*, Long Hill Road; *NB*, North Branford; *NH*, Notch Hill; *SM*, Summer Hill Road. (*B*) *FR*, Flatbrook Road; *HR*, Higganum Railroad; *HU*, Hurd State Park; *JA*, Jackson Road; *MJ*, Marlborough Johnson Road; *MQ*, Marlborough Quarry; *PO*, Ponset Road.

(*continued*)

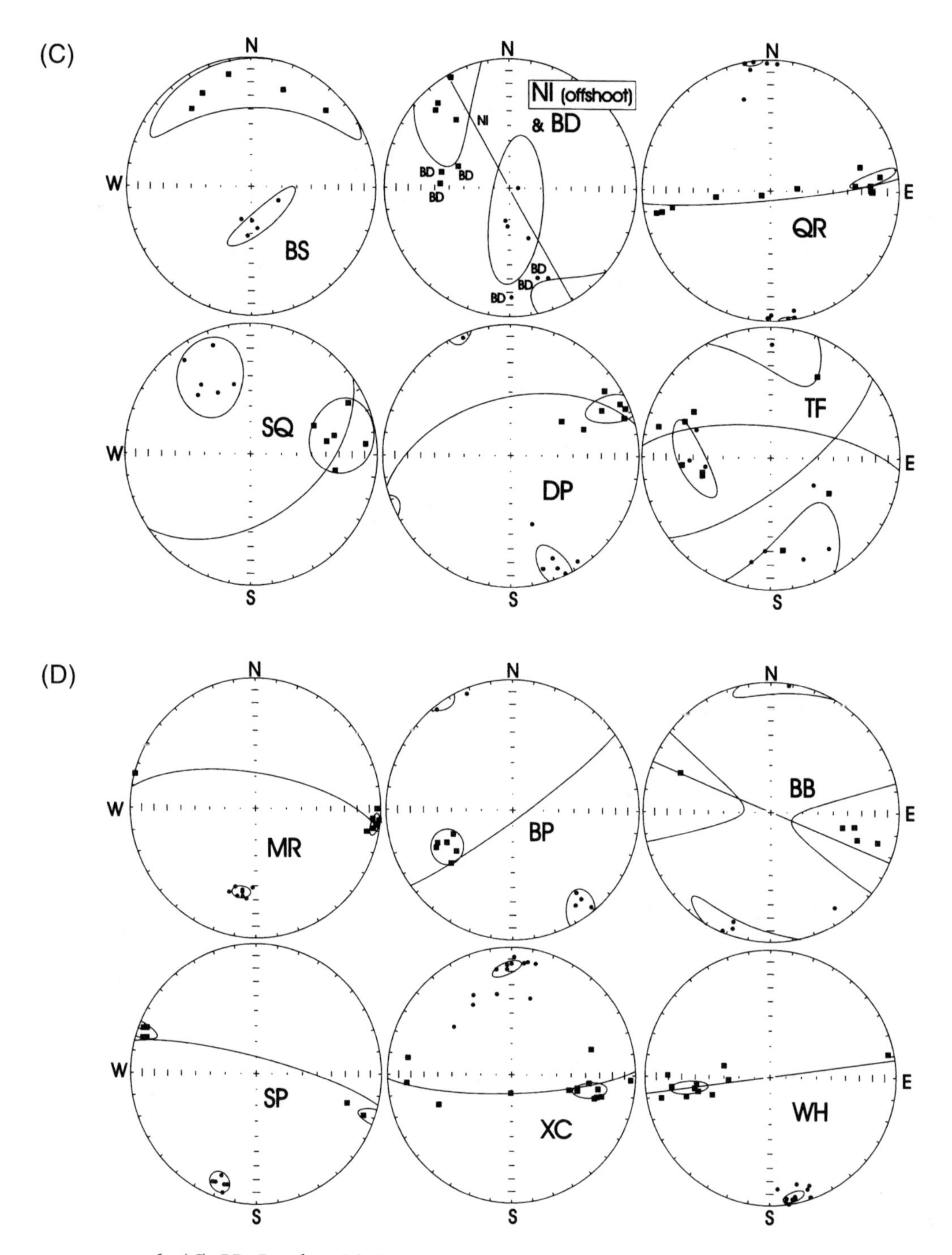

FIGURE 13.6 (*C*) *BD*, Bearden; *BS*, Bear Swamp; *DP*, Doyle Point; *NI*, Nipmuck State Forest; *QR*, Quanapoxet Reservoir; *SQ*, Sterling Quarry; *TF*, Tank Farm. (*D*) *BB*, Below Bijhouwer's; *BP*, Bijhouwer's Property; *MR*, Mountain Road; *SP*, Spruce Point; *WH*, Whitehead Island; *XC*, Christmas Cove.

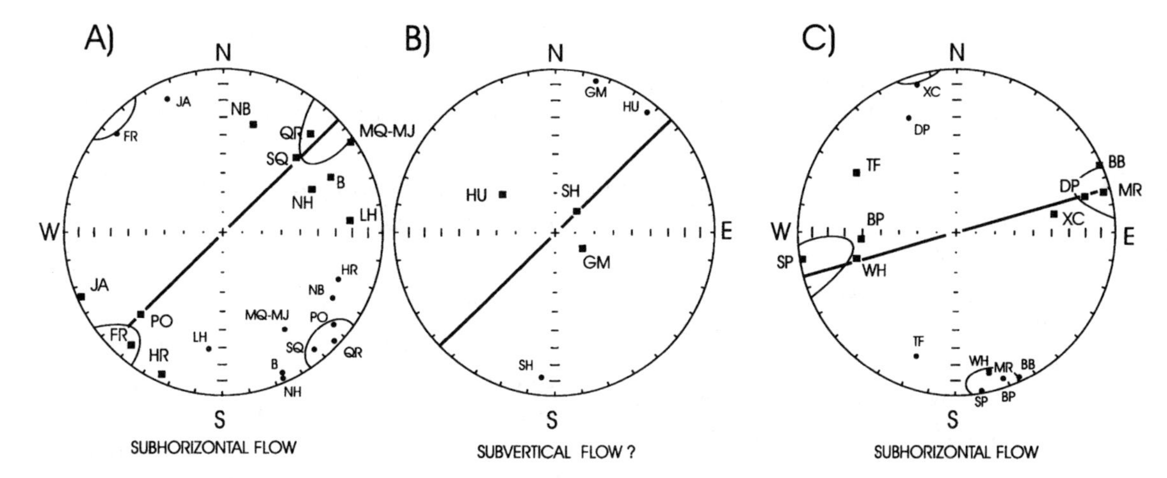

FIGURE 13.7 Summary of AMS data for the (*A* and *B*) Higganum-Holden and (*C*) Christmas Cove sites. AMS maxima are indicated by squares; AMS minima, by dots. Site data are rotated to a common strike (NE) and dip (vertical) for the Higganum-Holden sites and to a common strike (ENE) and dip (vertical) for the Christmas Cove sites.

Discussion

Christmas Cove Dike

At this time, 10 outcrops of the Christmas Cove dike have been located, two recently. AMS measurements were carried out on eight sites. Mean values of H, which provides a measure of the degree of anisotropy, range from 2.0 to 8.2%. The low value was found for the Christmas Cove type locality (XC); the highest value, at site SP. Ground-level magnetic surveys indicate that the dike is approximately twice as wide at its type locality than at SP. Therefore, H may be a function of dike width and may be indicative of different flow regimes or reflect differences in average grain size.

All eight Christmas Cove sites provided subhorizontal K_{max} axes (figures 13.6 and 13.7C). Their general trends are E–W, which conforms with the direction of the principal dike segments (figure 13.6). The K_{max} groupings at individual sites either cluster tightly (BP and MR) or exhibit some streaking (WH). In the latter, the westward-plunging K_{max} axes possess inclinations varying from 13 to 63°. The more steeply inclined K_{max} directions were encountered in cores collected near the dike's contacts, indicating the effect of friction. Site WH, the best-exposed outcrop of the Christmas Cove dike, contains several flow indicators such as basalt spurs that intruded obliquely into the granitic country rock and granitic slivers that were ripped from the contact and were carried along by the basaltic magma. They all indicate eastward flow, implying that the magma rose at low angles in that direction.

K_{min} vectors generally plot more or less at right angles to the dike plane. Site XC shows streaking, with K_{min} axes varying in trend from N14°E to N51°W. They appear to have rotated from directions more or less at right angles to the dike to more steeply inclined directions that approach the dike plane. The latter axes were obtained mainly from cores in or near the dike's center. This magnetic fabric therefore may be indicative of incomplete rotation during late-stage vertical compaction of the magma.

Higganum-Holden Dike

The Higganum-Holden dike is especially well exposed along its southwesternmost segment. Farther north, outcrops are more scattered and more poorly exposed. Good cores were obtained from 17 sites. These sites include subvertical dike segments, inclined segments, a sill, and a narrow dike offshoot.

Mean H values, a measure of the degree of magnetic anisotropy, vary from 0.9 to 7.3%. The lowest value was encountered in a sill, and the highest in a 1 m wide offshoot of the dike (site NI). Most H values range from 1.3 to 5.9%. Where the dike is widest (site PO), H is 2.6%; where it is relatively narrow (site QR), H is 5.9%. As in the Christmas Cove dike, H appears to be a function of dike width, hence representing different flow regimes during emplacement or average grain size differences. In 11 sites, the K_{max} vectors are subhorizontal and trend NE–SW parallel to the general direction of the dike (figures 13.6 and 13.7A). Three sites (GM, SH, and HU) possess subvertical K_{max} axes (figure 13.7B). GM is located in a major "dogleg," a N-NE-trending segment of the dike that connects two en echelon NE-trending segments. Sites SH and HU are located where individual dike segments abut against major NW-trending fault zones that had formed before dike emplacment. In these three cases, it appears as if the horizontally flowing magma encountered obstacles that forced it to rise (and to become more turbulent).

The distribution of K_{max} axes at site QR is streaked. Most vectors are subhorizontal, but three cores provided steeply inclined K_{max} directions. Of the latter, two were collected near the dike's contact and one close to a gneissic screen incorporated into the dike. Turbulence resulting from frictional processes near the dike walls probably led to different flow regimes and can account for this phenomenon.

K_{min} vectors cluster in a few sites (B, NB, SQ) but show streaked distributions in most. In virtually all streaks, the K_{min} axes appear to have rotated from positions at right angles to the dike plane to trends approaching those of the dike. A good example is provided by site NH (figure 13.6). In this dike segment, the K_{min} rotates from subhorizontal SE via low-angle southward-dipping directions to moderately plunging SW trends. Cores collected near the dike center exhibit the largest rotation away from the normal to the dike. Streaking, therefore, most probably reflects the effect of partial compaction at late stages in the cooling process. Differential compaction could and should have affected K_{max} directions as well. K_{max} cones of confidence indeed generally appear to be larger when K_{min} distributions are streaked.

An unusual distribution of AMS directions was obtained for site BS (figure 13.6C). Here cores were drilled from a subhorizontal glaciated surface, with no evidence for contacts or cooling columns. Considering the data for sill site JA (figure 13.6B), it seems possible that site BS also represents part of a sill, especially because its grain size is relatively coarse. Site JA is located relatively close (few tens of meters) from the Higganum-Holden feeder dike. Judging from the general trends of dike segments northeast and southwest of BS, this site may be as much as 50 m from where its feeder should be located (but is not exposed). Difference in distance to source area may explain the preference for subhorizontal NE–SW directions in site BS.

Low-Grade Metamorphism as Possible Reason for AMS Noise

The southwesternmost outcrop of the Higganum-Holden dike, B, has the highest noise level (figure 13.6). K_{max} directions plot in a zone parallel to the dike trend but vary considerably in dip. K_{min} are subhorizontal and at right angles to the dike contact. The K_{max} variation therefore cannot be attributed to differential rotation during compaction. Multiple injection could have caused this phenomenon, but no clear evidence has been found for this hypothesis. Scatter most likely resulted from low-grade metamorphism. During demagnetization of the oriented cores, a magnetic component emerged at temperatures of approximately 200°C (unpublished data). The direction of this magnetic component was similar to that of a chemical remanent overprint found in the Eastern North America flows of the Hartford basin. The latter was introduced during a period of relatively high heat flow in the Late Jurassic or Early Cretaceous or both. The thermochemical overprint decreases rapidly with increasing distance from the dike contacts. Its effect also decreases with increasing distance from the eastern border fault of the rift valley and is restricted to outcrops in the southwesternmost segment of the Higganum-Holden dike. Both the degree of overprinting (as reflected by Chemical Remnant Magnetization [CRM] intensity) and the scatter in AMS directions decrease north to northeastward. A correlation between strong CRM and scattered AMS in site B suggests that such overprints may explain AMS scatter in other sites.

Site PO of the Higganum-Holden Dike

Site PO was studied in much detail because it is the type locality for the Higganum-Holden dike and the outcrop from which the most reliable flow data were obtained petrographically by Philpotts and Asher (1994). A total of 28 AMS cores were collected. K_{min} axes generally trend at right angles to the inclined dike plane (figure 13.6) supportive of a flow origin, but the data distribution shows some streaking. The majority of samples provided subhorizontal NE–SW-trending K_{max} directions (figures 13.6–13.8) that suggest sub-

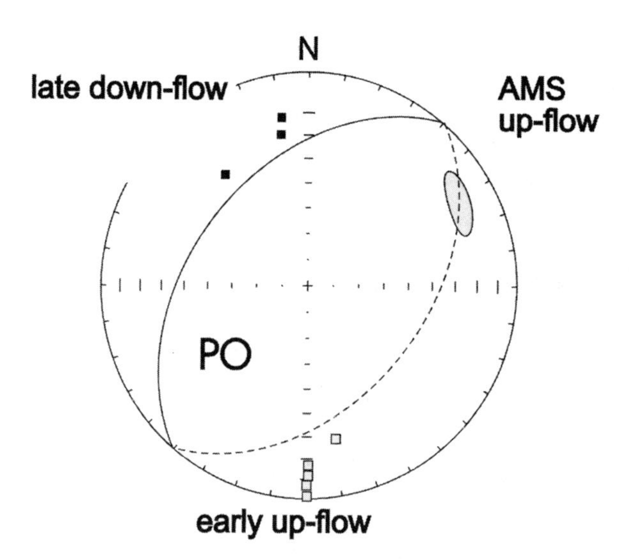

FIGURE 13.8 Stereonet plot (lower hemisphere) of flow directions obtained at PO site of Higganum-Holden dike. Solid and open squares (late down-flow and early up-flow) are Philpotts and Asher (1994) data plotted in the lower and the upper hemisphere, respectively. Our AMS up-flow data are represented by its 95% confidence ellipse and are plotted on the upper hemisphere.

horizontal flow. These results for the Higganum-Holden dike differ from the flow directions obtained by Philpotts and Asher (1994) and require further discussion. The dike at PO is wide and very well exposed along highway cuts and at several entrance and exit ramps. The exposed upper contact, well-developed columnar structures, and ground magnetic surveys indicate that the dike trends NE–SW and dips approximately 45° NW. Philpotts and Asher (1994) collected their petrographic evidence along the upper contact of the dike, which is characterized by at least five more or less parallel gneiss screens (planar xenoliths) separated by dikelets varying from a few centimeters to meters in thickness (figure 13.9). The petrographic data showed early subhorizontal flow toward the S and later backflow toward the N–NW.

Cores were drilled from the same spots sampled by Philpotts and Asher (1994) at approximately 10 cm intervals; cores also were drilled at approximately 1 m intervals from this upper contact to the center of the dike. The petrographic flow indicators do not parallel the dike plane or the local contact along which they were collected and therefore indicate imbrication or rotation or both. The petrographic data for S-directed upward flow plunge at lower angles than the data obtained for northward backflow, which implies rotation of the upflow fabric by the backflow event.

If the original flow was southward and parallel to the inclined fault plane, as suggested by Philpotts and Asher (1994), backflow would have rotated the K_{max} axes into a subhorizontal N–S direction. Four out of 28 AMS axes do indeed show such a trend (figure 13.6). All four were obtained in cores collected from dike segments between the gneiss screens—that is, from the same location where the petrographic evidence was collected. However, the remaining 24 K_{max} axes are subhorizontal and are located within the dike away from these contacts. Northward backflow could not have significantly affected axes with such trends. Therefore, neither the four deviating AMS axes nor the petrographic data of Philpotts and Asher (1994) appear to be representative for flow in the Higganum-Holden dike at this type locality. Perturbation due to the presence of xenolithic screens appears to have been responsible for the deviations.

En Echelon Segmentation of the Dikes

The AMS data for both the Higganum-Holden and Christmas Cove dikes predominantly show subhorizontal K_{max} directions. Field evidence, although sparse, suggests that lateral flow was to the NE and E. Because the Eastern North America dikes converge to the southwest, toward the common center of the Blake Plateau, a lateral northeastward (eastward) flow appears the most acceptable at this stage of our study.

How can a predominantly lateral flow be reconciled with the presence of en echelon segmentation of the dikes, which seems to imply vertical flow? Pollard (1987) showed that a rotation of the orientation of the extensional stress about an axis parallel to the direction of dike emplacement might produce segmentation and en echelon offsets in dike trend. Greenough and Hodych (1990) used a variation of this model and calculated that a 3 km offset between 15 km long dike segments (such as in the Caraquet dike in Canada) implied a rotation of approximately 10°. In addition, they used Pollard's model to illustrate how magma flow can be dominated by subhorizontal trajectories and can maintain only a small component of vertical motion to produce segmentation of a dike system.

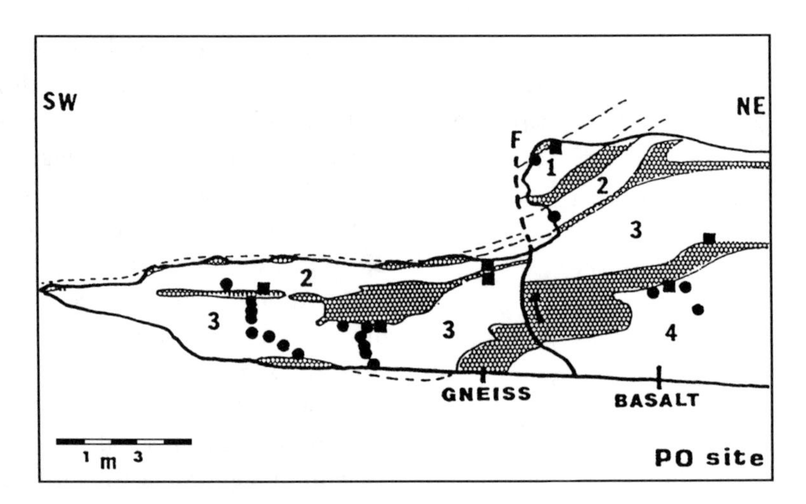

FIGURE 13.9 Cross section of outcrop along the northwestern side of the Higganum-Holden dike at site PO. Squares indicate samples collected by Philpotts and Asher (1994); dots, the locations of cores collected for this study. Diabase (basalt) is unornamented, and screens of granite are shown in a dense crosshatch pattern. Numbers locate different domains based on the distribution of granite screens; the dotted line marks NW contact of the dike; and the arrow indicates the sense of offset along a later fault.

Comparing their scenario with the Higganum-Holden dike, where offsets between dike segments are generally less than 3 km, suggests that the rotation of the extensional stress, if it indeed occurred, must have been no more than a few degrees.

AMS and flow data for dikes in New England and the Maritime Provinces suggest that the magmas moved predominantly along horizontal transects that are distributed over a cumulative distance of at least 1,000 km. This result supports a plume source in the Blake Plateau region rather than multiple sources underlying the rift basins (Philpotts 1992) or a mantle keel (Cummins, Arthur, and Ragland 1992).

Results from Other Dike Swarms

Studies of the magnetic fabric in other dike swarms have shown a predominance of evidence for subvertical or steeply inclined flow near the hypothetical plume source and for predominant lateral flow in distal segments. In the 1270 Ma Mackenzie swarm of northern Canada, subvertical flow dominated to 500 km from the focal region of the swarm, and lateral flow dominated from 600 km to the farthest dikes studied at 2,300 km (Ernst and Baragar 1992). Patterns similar to that of the Mackenzie swarm were encountered in the 180 Ma Botswana swarm of southern Africa (Ernst and Duncan 1995) and the 130 Ma Ponta Grossa swarm of South America (Raposa and Ernesto 1995). The general flow pattern of the Eastern North America swarm may resemble that of these intrusive systems.

Baragar et al. (1996) suggested that magmas are emplaced in different pulses during progressive rise of the plume head. Early more primitive magmas would not be driven far from the center of uplift, but later more evolved magmas, being emplaced from the more uplifted central region, would have a greater hydrostatic head and would be driven laterally over longer distances (especially in zones of crustal weakness).

Application of this model to the Eastern North America dike swarm pattern would imply that the older olivine-normative magmas would have intruded into a radial set of fractures crosscutting the Appalachian grain in relative close proximity to the Blake Plateau, whereas the younger quartz-normative (geochemically evolved) magmas traveled over long distances mainly parallel and at low angle to the regional tectonic grain. A third pulse may be represented by the N- to NNW-trending Charleston swarm, which is composed of LIL (large-ion lithophile elements) enriched quartz-normative dikes that clearly converge on the Blake Plateau focal region.

Implication for the Origin of Circa 200 Ma Atlantic Bordering Dikes

The model for a point source is consistent with the geochemical data. Puffer (1992) showed that flows in the "Newark" rift basins, which were presumably fed by the quartz-normative dikes, exhibit a remarkable degree of chemical homogeneity throughout from the Carolinas to Newfoundland. He concluded that this characteristic makes it very unlikely that localized processes unique to individual rift basins were the dominant factor in controlling magma generation.

The complementary dikes on the opposite side of the central Atlantic extend from western Africa northward into the Iberian Peninsula over a distance of approximately 2,000 km. Sebai et al. (1991) showed that these dikes and related flows possess similar trace-element features characterized by LREE (light rare earth element) enrichment and negative Nb anomalies. Th:La ratios for these igneous rocks in a zone extending from Spain to Mali, for instance, are remarkably constant (5.7 ± 1.2). Chemical homogeneity thus characterizes the Early Jurassic igneous rocks along both the eastern and western margins of the central Atlantic.

Puffer (1994) suggested that the magmas originated during prolonged periods of insulation and thermal accumulation below the entire Pangean lithosphere and intruded subvertically into zones of crustal extension during the earliest phase in the breakup of this supercontinent. The relatively enriched magmas then could have originated by partial melting in a mantle "keel," as suggested by Cummins, Arthur, and Ragland (1992).

However, it seems unlikely that the upper mantle below the eastern part of the Appalachian mountain range (which presumably contains imbricated remains of subducted lithosphere) and the upper mantle below the western part of the West African craton were (are) as homogeneous chemically as required by the remarkably constant data for the HTQ subprovince. It is less difficult to assume that a homogeneous LREE-enriched magma could have originated in the asthenosphere and that it rose in a single plume below the south-central Atlantic region and spread laterally over large distances within the weakened lithosphere above it.

Oyarzun et al. (1997) developed a comparable model. They believe that the tholeiitic magma originated in a central Atlantic plume and migrated N–NE by means of large-scale sublithospheric channeling in the tectonically weakened corridor that had developed along the "collapsed" southern branch of the Variscan foldbelt. Magmas headed for the highly thinned European realm, which Oyarzun et al. (1997) referred to as a large thin-spot–type domain. Our data are consistent with their general sense of magma flow.

The most volumetric and regionally extensive magmatism in the opening phase of the Atlantic occurred in a relatively short period around 200 Ma. Crustal extension (normal faulting) preceded this event by around 30 Ma, and true oceanic crust did not form until approximately 30 Ma after magmatism. The Early Jurassic magmatic pulse therefore should be regarded as a climax in the lithospheric separation process. It is our opinion that the rise of an upper-mantle plume below the present-day Blake Plateau did play a crucial role in the slow process of crustal divergence and that it was responsible for this magmatic climax.

Conclusions

AMS measurements were performed on 215 samples collected from 25 sites distributed along two dike segments of the 200 Ma Eastern North America swarm in New England. The data are characterized by AMS ratios typically of 1 to 8%, although narrow dike offshoots and samples very close to dike contacts can have values greater than 10%. The AMS fabric indicates predominantly lateral flow with minor local intervals of vertical flow.

Evidence for lateral flow has also been reported previously for dikes of this swarm in the Canadian Maritime Provinces. These results and the overall radiating pattern of the circum-Atlantic dikes support a plume source in the vicinity of the Blake Plateau.

Acknowledgments

We thank Art Hussey and Greg McHone for helping us locate outcrops of the Christmas Cove dike in Maine and Ken Buchan for ongoing discussions on magma flow in dikes and for comments on a draft version of the manuscript for this chapter. This chapter is Geological Survey of Canada contribution no. 1997188.

Literature Cited

Baragar, W. R. A., R. E. Ernst, L. Hulbert, and T. Peterson. 1996. Longitudinal petrochemical variation in the Mackenzie dike swarm, northwestern Canadian shield. *Journal of Petrology* 37:317–359.

Bertrand, H. 1991. The Mesozoic tholeiitic province of northwest Africa: A volcano-tectonic record of the early opening of the central Atlantic. In A. B. Kampunzu and R. T. Lubala, eds., *Magmatism in Extensional Structural Settings: The Phanerozoic African Plate,* pp. 147–191. New York: Springer.

Cañón-Tapia, E. 1996. Single-grain versus distribution anisotropy: A simple three-dimensional model. *Physics of the Earth and Planetary Interiors* 94:149–158.

Cañón-Tapia, E., G. P. L. Walker, and E. Herrero-Bervera. 1996. The internal structure of lava flows: Insights from AMS measurements. I. Near-vent a'a. *Journal of Volcanology and Geothermal Research* 70:21–36.

Cummins, L. E., J. D. Arthur, and P. C. Ragland. 1992. Classification and tectonic implications for early Mesozoic magma types of the circum-Atlantic. In J. H. Puffer and P. C. Ragland, eds., *Eastern North American Mesozoic Magmatism,* pp. 119–135. Geological Society of America Special Paper, no. 268. Boulder, Colo.: Geological Society of America.

de Boer, J. Z. 1992. Stress configurations during and following emplacement of ENA basalts in the northern Appalachians. In J. H. Puffer and P. C. Ragland, eds., *Eastern North American Mesozoic Magmatism,* pp. 361–378. Geological Society of America Special Paper, no. 268. Boulder, Colo.: Geological Society of America.

de Boer, J. Z., J. G. McHone, J. H. Puffer, P. C. Ragland, and D. Whittington. 1988. Mesozoic and Cenozoic magmatism. In R. E. Sheridan and J. A. Grow, eds., *The Atlantic Continental Margin,* pp. 217–241. Vol. I-2 of *The Geology of North America.* Boulder, Colo.: Geological Society of America.

Dostal, J., and T. D. Greenough. 1992. Geochemistry and petrogenesis of the early Mesozoic North Mountain Basalts of Nova Scotia, Canada. In J. H. Puffer and P. C. Ragland, eds., *Eastern North American Mesozoic Magmatism,* pp. 149–160. Geological Society of America Special Paper, no. 268. Boulder, Colo.: Geological Society of America.

Dunn, A. M., P. H. Reynolds, D. B. Clarke, and J. M. Ugidos. 1998. A comparison of the age and composition of the Shelburne dyke, Nova Scotia, and the Messejana dyke, Spain. *Canadian Journal of Earth Sciences* 35:1110–1115.

Dunn, T., and P. Stringer. 1990. Petrology and petrogenesis of the Ministers Island dike, southwest New Brunswick, Canada. *Contributions to Mineralogy and Petrology* 105:55–65.

Dunning, G., and J. P. Hodych. 1990. U/Pb zircon and baddeleyite ages for the Palisades and Gettysburg sills of the northeastern United States: Implications for the ages of the Triassic/Jurassic boundary. *Geology* 18: 795–798.

Ernst, R. E. 1994. Mapping the magma flow pattern in the Sudbury dyke swarm in Ontario using magnetic fabric analysis. In *Current Research 1994-E, Geological Survey of Canada,* pp. 183–192. Ottawa: Geological Survey of Canada.

Ernst, R. E., and W. R. A. Baragar. 1992. Evidence from magnetic fabric for the flow pattern of magma in the Mackenzie giant radiating dike swarm. *Nature* 356: 511–513.

Ernst, R. E., and K. L. Buchan. 1997. Giant radiating dyke swarms: Their use in identifying pre-Mesozoic large igneous provinces and mantle plumes. In J. J. Mahoney and M. F. Coffin, eds., *Large Igneous Provinces: Continental, Oceanic, and Planetary Flood Volcanism,* pp. 297–333. American Geophysical Union Geophysical Monograph, no. 100. Washington, D.C.: American Geophysical Union.

Ernst, R. E., and A. R. Duncan. 1995. Magma flow in the giant Botswana dyke swarm from analysis of magnetic fabric [abstract]. In A. Agnon and G. Baer, eds., *Program and Abstract for the Third International Dyke Conference,* p. 30. Jerusalem: Geological Survey of Israel, Israel Geological Society, Hebrew University (Jerusalem), Ben-Gurion University of the Negev (Be'er Sheva).

Ernst, R. E., J. W. Head, E. Parfitt, E. Grosfils, and L. Wilson. 1995. Giant radiating dyke swarms on Earth and Venus. *Earth-Science Reviews* 39:1–58.

Ernst, R. E., and G. W. Pearce. 1989. Averaging of anisotropy of magnetic susceptibility data. In F. P. Agterberg and G. F. Bonham-Carter, eds., *Statistical Applications in the Earth Sciences,* pp. 297–305. Geological Survey of Canada Paper, no. 89–9. Ottawa: Geological Survey of Canada.

Greenough, J. D., and J. P. Hodych. 1990. Evidence for lateral injection in the early Mesozoic dikes of eastern North America. In A. J. Parker, P. C. Rickwood and

D. H. Tucker, eds., *Mafic Dikes and Emplacement Mechanisms,* pp. 35–46. Rotterdam: Balkema.

Hargraves, R. B., D. Johnson, and C. Y. Chan. 1991. Distribution anisotropy: The cause of AMS in igneous rocks? *Geophysical Research Letters* 18:2193–2196.

Hext, G. R. 1963. The estimation of second-order tensors, with related tests and designs. *Biometrika* 50:353–373.

Hill, R. I. 1991. Starting plumes and continental breakup. *Earth and Planetary Science Letters* 104:398–416.

Hussey, A. M. 1971. *Geologic Map and Cross-Sections of the Orrs Island 7.5 Minute Quadrangle and Adjacent Area, Maine.* Maine Geologic Survey Map GM-2. Augusta: Maine Geologic Survey.

Jelinek, V. 1978. Statistical processing of anisotropy of magnetic susceptibility measured on groups of sediments. *Studia Geophysica et Geodaetika* 22:50–62.

Jelinek, V. 1981. Characterization of the magnetic fabric of rocks. *Tectonophysics* 79:63–67.

Knight, M. D., and G. P. L. Walker. 1988. Magma flow directions in dikes of the Koolau Complex, Oahu, determined from magnetic fabric studies. *Journal of Geophysical Research* 93:4301–4319.

Koza, D. 1976. Petrology of the Higganum diabase dike in Connecticut and Massachusetts. M.S. thesis, University of Connecticut.

Lienert, B. R. 1991. Monte Carlo simulation of errors in the anisotropy of magnetic susceptibility: A second-rank symmetric tensor. *Journal of Geophysical Research* 96:19539–19544.

Lindsey, A. G. 1995. Emplacement mechanism and deformation of the Higganum-Holden dike in Connecticut and Massachusetts and the Christmas Cove dike in Maine. B.A. thesis, Wesleyan University.

Manspeizer, W., and H. L. Cousminer. 1988. Late Triassic–Early Jurassic synrift basins of the U.S. Atlantic margin. In R. E. Sheridan and J. A. Grow, eds., *The Atlantic Continental Margin,* pp. 197–216. Vol. I-2 of *The Geology of North America.* Boulder, Colo.: Geological Society of America.

Marzoli, A., P. R. Renne, E. M. Piccirillo, M. Ernesto, G. Bellieni, and A. de Min. 1999. Extensive 200-million-year-old continental flood basalts of the Central Atlantic Magmatic Province. *Science* 284:616–618.

May, P. R. 1971. Pattern of Triassic–Jurassic diabase dikes around the North Atlantic in context of predrift position of the continents. *Geological Society of America Bulletin* 82:1285–1292.

McEnroe, S. A. 1989. *Paleomagnetism and Geochemistry of Mesozoic Dikes and Sills in West-Central Massachusetts.* Department of Geology and Geography Contribution, no. 64. Amherst: University of Massachusetts.

McEnroe, S. A. 1993. Paleomagnetism of Mesozoic intrusions in New England: Implications for the North American apparent polar wander path. Ph.D. diss., University of Massachusetts.

McHone, J. G. 1996. Broad-terrane Jurassic flood basalts across northeastern North America, *Geology* 24:319–322.

McHone, J. G., and J. R. Butler. 1984. Mesozoic igneous provinces of New England and the opening of the North Atlantic Ocean. *Geological Society of America Bulletin* 95:757–765.

McHone, J. G., D. P. West, A. M. Hussey, and H. W. McHone. 1995. The Christmas Cove dike, coastal Maine: Petrology and regional significance. *Geological Society of America, Abstracts with Programs* 27:67–68.

Oliveira, E. P., J. Tarney, and X. J. João. 1990. Geochemistry of the Mesozoic Amapa and Jari dyke swarms, northern Brazil: Plume-related magmatism during opening of the central Atlantic. In A. J. Parker, P. C. Rickwood, and D. H. Tucker, eds., *Mafic Dykes and Emplacement Mechanisms,* pp. 173–183. Rotterdam: Balkema.

Owens, W. H. 1974. Mathematical model studies on factors affecting the magnetic anisotropy of deformed rocks. *Tectonophysics* 24:115–131.

Oyarzun, R., M. Doblas, J. L. Lopez-Ruiz, and J. M. Cebria. 1997. Opening of the central Atlantic and asymmetric mantle upwelling phenomena: Implications for long lived magmatism in western North Africa and Europe. *Geology* 8:727–730.

Papezik, V. S., J. D. Greenough, J. A. Colwell, and T. J. Mallinson. 1988. North Mountain Basalt from Digby, Nova Scotia: Models for a fissure eruption from stratigraphy and petrochemistry. *Canadian Journal of Earth Sciences* 25:74–83.

Park, J. K., E. I. Tanczyk, and A. Desbarats. 1988. Magnetic fabric and its significance in the 1400 Ma Mealy diabase dykes of Labrador, Canada. *Journal of Geophysical Research* 93:13689–13704.

Pe-Piper, G., L. F. Jansa, and R. S. J. Lambert. 1992. Early Mesozoic magmatism on the eastern Canadian margin: Petrogenetic and tectonic significance. In J. H. Puffer and P. C. Ragland, eds., *Eastern North American Mesozoic Magmatism,* pp. 13–36. Geological Society of America Special Paper, no. 268. Boulder, Colo.: Geological Society of America.

Philpotts, A. R. 1992. A model for emplacement of magma in the Mesozoic Hartford basin. In J. H. Puffer and P. C. Ragland, eds., *Eastern North American Mesozoic Magmatism,* pp. 137–148. Geological Society of America Special Paper, no. 268. Boulder, Colo.: Geological Society of America.

Philpotts, A. R., and P. M. Asher. 1994. Magnetic flow-direction indicators in a giant diabase feeder dike, Connecticut. *Geology* 22:363–366.

Pollard, D. D. 1987. Elementary fracture mechanics applied to the structural interpretation of dykes. In H. C. Halls and W. F. Fahrig, eds., *Mafic Dyke Swarms,* pp. 5–24. Geological Association of Canada Special Paper, no. 34. Ottawa: Geological Association of Canada.

Puffer, J. H. 1992. Eastern North American flood basalts in the context of the incipient breakup of Pangea. In J. H. Puffer and P. C. Ragland, eds., *Eastern North American Mesozoic Magmatism,* pp. 95–118. Geological Society of America Special Paper, no. 268. Boulder, Colo.: Geological Society of America.

Puffer, J. H. 1994. Initial and secondary Pangean basalts. In B. Beauchamp, A. F. Embry, and D. Glass, eds., *Pangea: Global Environments and Resources,* pp. 85–95. Canadian Society of Petroleum Geologists Memoir, no. 17. Calgary: Canadian Society of Petroleum Geologists.

Ragland, P. C., L. E. Cummins, and J. D. Arthur. 1992. Compositional patterns for early Mesozoic diabases from South Carolina to central Virginia. In J. H. Puffer and P. C. Ragland, eds., *Eastern North American Mesozoic Magmatism,* pp. 309–332. Geological Society of America Special Paper, no. 268. Boulder, Colo.: Geological Society of America.

Raposa, M. I. B., and M. Ernesto. 1995. Anisotropy of magnetic susceptibility in the Ponta Grossa dyke swarm (Brazil) and its relationship with magma flow direction. *Physics of the Earth and Planetary Interiors* 87:183–196.

Rochette, P., M. Jackson, and C. Aubourg. 1992. Rock magnetism and the interpretation of anisotropy of magnetic susceptibility. *Reviews of Geophysics* 30: 209–226.

Sebai, A., G. Feraud, H. Bertrand, and J. Hanes. 1991. $^{40}Ar/^{39}Ar$ dating and geochemistry of tholeiitic magmatism related to the early opening of the central Atlantic rift. *Earth and Planetary Science Letters* 104:455–472.

Stephenson, A. 1994. Distribution anisotropy: Two simple models for magnetic lineation and foliation. *Physics of the Earth and Planetary Interiors* 82:49–53.

Sundeen, D. A., and M. C. Huff. 1992. Petrography, petrology, and K-Ar geochronology of hypabyssal mafic and silicic Mesozoic igneous rocks in southeastern New Hampshire. In J. H. Puffer and P. C. Ragland, eds., *Eastern North American Mesozoic Magmatism,* pp. 75–94. Geological Society of America Special Paper, no. 268. Boulder, Colo.: Geological Society of America.

Sutter, J. F. 1988. Innovative approaches to the dating of igneous events in the early Mesozoic basins of the eastern United States. In A. J. Froelich and G. R. Robinson Jr., eds., *Studies of the Early Mesozoic Basins of the Eastern United States,* pp. 194–200. U.S. Geological Survey Bulletin, no. 1776. Washington, D.C.: Government Printing Office.

Swanson, M. T. 1992. Structural sequence and tectonic significance of Mesozoic dikes in southern coastal Maine. In J. H. Puffer and P. C. Ragland, eds., *Eastern North American Mesozoic Magmatism,* pp. 37–62. Geological Society of America Special Paper, no. 268. Boulder, Colo.: Geological Society of America.

Tarling, D. H., and F. Hrouda. 1993. *The Magnetic Anisotropy of Rocks.* London: Chapman and Hall.

Weigand, P. W., and P. C. Ragland. 1970. Geochemistry of Mesozoic dolerite dikes from eastern North America. *Contributions to Mineralogy and Petrology* 29:195–214.

White, R. S., and D. McKenzie. 1989. Magmatism at rift zones: The generation of volcanic continental margins and flood basalts. *Journal of Geophysical Research* 94:7685–7729.

CONTRIBUTORS

Rolf V. Ackermann
ExxonMobil Upstream Research Company
Mailstop SW-526
3120 Buffalo Speedway
Houston, Texas 77252

Robert J. Altamura
Department of Geosciences
Pennsylvania State University
University Park, Pennsylvania 16802
and
College of Earth and Mineral Sciences
Pennsylvania State University
DuBois, Pennsylvania 15801

Colleen Barton
Department of Geophysics
Stanford University
Stanford, California 94093

Michael Caputi
P.O. Box 61933
Shell Offshore
New Orleans, Louisiana 70161

Jelle Zeilinga de Boer
Department of Earth and Environmental Sciences
Wesleyan University
Middletown, Connecticut 06459

Richard E. Ernst
Geological Survey of Canada
601 Booth Street
Ottawa, Ontario K1A 0E8
Canada

David Goldberg
Lamont-Doherty Earth Observatory
Columbia University
Palisades, New York 10964

Lois A. Johnson
Department of Geological Sciences
Rutgers University
Piscataway, New Jersey 08854

Dennis V. Kent
Department of Geological Sciences
Rutgers University
Piscataway, New Jersey 08854
and
Lamont-Doherty Earth Observatory
Columbia University
Palisades, New York 10964

Peter M. LeTourneau
Lamont-Doherty Earth Observatory
Columbia University
Palisades, New York 10964

Andrew G. Lindsey
Department of Earth and Environmental Sciences
Wesleyan University
Middletown, Connecticut 06459

Tony Lupo
Department of Earth and Environmental Sciences
New Mexico Institute of Technology
Socorro, New Mexico 87801
currently
Phillips Petroleum Company
4th and Keeler Ave.
Bartlesville, Oklahoma 74004

MaryAnn Love Malinconico
Lamont-Doherty Earth Observatory
Columbia University
Palisades, New York 10964

J. Gregory McHone
Department of Geology and Geophysics
University of Connecticut
Storrs, Connecticut 06269

Giovanni Muttoni
Department of Earth Sciences
University of Milan
Via Mangiagalli 34
20133 Milan
Italy

Paul E. Olsen
Lamont-Doherty Earth Observatory
Columbia University
Palisades, New York 10964

William C. Parker
Department of Geological Sciences
Florida State University
Tallahassee, Florida 32306

Lina C. Patiño
Department of Geology
Michigan State University
East Lansing, Michigan 48824

John H. Puffer
Department of Earth and Environmental Sciences
Rutgers University
Newark, New Jersey 07102

Paul C. Ragland
Department of Geological Sciences
National High Magnetic Field Laboratory
Florida State University
Tallahassee, Florida 32306

Vincent J. M. Salters
Department of Geological Sciences
National High Magnetic Field Laboratory
Florida State University
Tallahassee, Florida 32306

Roy W. Schlische
Department of Geological Sciences
Rutgers University
Piscataway, New Jersey 08854

Leonardo Seeber
Lamont-Doherty Earth Observatory
Columbia University
Palisades, New York 10964

Martha Oliver Withjack
Department of Geological Sciences
Rutgers University
Piscataway, New Jersey 08854

INDEX